Short Courses

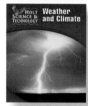

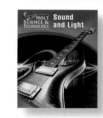

Teacher Edition WALK-THROUGH

Student Edition CONTENTS IN BRIEF

HOLT, RINEHART AND WINSTON
A Harcourt Education Company
Orlando • **Austin** • New York • San Diego • Toronto • London

Designed to meet the needs of all students

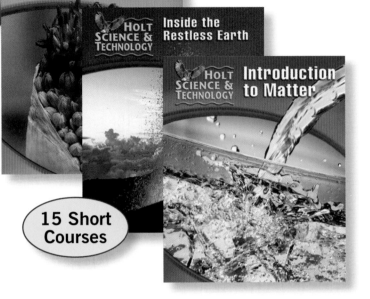

15 Short Courses

Holt Science & Technology: Short Course Series allows you to match your curriculum by choosing from 15 books covering life, earth, and physical sciences. The program reflects current curriculum developments and includes the strongest skills-development strand of any middle school science series. Students of all abilities will develop skills that they can use both in science as well as in other courses.

STUDENTS OF ALL ABILITIES RECEIVE THE READING HELP AND TAILORED INSTRUCTION THEY NEED.

- The *Student Edition* is accessible with a clean, easy-to-follow design and highlighted vocabulary words.
- Inclusion strategies and different learning styles help support all learners.
- Comprehensive Section and Chapter Reviews and Standardized Test Preparation allow students to practice their test-taking skills.
- Reading Comprehension Guide and Guided Reading Audio CDs help students better understand the content.

CROSS-DISCIPLINARY CONNECTIONS LET STUDENTS SEE HOW SCIENCE RELATES TO OTHER DISCIPLINES.

- Mathematics, reading, and writing skills are integrated throughout the program.
- Cross-discipline Connection To features show students how science relates to language arts, social studies, and other sciences.

A FLEXIBLE LABORATORY PROGRAM HELPS STUDENTS BUILD IMPORTANT INQUIRY AND CRITICAL-THINKING SKILLS.

- The laboratory program includes labs in each chapter, labs in the **LabBook** at the end of the text, six different lab books, and **Video Labs.**
- All labs are teacher-tested and rated by difficulty in the *Teacher Edition,* so you can be sure the labs will be appropriate for your students.
- A variety of labs, from **Inquiry Labs** to **Skills Practice Labs,** helps you meet the needs of your curriculum and work within the time constraints of your teaching schedule.

INTEGRATED TECHNOLOGY AND ONLINE RESOURCES EXPAND LEARNING BEYOND CLASSROOM WALLS.

- An **Enhanced Online Edition** or **CD-ROM Version** of the student text lightens your students' load.

- **SciLinks,** a Web service developed and maintained by the National Science Teachers Association (NSTA), contains current prescreened links directly related to the textbook.

- **Brain Food Video Quizzes** on videotape and DVD are game-show style quizzes that assess students' progress and motivate them to study.

- The **One-stop Planner® CD-ROM** with **Exam View® Test Generator** contains all of the resources you need including an *Interactive Teacher Edition,* worksheets, customizable lesson plans, **Holt Calendar Planner,** a powerful test generator, **Lab Materials QuickList Software,** and more.

- Spanish Resources include **Guided Reading Audio CD** in Spanish.

CHAPTER RESOURCE FILES FOR

Inside the Restless Earth

Skills Worksheets
- Directed Reading A
- Directed Reading B
- Vocabulary & Notes
- Section Reviews
- Chapter Review
- Reinforcement
- Critical Thinking

Assessments
- Section Quizzes
- Chapter Test A
- Chapter Test B
- Chapter Test C
- Performance-Based Assessment
- Standardized Test Preparation

Labs and Activities
- Datasheets for In-Text Labs
- Datasheets for Quick Labs
- Datasheets for LabBook
- Vocabulary Activity
- SciLinks' Activity

Teacher Resources
- Teacher Notes for Performance-Based Assessment
- Lab Notes and Answers
- Answer Keys
- Lesson Plans
- Test Item Listing for ExamView® Test Generator
- Teaching Transparencies
- Chapter Starter Transparencies
- Bellringer Transparencies
- Concept Mapping Transparencies

HOLT CIENCIAS Y TECNOLOGÍA

Guided Reading Audio CD Program

Direct read of the student text

INTRODUCCIÓN A LA MATERIA

K

Life Science

<table>
<tr><th></th><th>A MICROORGANISMS, FUNGI, AND PLANTS</th><th>B ANIMALS</th></tr>
<tr>
<td>CHAPTER 1</td>
<td>

It's Alive!! Or, Is It?
- Characteristics of living things
- Homeostasis
- Heredity and DNA
- Producers, consumers, and decomposers
- Biomolecules

</td>
<td>

Animals and Behavior
- Characteristics of animals
- Classification of animals
- Animal behavior
- Hibernation and estivation
- The biological clock
- Animal communication
- Living in groups

</td>
</tr>
<tr>
<td>CHAPTER 2</td>
<td>

Bacteria and Viruses
- Binary fission
- Characteristics of bacteria
- Nitrogen-fixing bacteria
- Antibiotics
- Pathogenic bacteria
- Characteristics of viruses
- Lytic cycle

</td>
<td>

Invertebrates
- General characteristics of invertebrates
- Types of symmetry
- Characteristics of sponges, cnidarians, arthropods, and echinoderms
- Flatworms versus roundworms
- Types of circulatory systems

</td>
</tr>
<tr>
<td>CHAPTER 3</td>
<td>

Protists and Fungi
- Characteristics of protists
- Types of algae
- Types of protozoa
- Protist reproduction
- Characteristics of fungi and lichens

</td>
<td>

Fishes, Amphibians, and Reptiles
- Characteristics of vertebrates
- Structure and kinds of fishes
- Development of lungs
- Structure and kinds of amphibians and reptiles
- Function of the amniotic egg

</td>
</tr>
<tr>
<td>CHAPTER 4</td>
<td>

Introduction to Plants
- Characteristics of plants and seeds
- Reproduction and classification
- Angiosperms versus gymnosperms
- Monocots versus dicots
- Structure and functions of roots, stems, leaves, and flowers

</td>
<td>

Birds and Mammals
- Structure and kinds of birds
- Types of feathers
- Adaptations for flight
- Structure and kinds of mammals
- Function of the placenta

</td>
</tr>
<tr>
<td>CHAPTER 5</td>
<td>

Plant Processes
- Pollination and fertilization
- Dormancy
- Photosynthesis
- Plant tropisms
- Seasonal responses of plants

</td>
<td></td>
</tr>
<tr>
<td>CHAPTER 6</td>
<td></td>
<td></td>
</tr>
<tr>
<td>CHAPTER 7</td>
<td></td>
<td></td>
</tr>
</table>

PROGRAM SCOPE AND SEQUENCE

Selecting the right books for your course is easy. Just review the topics presented in each book to determine the best match to your district curriculum.

C CELLS, HEREDITY, & CLASSIFICATION

Cells: The Basic Units of Life
- Cells, tissues, and organs
- Populations, communities, and ecosystems
- Cell theory
- Surface-to-volume ratio
- Prokaryotic versus eukaryotic cells
- Cell organelles

The Cell in Action
- Diffusion and osmosis
- Passive versus active transport
- Endocytosis versus exocytosis
- Photosynthesis
- Cellular respiration and fermentation
- Cell cycle

Heredity
- Dominant versus recessive traits
- Genes and alleles
- Genotype, phenotype, the Punnett square and probability
- Meiosis
- Determination of sex

Genes and Gene Technology
- Structure of DNA
- Protein synthesis
- Mutations
- Heredity disorders and genetic counseling

The Evolution of Living Things
- Adaptations and species
- Evidence for evolution
- Darwin's work and natural selection
- Formation of new species

The History of Life on Earth
- Geologic time scale and extinctions
- Plate tectonics
- Human evolution

Classification
- Levels of classification
- Cladistic diagrams
- Dichotomous keys
- Characteristics of the six kingdoms

D HUMAN BODY SYSTEMS & HEALTH

Body Organization and Structure
- Homeostasis
- Types of tissue
- Organ systems
- Structure and function of the skeletal system, muscular system, and integumentary system

Circulation and Respiration
- Structure and function of the cardiovascular system, lymphatic system, and respiratory system
- Respiratory disorders

The Digestive and Urinary Systems
- Structure and function of the digestive system
- Structure and function of the urinary system

Communication and Control
- Structure and function of the nervous system and endocrine system
- The senses
- Structure and function of the eye and ear

Reproduction and Development
- Asexual versus sexual reproduction
- Internal versus external fertilization
- Structure and function of the human male and female reproductive systems
- Fertilization, placental development, and embryo growth
- Stages of human life

Body Defenses and Disease
- Types of diseases
- Vaccines and immunity
- Structure and function of the immune system
- Autoimmune diseases, cancer, and AIDS

Staying Healthy
- Nutrition and reading food labels
- Alcohol and drug effects on the body
- Hygiene, exercise, and first aid

E ENVIRONMENTAL SCIENCE

Interactions of Living Things
- Biotic versus abiotic parts of the environment
- Producers, consumers, and decomposers
- Food chains and food webs
- Factors limiting population growth
- Predator-prey relationships
- Symbiosis and coevolution

Cycles in Nature
- Water cycle
- Carbon cycle
- Nitrogen cycle
- Ecological succession

The Earth's Ecosystems
- Kinds of land and water biomes
- Marine ecosystems
- Freshwater ecosystems

Environmental Problems and Solutions
- Types of pollutants
- Types of resources
- Conservation practices
- Species protection

Energy Resources
- Types of resources
- Energy resources and pollution
- Alternative energy resources

Earth Science

H WATER ON EARTH

The Flow of Fresh Water
- Water cycle
- River systems
- Stream erosion
- Life cycle of rivers
- Deposition
- Aquifers, springs, and wells
- Ground water
- Water treatment and pollution

Exploring the Oceans
- Properties and characteristics of the oceans
- Features of the ocean floor
- Ocean ecology
- Ocean resources and pollution

The Movement of Ocean Water
- Types of currents
- Characteristics of waves
- Types of ocean waves
- Tides

I WEATHER AND CLIMATE

The Atmosphere
- Structure of the atmosphere
- Air pressure
- Radiation, convection, and conduction
- Greenhouse effect and global warming
- Characteristics of winds
- Types of winds
- Air pollution

Understanding Weather
- Water cycle
- Humidity
- Types of clouds
- Types of precipitation
- Air masses and fronts
- Storms, tornadoes, and hurricanes
- Weather forecasting
- Weather maps

Climate
- Weather versus climate
- Seasons and latitude
- Prevailing winds
- Earth's biomes
- Earth's climate zones
- Ice ages
- Global warming
- Greenhouse effect

J ASTRONOMY

Studying Space
- Astronomy
- Keeping time
- Types of telescope
- Radioastronomy
- Mapping the stars
- Scales of the universe

Stars, Galaxies, and the Universe
- Composition of stars
- Classification of stars
- Star brightness, distance, and motions
- H-R diagram
- Life cycle of stars
- Types of galaxies
- Theories on the formation of the universe

Formation of the Solar System
- Birth of the solar system
- Structure of the sun
- Fusion
- Earth's structure and atmosphere
- Planetary motion
- Newton's Law of Universal Gravitation

A Family of Planets
- Properties and characteristics of the planets
- Properties and characteristics of moons
- Comets, asteroids, and meteoroids

Exploring Space
- Rocketry and artificial satellites
- Types of Earth orbit
- Space probes and space exploration

Physical Science

		K INTRODUCTION TO MATTER	**L** INTERACTIONS OF MATTER
CHAPTER	**1**	**The Properties of Matter** • Definition of matter • Mass and weight • Physical and chemical properties • Physical and chemical change • Density	**Chemical Bonding** • Types of chemical bonds • Valence electrons • Ions versus molecules • Crystal lattice
CHAPTER	**2**	**States of Matter** • States of matter and their properties • Boyle's and Charles's laws • Changes of state	**Chemical Reactions** • Writing chemical formulas and equations • Law of conservation of mass • Types of reactions • Endothermic versus exothermic reactions • Law of conservation of energy • Activation energy • Catalysts and inhibitors
CHAPTER	**3**	**Elements, Compounds, and Mixtures** • Elements and compounds • Metals, nonmetals, and metalloids (semiconductors) • Properties of mixtures • Properties of solutions, suspensions, and colloids	**Chemical Compounds** • Ionic versus covalent compounds • Acids, bases, and salts • pH • Organic compounds • Biomolecules
CHAPTER	**4**	**Introduction to Atoms** • Atomic theory • Atomic model and structure • Isotopes • Atomic mass and mass number	**Atomic Energy** • Properties of radioactive substances • Types of decay • Half-life • Fission, fusion, and chain reactions
CHAPTER	**5**	**The Periodic Table** • Structure of the periodic table • Periodic law • Properties of alkali metals, alkaline-earth metals, halogens, and noble gases	
CHAPTER	**6**		

M FORCES, MOTION, AND ENERGY

Matter in Motion
- Speed, velocity, and acceleration
- Measuring force
- Friction
- Mass versus weight

Forces in Motion
- Terminal velocity and free fall
- Projectile motion
- Inertia
- Momentum

Forces in Fluids
- Properties in fluids
- Atmospheric pressure
- Density
- Pascal's principle
- Buoyant force
- Archimedes' principle
- Bernoulli's principle

Work and Machines
- Measuring work
- Measuring power
- Types of machines
- Mechanical advantage
- Mechanical efficiency

Energy and Energy Resources
- Forms of energy
- Energy conversions
- Law of conservation of energy
- Energy resources

Heat and Heat Technology
- Heat versus temperature
- Thermal expansion
- Absolute zero
- Conduction, convection, radiation
- Conductors versus insulators
- Specific heat capacity
- Changes of state
- Heat engines
- Thermal pollution

N ELECTRICITY AND MAGNETISM

Introduction to Electricity
- Law of electric charges
- Conduction versus induction
- Static electricity
- Potential difference
- Cells, batteries, and photocells
- Thermocouples
- Voltage, current, and resistance
- Electric power
- Types of circuits

Electromagnetism
- Properties of magnets
- Magnetic force
- Electromagnetism
- Solenoids and electric motors
- Electromagnetic induction
- Generators and transformers

Electronic Technology
- Properties of semiconductors
- Integrated circuits
- Diodes and transistors
- Analog versus digital signals
- Microprocessors
- Features of computers

O SOUND AND LIGHT

The Energy of Waves
- Properties of waves
- Types of waves
- Reflection and refraction
- Diffraction and interference
- Standing waves and resonance

The Nature of Sound
- Properties of sound waves
- Structure of the human ear
- Pitch and the Doppler effect
- Infrasonic versus ultrasonic sound
- Sound reflection and echolocation
- Sound barrier
- Interference, resonance, diffraction, and standing waves
- Sound quality of instruments

The Nature of Light
- Electromagnetic waves
- Electromagnetic spectrum
- Law of reflection
- Absorption and scattering
- Reflection and refraction
- Diffraction and interference

Light and Our World
- Luminosity
- Types of lighting
- Types of mirrors and lenses
- Focal point
- Structure of the human eye
- Lasers and holograms

Program resources make teaching and learning easier.

CHAPTER RESOURCES

A *Chapter Resources book* accompanies each of the 15 *Short Courses*. Here you'll find everything you need to make sure your students are getting the most out of learning science—all in one book.

Skills Worksheets
- Directed Reading A: Basic
- Directed Reading B: Special Needs
- Vocabulary and Chapter Summary
- Section Reviews
- Chapter Reviews
- Reinforcement
- Critical Thinking

Labs & Activities
- Datasheets for Chapter Labs
- Datasheets for Quick Labs
- Datasheets for LabBook
- Vocabulary Activity
- SciLinks® Activity

Assessments
- Section Quizzes
- Chapter Tests A: General
- Chapter Tests B: Advanced
- Chapter Tests C: Special Needs
- Performance-Based Assessments
- Standardized Test Preparation

Teacher Resources
- Lab Notes and Answers
- Teacher Notes for Performance-Based Assessment
- Answer Keys
- Lesson Plans
- Test Item Listing for ExamView® Test Generator
- Full-color Teaching Transparencies, plus section Bellringers, Concept Mapping, and Chapter Starter Transparencies.

SPANISH RESOURCES

Spanish materials are available for each *Short Course:*

- *Student Edition*
- *Spanish Resources* booklet contains worksheets and assessments translated into Spanish with an English Answer Key.
- Guided Reading Audio CD Program

ONLINE RESOURCES

- *Enhanced Online Editions* engage students and assist teachers with a host of interactive features that are available anytime and anywhere you can connect to the Internet.
- CNNStudentNews.com provides award-winning news and information for both teachers and students.
- SciLinks—a Web service developed and maintained by the National Science Teachers Association—links you and your students to up-to-date online resources directly related to chapter topics.
- go.hrw.com links you and your students to online chapter activities and resources.
- Current Science articles relate to students' lives.

ADDITIONAL LAB AND SKILLS RESOURCES

- *Calculator-Based Labs* incorporates scientific instruments, offering students insight into modern scientific investigation.
- *EcoLabs & Field Activities* develops awareness of the natural world.
- *Holt Science Skills Workshop: Reading in the Content Area* contains exercises that target reading skills key.
- *Inquiry Labs* taps students' natural curiosity and creativity with a focus on the process of discovery.
- *Labs You Can Eat* safely incorporates edible items into the classroom.
- *Long-Term Projects & Research Ideas* extends and enriches lessons.
- *Math Skills for Science* provides additional explanations, examples, and math problems so students can develop their skills.
- *Science Skills Worksheets* helps your students hone important learning skills.
- *Whiz-Bang Demonstrations* gets your students' attention at the beginning of a lesson.

ADDITIONAL RESOURCES

- *Assessment Checklists & Rubrics* gives you guidelines for evaluating students' progress.
- *Holt Anthology of Science Fiction* sparks your students' imaginations with thought-provoking stories.
- *Holt Science Posters* visually reinforces scientific concepts and themes with seven colorful posters including **The Periodic Table of the Elements.**

- *Professional Reference for Teachers* contains professional articles that discuss a variety of topics, such as classroom management.
- *Program Introduction Resource File* explains the program and its features and provides several additional references, including lab safety, scoring rubrics, and more.
- *Science Fair Guide* gives teachers, students, and parents tips for planning and assisting in a science fair.
- *Science Puzzlers, Twisters & Teasers* activities challenge students to think about science concepts in different ways.

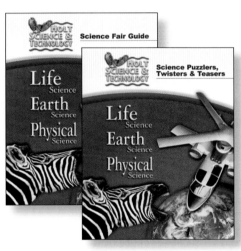

TECHNOLOGY RESOURCES

- *CNN Presents Science in the News: Video Library* helps students see the impact of science on their everyday lives with actual news video clips.
  - Multicultural Connections
  - Science, Technology & Society
  - Scientists in Action
  - Eye on the Environment
- *Guided Reading Audio CD Program*, available in English and Spanish, provides students with a direct read of each section.
- *HRW Earth Science Videotape* takes your students on a geology "field trip" with full-motion video.
- *Interactive Explorations CD-ROM Program* develops students' inquiry and decision-making skills as they investigate science phenomena in a virtual lab setting.

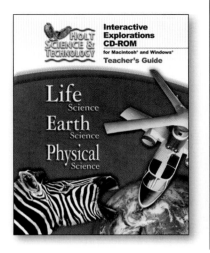

- *One-Stop Planner CD-ROM®* organizes everything you need on one disc, including printable worksheets, customizable lesson plans, a powerful test generator, **PowerPoint® LectureNotes, Lab Materials QuickList Software, Holt Calendar Planner, Interactive Teacher Edition,** and more.
- *Science Tutor CD-ROMs* help students practice what they learn with immediate feedback.
- *Lab Videos* make it easier to integrate more experiments into your lessons without the preparation time and costs. Available on DVD and VHS.
- **Brain Food Video Quizzes** are game-show style quizzes that assess students' progress. Available on DVD and VHS.
- *Visual Concepts CD-ROMs* include graphics, animations, and movie clips that demonstrate key chapter concepts.

Science and Math Worksheets

The **Holt Science & Technology** program helps you meet the needs of a wide variety of students, regardless of their skill level. The following pages provide examples of the worksheets available to improve your students' science and math skills whether they already have a strong science and math background or are weak in these areas. Samples of assessment checklists and rubrics are also provided.

In addition to the skills worksheets represented here, **Holt Science & Technology** provides a variety of worksheets that are correlated directly with each chapter of the program. Representations of these worksheets are found at the beginning of each chapter in this *Teacher Edition*.

Many worksheets are also available on the Holt Web site. The address is **go.hrw.com**.

Science Skills Worksheets: Thinking Skills

BEING FLEXIBLE

USING YOUR SENSES

THINKING OBJECTIVELY

UNDERSTANDING BIAS

USING LOGIC

BOOSTING YOUR MEMORY

IMPROVING YOUR STUDY HABITS

READING A SCIENCE TEXTBOOK

Science Skills Worksheets: Experimenting Skills

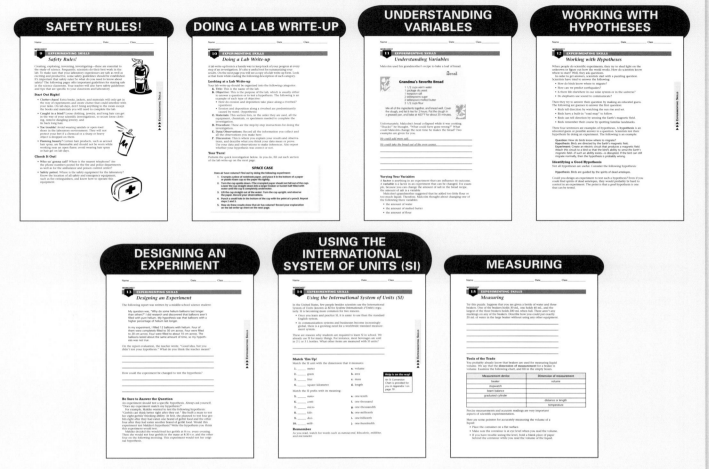

SAFETY RULES!

DOING A LAB WRITE-UP

UNDERSTANDING VARIABLES

WORKING WITH HYPOTHESES

DESIGNING AN EXPERIMENT

USING THE INTERNATIONAL SYSTEM OF UNITS (SI)

MEASURING

Science Skills Worksheets: Researching Skills

CHOOSING YOUR TOPIC

ORGANIZING YOUR RESEARCH

FINDING USEFUL SOURCES

RESEARCHING ON THE WEB

Science Skills Worksheets: Researching Skills (continued)

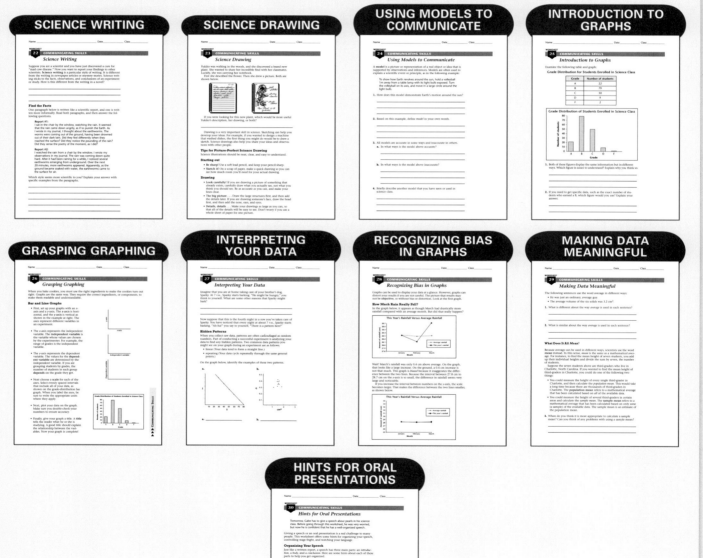

IDENTIFYING BIAS

TAKING NOTES

Science Skills Worksheets: Communicating Skills

SCIENCE WRITING

SCIENCE DRAWING

USING MODELS TO COMMUNICATE

INTRODUCTION TO GRAPHS

GRASPING GRAPHING

INTERPRETING YOUR DATA

RECOGNIZING BIAS IN GRAPHS

MAKING DATA MEANINGFUL

HINTS FOR ORAL PRESENTATIONS

Math Skills for Science

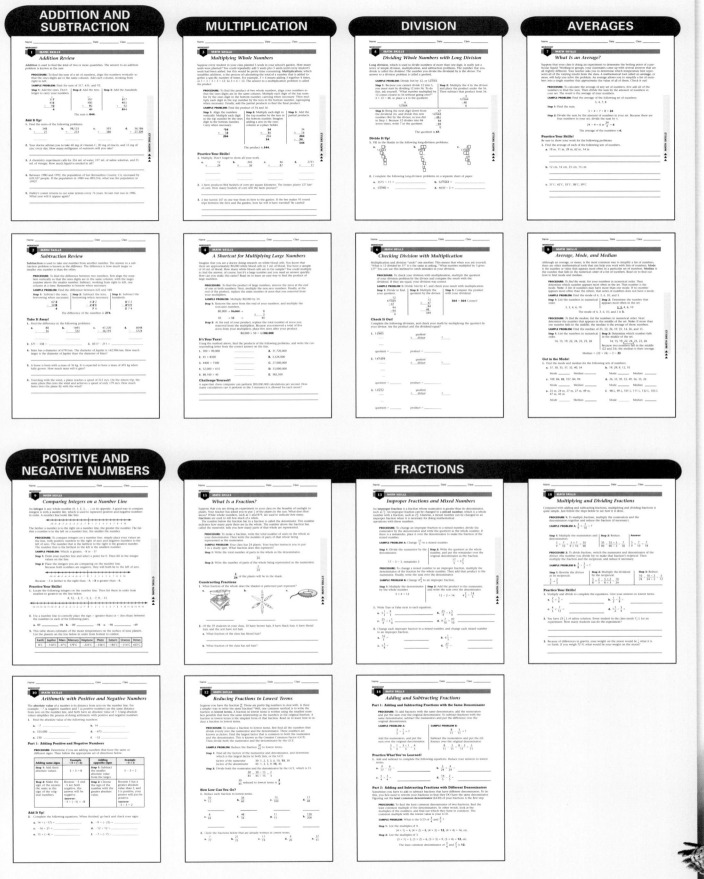

Math Skills for Science (continued)

RATIOS AND PROPORTIONS

16 MATH SKILLS — What Is a Ratio?

Imagine that you are planning a science experiment for your class and you want to make sure you have enough beakers for everyone. What do you do? Well, you could simply count the total number of beakers you have and compare it with the number of students in your class. You want to make it balanced, but you just made a ratio! A **ratio** is a comparison between numbers, and can be written in words (3 to 7), as a fraction (3/7), or with a colon (3:7).

PROCEDURE: To find the ratio between two quantities, show the two quantities as a fraction, and then reduce. The result is the ratio.

SAMPLE PROBLEM: Find the ratio of thermometers to students if you have 36 thermometers and 48 students in your class.

Step 1: Make the ratio:
36 thermometers
48 students

Step 2: Reduce.
$\frac{36}{48} = \frac{36 \div 12}{48 \div 12} = \frac{3}{4}$

The ratio of thermometers to students is 3 to 4, 3/4, or 3:4.

Wildflower Research Results

Field	Average number of flowers (per 10 m²)	Number of species	Species currently flowering
1	51	12	9
2	17	11	7
3	22	21	14

Analyze Your Data!

1. What is the ratio between the currently flowering species and the total number of species of flowers in Field 1?

2. What is the ratio between the number of species currently flowering in Field 1 and Field 2 and the number of species currently flowering in Field 3?

3. What is the ratio between the number of species currently flowering and the total number of flowers in all three fields?

17 MATH SKILLS — Using Proportions and Cross-Multiplication

Ratios are a powerful tool in science and math. But in order to take full advantage of them, we have to do more than just calculate ratios—we have to put them to work! For example, if you have these bacteria specimens for every student in your class, you know that you will have a ratio of 3 to 1, 3, or 3:1. But this ratio does not tell you the total number of specimens. To find that, you need to use a proportion.

A **proportion** is a statement of equality between two ratios. This means that the numerator of one ratio multiplied by the denominator of the other ratio is equal to the product of the other numerator and denominator. An example looks like this:

$\frac{3 \times 4}{4 \times 4} = \frac{12}{16}$

$3 \times 4 = 12$
$12 = 12$

Notice that we are multiplying across the equal sign in your proportion. This process is called cross-multiplication. Cross-multiplication is useful because if you know three of the quantities in a proportion, you can find the fourth.

PROCEDURE: To find an unknown quantity in a proportion, set up the numbers you know in equal ratios. Leave the place for the quantity you do not know empty for now. Cross-multiply the known numerator of one ratio with the known denominator of the other. Then divide this product by your remaining known quantity. The quotient is your answer.

SAMPLE PROBLEM: Find the missing number in this proportion:

Try It Yourself!

1. Find the unknown quantities in the following proportions:

DECIMALS

18 MATH SKILLS — Decimals and Fractions

Many numbers you will use in science class and other places will be decimal numbers. Like fractions, **decimals** are used to show how much, or what part, of a whole. A decimal point (.) separates the whole number part of a decimal number on the left from the fraction part on the right. The value of a decimal number is determined by its place value. The chart on the right shows the place values for the decimal system. The first place after the decimal point shows parts of ten, or tenths, the second place shows hundredths, and so on. For example, 3.74 is the same as 3 + 7/10 + 4/100. Any fraction can be changed into a decimal number, and vice versa.

PROCEDURE: To change a fraction into a decimal, divide the numerator of the fraction by the denominator. If you have a mixed number (a whole number with a fraction), put the whole-number part of your number before the decimal point.

SAMPLE PROBLEM: Change 1/20 into a decimal number.

19 MATH SKILLS — Arithmetic with Decimals

How much would you expect to pay if you were buying a bag of chips for 30 cents and a cola for 75 cents? $1.25, right? Well, if you know that one, you already know how to add decimals. Doing arithmetic with decimals is a lot like doing arithmetic with whole numbers. Read on to see how it's done.

Part 1: Adding and Subtracting Decimals

PROCEDURE: To add or subtract decimals, line up your numbers vertically so that the decimal points line up. Then add or subtract the columns from right to left, carrying or borrowing numbers when necessary.

SAMPLE PROBLEM: Add the following numbers: 3.1415 and 2.96.

PERCENTAGES

20 MATH SKILLS — Parts of 100: Calculating Percentages

Let's say you scored 83 percent (%) on your last science test. Does that mean you got 83 questions right? Probably not. The score on your test is expressed as a percentage. The word percent comes from Latin words meaning "parts of a 100," and that's exactly what a percentage is. A **percentage** is a ratio that compares a number with 100. Read on to learn how to find a percentage of a number.

21 MATH SKILLS — Percentages, Fractions, and Decimals

22 MATH SKILLS — Working with Percentages and Proportions

When working with percentages, it is often helpful to think of them in terms of ratios and proportions. For instance, if someone asks you, "What is 10% of 40?" you could simply change 10% into a decimal (0.1) and multiply it by 40 to get 4. But what if you were asked, "5% of what number is 10?" That's a little trickier. To do this calculation, it is convenient to use a proportion.

POWERS OF 10

23 MATH SKILLS — Counting the Zeros

A **power of 10** is a number that can have 10 as its only factors. For instance, (10 × 10) = 100 and (10 × 10 × 10) = 10,000 are both powers of 10. Multiplying and dividing by powers of 10 is as easy as counting the zeros and moving your decimal point the same number of places.

Part 1: Multiplying by Powers of 10

24 MATH SKILLS — Creating Exponents

Imagine that you are writing a paper for your science class and need to write very large numbers, such as 10,000,000,000,000. Your fingers would get pretty tired writing all those zeros. However, there is a simpler way to express these large powers of 10.

SCIENTIFIC NOTATION

25 MATH SKILLS — What Is Scientific Notation?

Sometimes scientific calculations result in very large numbers, like 918,700,000,000,000, or in very small numbers, such as 0.00000000078. **Scientific notation** is a short way of representing such numbers without writing all the place-holding zeros. In scientific notation, we write the number as a product of two factors: the first is a number between 1 and 10, and the second is a power of ten, written as 10^something.

26 MATH SKILLS — Multiplying and Dividing in Scientific Notation

Part 1: Multiplying in Scientific Notation

SI MEASUREMENT AND CONVERSION

27 MATH SKILLS — What Is SI?

To make sharing information easier, most of the world uses the SI system of measurement. SI, which stands for Système International, is a standard for measuring mass, length, volume, and other quantities.

Quantity	Unit	Symbol
length	meter	m
volume	liter	L
mass	gram	g

Prefix	Powers of 10	Symbol	Example
kilo-	1000	(10³)	k	kilogram (kg)
hecto-	100	(10²)	h	hectoliter (hL)
deca-	10	(10¹)	da	decameter (dam)
	1	(10⁰)		meter (m), gram (g), liter (L)
deci-	0.1	(10⁻¹)	d	decigram (dg)
centi-	0.01	(10⁻²)	c	centimeter (cm)
milli-	0.001	(10⁻³)	m	milliliter (mL)

28 MATH SKILLS — A Formula for SI Catch-up

Scientists use the SI system all the time, but most people in the United States still use non-SI units. So what do you do if you data in non-SI units and you have to convert the data into SI units, or vice versa? Here no fear! Conversion charts, like the one shown below, can help you to accomplish the task with ease.

SI Conversion Chart

If you know	Multiply by	To find
inches (in.)	2.54	centimeters (cm)
feet (ft)	30.50	centimeters (cm)
yards (yd)	0.91	meters (m)
miles (mi)	1.61	kilometers (km)
ounces (oz)	28.35	grams (g)
pounds (lb)	0.45	kilograms (kg)
fluid ounces (fl oz)	29.57	milliliters (mL)
cups (c)	0.24	liters (L)
pints (pt)	0.47	liters (L)
quarts (qt)	0.94	liters (L)
gallons (gal)	3.79	liters (L)

Math Skills for Science (continued)

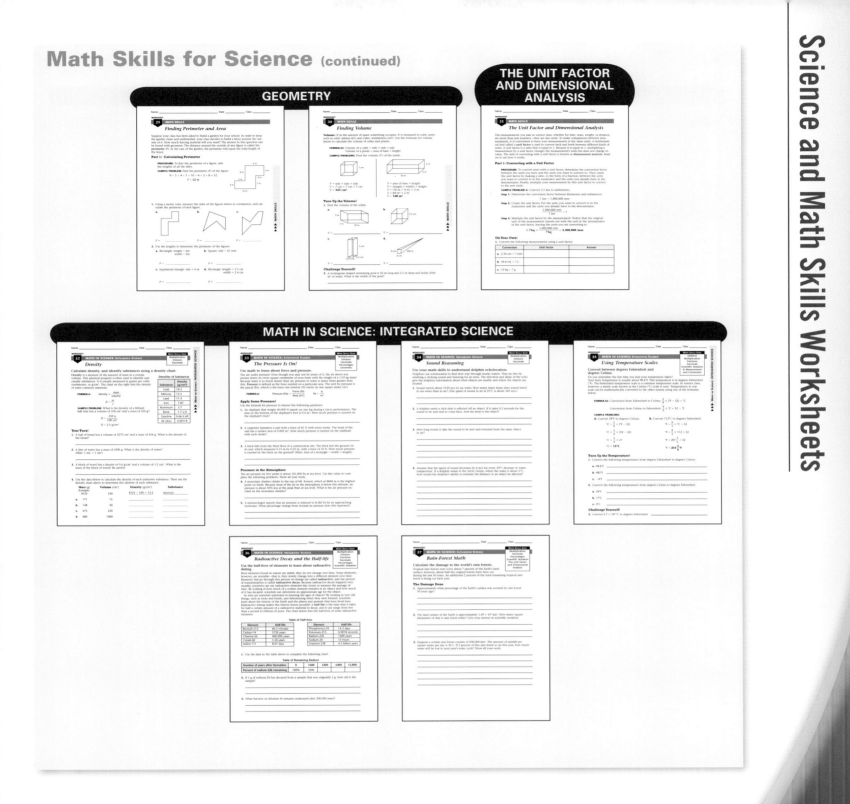

GEOMETRY

THE UNIT FACTOR AND DIMENSIONAL ANALYSIS

MATH IN SCIENCE: INTEGRATED SCIENCE

Math Skills for Science (continued)

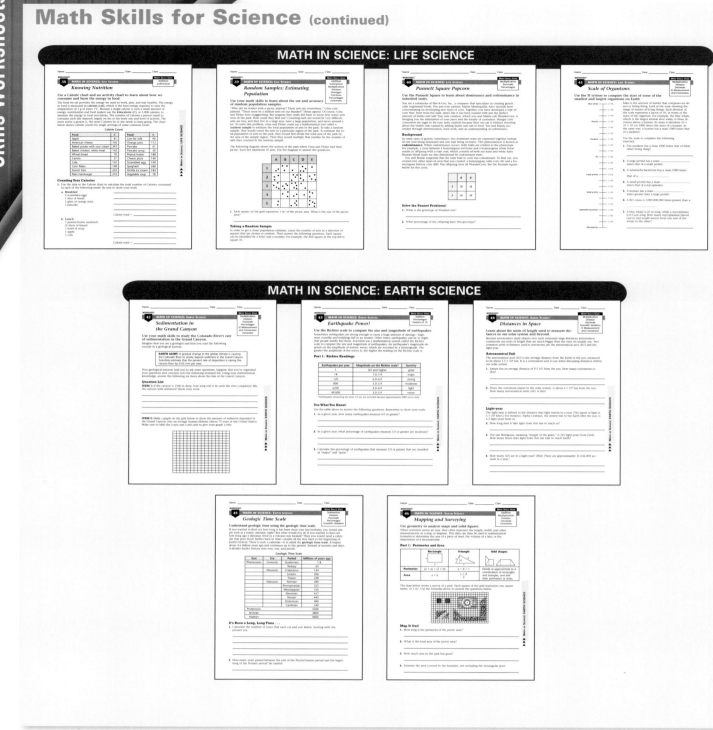

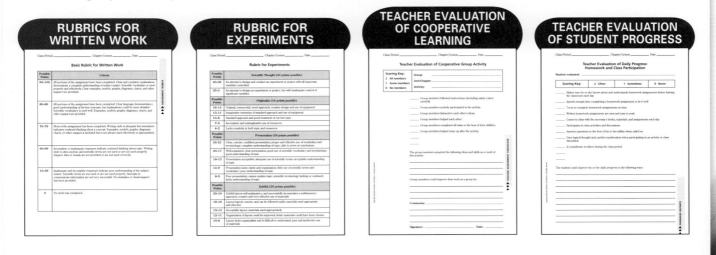

MATH IN SCIENCE: PHYSICAL SCIENCE

Assessment Checklist & Rubrics

The following is just a sample of over 50 checklists and rubrics contained in this booklet.

National Science Education Standards

The following lists show the chapter correlation of *Holt Science & Technology: Astronomy* with the *National Science Education Standards* (grades 5–8).

Unifying Concepts and Processes

Standard	Chapter Correlation	
Systems, order, and organization Code: UCP 1	Chapter 2 Chapter 3 Chapter 4	2.1, 2.2, 2.3, 2.4 3.1, 3.4 4.1, 4.2, 4.3, 4.4
Evidence, models, and explanation Code: UCP 2	Chapter 1 Chapter 2 Chapter 3 Chapter 4 Chapter 5	1.1 2.1, 2.2, 2.4 3.1, 3.3, 3.4 4.1, 4.4 5.1, 5.2
Change, constancy, and measurement Code: UCP 3	Chapter 1 Chapter 2 Chapter 3 Chapter 4 Chapter 5	1.3 2.1, 2.2, 2.4 3.4 4.1, 4.2, 4.3, 4.4 5.1, 5.2
Evolution and equilibrium Code: UCP 4	Chapter 3	3.1, 3.3
Form and function Code: UCP 5	Chapter 2 Chapter 5	2.2, 2.3, 2.4 5.3, 5.4

Science as Inquiry

Standard	Chapter Correlation	
Abilities necessary to do scientific inquiry Code: SAI 1	Chapter 1 Chapter 2 Chapter 3 Chapter 4 Chapter 5	1.3 2.1, 2.2, 2.3, 2.4 3.3, 3.4 4.1, 4.2, 4.3, 4.4 5.2
Understandings about scientific inquiry Code: SAI 2	Chapter 1 Chapter 2 Chapter 3	1.1 2.1 3.4

Science and Technology

Standard	Chapter Correlation	
Abilities of technological design Code: ST 1	Chapter 2	2.3
	Chapter 4	4.1, 4.4
Understandings about science and technology Code: ST 2	Chapter 1	1.3
	Chapter 2	2.3
	Chapter 3	3.1, 3.2, 3.4
	Chapter 4	4.1, 4.2, 4.3
	Chapter 5	5.2, 5.3, 5.4

Science in Personal Perspectives

Standard	Chapter Correlation	
Science and technology in society Code: SPSP 5	Chapter 1	1.1, 1.3
	Chapter 2	2.1, 2.3
	Chapter 3	3.1, 3.2, 3.4
	Chapter 4	4.1, 4.2, 4.3
	Chapter 5	5.2, 5.4

History and Nature of Science

Standard	Chapter Correlation	
Science as human endeavor Code: HNS 1	Chapter 1	1.1, 1.3
	Chapter 2	2.2, 2.3
	Chapter 3	3.2, 3.4
	Chapter 4	4.1, 4.2, 4.3, 4.4
	Chapter 5	5.1, 5.4
Nature of science Code: HNS 2	Chapter 1	1.1, 1.3
	Chapter 2	2.2, 2.3
	Chapter 3	3.1, 3.2, 3.4
History of science Code: HNS 3	Chapter 1	1.1
	Chapter 3	3.1, 3.2, 3.4
	Chapter 4	4.1, 4.2, 4.3, 4.4
	Chapter 5	5.1, 5.4

Earth Science Content Standards

Structure of the Earth System

Standard	Chapter Correlation	
The solid earth is layered with a lithosphere; hot, convecting mantle; and dense metallic core. Code: ES 1a	**Chapter 4**	4.4
Land forms are the result of a combination of constructive and destructive forces. Constructive forces include crustal deformation, volcanic eruption, and deposition of sediment, while destructive forces include weathering and erosion. Code: ES 1c	**Chapter 4**	4.1, 4.2, 4.3

Earth's History

Standard	Chapter Correlation	
Fossils provide important evidence of how life and environmental conditions have changed. Code: ES 2b	**Chapter 3**	3.3

Earth in the Solar System

Standard	Chapter Correlation	
The earth is the third planet from the sun in a system that includes the moon, the sun, eight other planets and their moons, and smaller objects, such as asteroids and comets. The sun, an average star, is the central and largest body in the solar system. Code: ES 3a	**Chapter 1** **Chapter 3** **Chapter 4**	1.1 3.1, 3.2 4.1, 4.2, 4.3, 4.4
Most objects in the solar system are in regular and predictable motion. Those motions explain such phenomena as the day, the year, phases of the moon, and eclipses. Code: ES 3b	**Chapter 1** **Chapter 3** **Chapter 4**	1.1 3.1, 3.4 4.1, 4.2, 4.3, 4.4
Gravity is the force that keeps planets in orbit around the sun and governs the rest of the motion in the solar system. Gravity alone holds us to the earth's surface and explains the phenomena of the tides. Code: ES 3c	**Chapter 1** **Chapter 3** **Chapter 4**	1.1 3.1 4.1, 4.4

HOLT SCIENCE & TECHNOLOGY

Astronomy

HOLT, RINEHART AND WINSTON

A Harcourt Education Company

Orlando • **Austin** • New York • San Diego • Toronto • London

Acknowledgments

Contributing Authors

Mary Kay Hemenway, Ph.D.
Research Associate and Senior Lecturer
Department of Astronomy
The University of Texas at Austin
Austin, Texas

Karen J. Meech, Ph.D.
Astronomer
Institute for Astronomy
University of Hawaii
Honolulu, Hawaii

Inclusion Specialist

Karen Clay
Inclusion Specialist Consultant
Boston, Massachusetts

Safety Reviewer

Jack Gerlovich, Ph.D.
Associate Professor
School of Education
Drake University
Des Moines, Iowa

Academic Reviewers

Dan Bruton, Ph.D.
Associate Professor
Department of Physics and Astronomy
Stephen F. Austin State University
Nacogdoches, Texas

Wesley N. Colley, Ph.D.
Lecturer
Department of Astronomy
University of Virginia
Charlottesville, Virginia

Mary Kay Hemenway, Ph.D.
Research Associate and Senior Lecturer
Astronomy Department
The University of Texas
Austin, Texas

Sten Odenwald, Ph.D.
Astronomer
NASA Goddard Space Flight Center and Raytheon ITSS
Greenbelt, Maryland

Teacher Reviewers

Diedre S. Adams
Physical Science Instructor
Science Department
West Vigo Middle School
West Terre Haute, Indiana

Laura Buchanan
Science Teacher and Department Chairperson
Corkran Middle School
Glen Burnie, Maryland

Randy Dye, M.S.
Middle School Science Department Head
Earth Science
Wood Middle School
Waynesville School District #6, Missouri

Meredith Hanson
Science Teacher
Westside Middle School
Rocky Face, Georgia

Laura Kitselman
Science Teacher and Coordinator
Loudoun Country Day School
Leesburg, Virginia

Astronomy

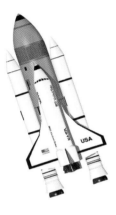

Contents   **v**

Skills Development

MATH PRACTICE

Connection to . . .

Science in Action

How to Use Your Textbook

Your Roadmap for Success with Holt Science and Technology

Reading Warm-Up

A Reading Warm-Up at the beginning of every section provides you with the section's objectives and key terms. The objectives tell you what you'll need to know after you finish reading the section.

Key terms are listed for each section. Learn the definitions of these terms because you will most likely be tested on them. Each key term is highlighted in the text and is defined at point of use and in the margin. You can also use the glossary to locate definitions quickly.

STUDY TIP Reread the objectives and the definitions to the key terms when studying for a test to be sure you know the material.

Get Organized

A Reading Strategy at the beginning of every section provides tips to help you organize and remember the information covered in the section. Keep a science notebook so that you are ready to take notes when your teacher reviews the material in class. Keep your assignments in this notebook so that you can review them when studying for the chapter test.

SECTION 3

The Earth Takes Shape

In many ways, Earth seems to be a perfect place for life.

We live on the third planet from the sun. The Earth, shown in **Figure 1,** is mostly made of rock, and nearly three-fourths of its surface is covered with water. It is surrounded by a protective atmosphere of mostly nitrogen and oxygen and smaller amounts of other gases. But Earth has not always been such an oasis in the solar system.

Formation of the Solid Earth

The Earth formed as planetesimals in the solar system collided and combined. From what scientists can tell, the Earth formed within the first 10 million years of the collapse of the solar nebula!

The Effects of Gravity

When a young planet is still small, it can have an irregular shape, somewhat like a potato. But as the planet gains more matter, the force of gravity increases. When a rocky planet, such as Earth, reaches a diameter of about 350 km, the force of gravity becomes greater than the strength of the rock. As the Earth grew to this size, the rock at its center was crushed by gravity and the planet started to become round.

The Effects of Heat

As the Earth was changing shape, it was also heating up. Planetesimals continued to collide with the Earth, and the energy of their motion heated the planet. Radioactive material, which was present in the Earth as it formed, also heated the young planet. After Earth reached a certain size, the temperature rose faster than the interior could cool, and the rocky material inside began to melt. Today, the Earth is still cooling from the energy that was generated when it formed. Volcanoes, earthquakes, and hot springs are effects of this energy trapped inside the Earth. As you will learn later, the effects of heat and gravity also helped form the Earth's layers when the Earth was very young.

Reading Check What factors heated the Earth during its early formation? (*See the Appendix for answers to Reading Checks.*)

Figure 1 *When Earth is seen from space, one of its unique features—the presence of water—is apparent.*

READING WARM-UP

Objectives
- Describe the formation of the solid Earth.
- Describe the structure of the Earth.
- Explain the development of Earth's atmosphere and the influence of early life on the atmosphere.
- Describe how the Earth's oceans and continents formed.

Terms to Learn
crust
mantle
core

READING STRATEGY

Discussion Read this section silently. Write down questions that you have about this section. Discuss your questions in a small group.

624   Chapter 20   Formation of the Solar System

Be Resourceful—Use the Web

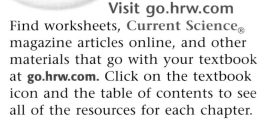

Internet Connect boxes in your textbook take you to resources that you can use for science projects, reports, and research papers. Go to scilinks.org, and type in the SciLinks code to get information on a topic.

Visit go.hrw.com Find worksheets, **Current Science**® magazine articles online, and other materials that go with your textbook at **go.hrw.com.** Click on the textbook icon and the table of contents to see all of the resources for each chapter.

How the Earth's Layers Formed

Have you ever watched the oil separate from vinegar in a bottle of salad dressing? The vinegar sinks because it is denser than oil. The Earth's layers formed in much the same way. As rocks melted, denser elements, such as nickel and iron, sank to the center of the Earth and formed the core. Less dense elements floated to the surface and became the crust. This process is shown in **Figure 2**.

The **crust** is the thin, outermost layer of the Earth. It is 5 to 100 km thick. Crustal rock is made of elements that have low densities, such as oxygen, silicon, and aluminum. The **mantle** is the layer of Earth beneath the crust. It extends 2,900 km below the surface. Mantle rock is made of elements such as magnesium and iron and is denser than crustal rock. The **core** is the central part of the Earth below the mantle. It contains the densest elements (nickel and iron) and extends to the center of the Earth—almost 6,400 km below the surface.

crust the thin and solid outermost layer of the Earth above the mantle

mantle the layer of rock between the Earth's crust and core

core the central part of the Earth below the mantle

Figure 2 The Formation of Earth's Layers

❶ All elements in the early Earth are randomly mixed.

❷ Rocks melt, and denser elements sink toward the center. Less dense elements rise and form layers.

❸ According to composition, the Earth is divided into three layers: the crust, the mantle, and the core.

Crust

Mantle

The Growth of Continents

After a while, some of the rocks were light enough to pile up on the surface. These rocks were the beginning of the earliest continents. The continents gradually thickened and slowly rose above the surface of the ocean. These scattered young continents did not stay in the same place, however. The slow transfer of thermal energy in the mantle pushed them around. Approximately 2.5 billion years ago, continents really started to grow. And by 1.5 billion years ago, the upper mantle had cooled and had become denser and heavier. At this time, it was easier for the cooler parts of the mantle to sink. These conditions made it easier for the continents to move in the same way that they do today.

INTERNET ACTIVITY

For another activity related to this chapter, go to **go.hrw.com** and type in the keyword **HZ5SOLW**.

SECTION Review

Summary

- The effects of gravity and heat created the shape and structure of Earth.
- The Earth is divided into three main layers based on composition: the crust, mantle, and core.
- The presence of life dramatically changed Earth's atmosphere by adding free oxygen.
- Earth's oceans formed shortly after the Earth did, when it had cooled off enough for rain to fall. Continents formed when lighter materials gathered on the surface and rose above sea level.

Using Key Terms

1. Use each of the following terms in a separate sentence: *crust*, *mantle*, and *core*.

Understanding Key Ideas

2. Earth's first atmosphere was mostly made of
   a. nitrogen and oxygen.
   b. chlorine, nitrogen, and sulfur.
   c. carbon dioxide and water vapor.
   d. water vapor and oxygen.
3. Describe the structure of the Earth.
4. Why did the Earth separate into distinct layers?
5. Describe the development of Earth's atmosphere. How did life affect Earth's atmosphere?
6. Explain how Earth's oceans and continents formed.

Critical Thinking

7. **Applying Concepts** How did the effects of gravity help shape the Earth?
8. **Making Inferences** How would the removal of forests affect the Earth's atmosphere?

Interpreting Graphics

Use the illustration below to answer the questions that follow.

9. Which of the layers is composed mostly of the elements magnesium and iron?
10. Which of the layers is composed mostly of the elements iron and nickel?

SciLINKS Developed and maintained by the National Science Teachers Association

For a variety of links related to this chapter, go to **www.scilinks.org**

Topic: The Layers of the Earth; The Oceans
SciLinks code: HSM0862; HSM1069

629

Use the Illustrations and Photos

Art shows complex ideas and processes. Learn to analyze the art so that you better understand the material you read in the text.

Tables and graphs display important information in an organized way to help you see relationships.

A picture is worth a thousand words. Look at the photographs to see relevant examples of science concepts that you are reading about.

Answer the Section Reviews

Section Reviews test your knowledge of the main points of the section. Critical Thinking items challenge you to think about the material in greater depth and to find connections that you infer from the text.

STUDY TIP When you can't answer a question, reread the section. The answer is usually there.

Do Your Homework

Your teacher may assign worksheets to help you understand and remember the material in the chapter.

STUDY TIP Don't try to answer the questions without reading the text and reviewing your class notes. A little preparation up front will make your homework assignments a lot easier. Answering the items in the Chapter Review will help prepare you for the chapter test.

Holt Online Learning

Visit Holt Online Learning

If your teacher gives you a special password to log onto the Holt Online Learning site, you'll find your complete textbook on the Web. In addition, you'll find some great learning tools and practice quizzes. You'll be able to see how well you know the material from your textbook.

CNN Student News

Visit CNN Student News

You'll find up-to-date events in science at **cnnstudentnews.com**.

SAFETY FIRST!

Exploring, inventing, and investigating are essential to the study of science. However, these activities can also be dangerous. To make sure that your experiments and explorations are safe, you must be aware of a variety of safety guidelines. You have probably heard of the saying, "It is better to be safe than sorry." This is particularly true in a science classroom where experiments and explorations are being performed. Being uninformed and careless can result in serious injuries. Don't take chances with your own safety or with anyone else's.

The following pages describe important guidelines for staying safe in the science classroom. Your teacher may also have safety guidelines and tips that are specific to your classroom and laboratory. Take the time to be safe.

Safety Rules!

Start Out Right

Always get your teacher's permission before attempting any laboratory exploration. Read the procedures carefully, and pay particular attention to safety information and caution statements. If you are unsure about what a safety symbol means, look it up or ask your teacher. You cannot be too careful when it comes to safety. If an accident does occur, inform your teacher immediately regardless of how minor you think the accident is.

Safety Symbols

All of the experiments and investigations in this book and their related worksheets include important safety symbols to alert you to particular safety concerns. Become familiar with these symbols so that when you see them, you will know what they mean and what to do. It is important that you read this entire safety section to learn about specific dangers in the laboratory.

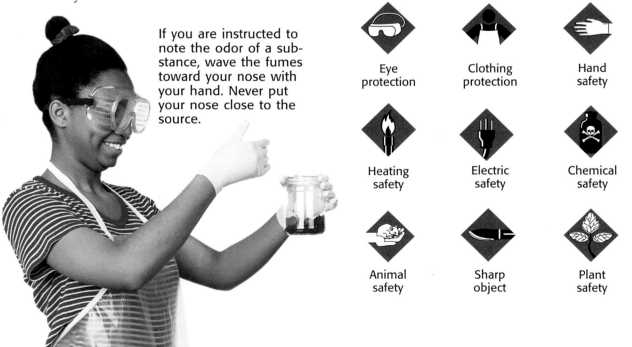

If you are instructed to note the odor of a substance, wave the fumes toward your nose with your hand. Never put your nose close to the source.

Eye protection

Clothing protection

Hand safety

Heating safety

Electric safety

Chemical safety

Animal safety

Sharp object

Plant safety

Eye Safety

Wear safety goggles when working around chemicals, acids, bases, or any type of flame or heating device. Wear safety goggles any time there is even the slightest chance that harm could come to your eyes. If any substance gets into your eyes, notify your teacher immediately and flush your eyes with running water for at least 15 minutes. Treat any unknown chemical as if it were a dangerous chemical. Never look directly into the sun. Doing so could cause permanent blindness.

Avoid wearing contact lenses in a laboratory situation. Even if you are wearing safety goggles, chemicals can get between the contact lenses and your eyes. If your doctor requires that you wear contact lenses instead of glasses, wear eye-cup safety goggles in the lab.

Safety Equipment

Know the locations of the nearest fire alarms and any other safety equipment, such as fire blankets and eyewash fountains, as identified by your teacher, and know the procedures for using the equipment.

Neatness

Keep your work area free of all unnecessary books and papers. Tie back long hair, and secure loose sleeves or other loose articles of clothing, such as ties and bows. Remove dangling jewelry. Don't wear open-toed shoes or sandals in the laboratory. Never eat, drink, or apply cosmetics in a laboratory setting. Food, drink, and cosmetics can easily become contaminated with dangerous materials.

Certain hair products (such as aerosol hair spray) are flammable and should not be worn while working near an open flame. Avoid wearing hair spray or hair gel on lab days.

Sharp/Pointed Objects

Use knives and other sharp instruments with extreme care. Never cut objects while holding them in your hands. Place objects on a suitable work surface for cutting.

Be extra careful when using any glassware. When adding a heavy object to a graduated cylinder, tilt the cylinder so that the object slides slowly to the bottom.

Heat

Wear safety goggles when using a heating device or a flame. Whenever possible, use an electric hot plate as a heat source instead of using an open flame. When heating materials in a test tube, always angle the test tube away from yourself and others. To avoid burns, wear heat-resistant gloves whenever instructed to do so.

Electricity

Be careful with electrical cords. When using a microscope with a lamp, do not place the cord where it could trip someone. Do not let cords hang over a table edge in a way that could cause equipment to fall if the cord is accidentally pulled. Do not use equipment with damaged cords. Be sure that your hands are dry and that the electrical equipment is in the "off" position before plugging it in. Turn off and unplug electrical equipment when you are finished.

Chemicals

Wear safety goggles when handling any potentially dangerous chemicals, acids, or bases. If a chemical is unknown, handle it as you would a dangerous chemical. Wear an apron and protective gloves when you work with acids or bases or whenever you are told to do so. If a spill gets on your skin or clothing, rinse it off immediately with water for at least 5 minutes while calling to your teacher.

Never mix chemicals unless your teacher tells you to do so. Never taste, touch, or smell chemicals unless you are specifically directed to do so. Before working with a flammable liquid or gas, check for the presence of any source of flame, spark, or heat.

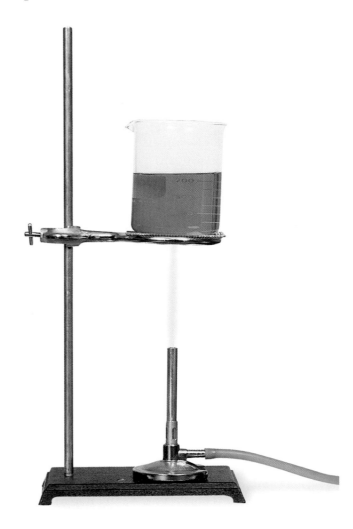

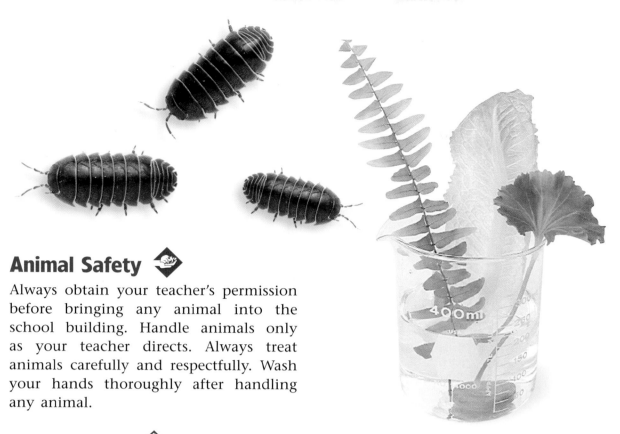

Animal Safety

Always obtain your teacher's permission before bringing any animal into the school building. Handle animals only as your teacher directs. Always treat animals carefully and respectfully. Wash your hands thoroughly after handling any animal.

Plant Safety

Do not eat any part of a plant or plant seed used in the laboratory. Wash your hands thoroughly after handling any part of a plant. When in nature, do not pick any wild plants unless your teacher instructs you to do so.

Glassware

Examine all glassware before use. Be sure that glassware is clean and free of chips and cracks. Report damaged glassware to your teacher. Glass containers used for heating should be made of heat-resistant glass.

Studying Space
Chapter Planning Guide

Compression guide:
To shorten instruction because of time limitations, omit the Chapter Lab.

OBJECTIVES	LABS, DEMONSTRATIONS, AND ACTIVITIES	TECHNOLOGY RESOURCES
PACING • 90 min pp. 2–7 **Chapter Opener**	**SE Start-up Activity**, p. 3 `GENERAL`	**OSP Parent Letter** ■ `GENERAL` **CD Student Edition on CD-ROM** **CD Guided Reading Audio CD** ■ **TR Chapter Starter Transparency*** **VID Brain Food Video Quiz**
Section 1 Astronomy: The Original Science • Identify the units of a calendar. • Describe two early ideas about the structure of the universe. • Describe the contributions of Brahe, Kepler, Galileo, Newton, and Hubble to modern astronomy.	**TE Activity** Naming the Months, p. 4 `GENERAL` **TE Group Activity** Astronomy Debate, p. 5 `ADVANCED` **TE Connection Activity** Real World, p. 7 `BASIC`	**CRF Lesson Plans*** **TR Bellringer Transparency***
PACING • 90 min pp. 8–13 **Section 2 Telescopes** • Compare refracting telescopes with reflecting telescopes. • Explain how the atmosphere limits astronomical observations, and explain how astronomers overcome these limitations. • List the types of electromagnetic radiation that astronomers use to study objects in space.	**TE Activity** Making a Waterdrop Lens, p. 8 `GENERAL` **TE Demonstration** Mystery of the Floating Penny, p. 9 `GENERAL` **TE Demonstration** Electromagnetic Spectrum, p. 11 `BASIC` **SE Connection to Physics** Detecting Infrared Radiation, p. 12 `GENERAL` **TE Group Activity** Reflecting Telescopes, p. 12 `GENERAL` **SE Skills Practice Lab** Through the Looking Glass, p. 22 ◆ `GENERAL` **CRF Datasheet for Chapter Lab*** **LB Whiz-Bang Demonstrations** Refraction Action* ◆ `GENERAL`	**CRF Lesson Plans*** **TR Bellringer Transparency*** **TR** Refracting and Reflecting Telescopes* **TR LINK TO PHYSICAL SCIENCE** How Your Eyes Work; How a Camera Works* **TR** The Electromagnetic Spectrum* **CRF SciLinks Activity*** `GENERAL` **VID Lab Videos for Earth Science**
PACING • 45 min pp. 14–21 **Section 3 Mapping the Stars** • Explain how constellations are used to organize the night sky. • Describe how the altitude of a star is measured. • Explain how the celestial sphere is used to describe the location of objects in the sky. • Compare size and scale in the universe, and explain how red shift indicates that the universe is expanding.	**SE Quick Lab** Using a Sky Map, p. 15 `GENERAL` **CRF Datasheet for Quick Lab*** **TE Group Activity** Classroom Planetarium, p. 15 `ADVANCED` **TE Activity** A Compass on Your Wrist, p. 16 `GENERAL` **TE Connection Activity** Real World, p. 16 `BASIC` **TE Group Activity** Space Science Exploration, p. 16 `GENERAL` **TE Activity** SpaceLog, p. 17 `GENERAL` **TE Connection Activity** Math, p. 18 `GENERAL` **TE Connection Activity** Math, p. 19 `GENERAL` **SE Science in Action** Math, Social Studies, and Language Arts Activities, pp. 28–29 `GENERAL` **SE Skills Practice Lab** The Sun's Yearly Trip Through the Zodiac, p. 164 ◆ `GENERAL` **CRF Datasheet for LabBook*** **LB Inquiry Labs** Constellation Prize* ◆ `ADVANCED` **LB Long-Term Projects & Research Ideas** Celestial Inspiration* `ADVANCED`	**CRF Lesson Plans*** **TR Bellringer Transparency*** **TR** Spring Constellations in the Northern Hemisphere* **TR** Zenith, Altitude, and Horizon* **TR** The Celestial Sphere* **TR** From Home Plate to 10 Million Light Years Away* **SE Internet Activity**, p. 18 `GENERAL`

PACING • 90 min

CHAPTER REVIEW, ASSESSMENT, AND STANDARDIZED TEST PREPARATION

CRF Vocabulary Activity* `GENERAL`
SE Chapter Review, pp. 24–25 `GENERAL`
CRF Chapter Review* ■ `GENERAL`
CRF Chapter Tests A* ■ `GENERAL`, **B*** `ADVANCED`, **C*** `SPECIAL NEEDS`
SE Standardized Test Preparation, pp. 26–27 `GENERAL`
CRF Standardized Test Preparation* `GENERAL`
CRF Performance-Based Assessment* `GENERAL`
OSP Test Generator `GENERAL`
CRF Test Item Listing* `GENERAL`

Online and Technology Resources

go.hrw.com

Visit **go.hrw.com** for a variety of free resources related to this textbook. Enter the keyword **HZ5OBS**.

Holt Online Learning

Students can access interactive problem-solving help and active visual concept development with the *Holt Science and Technology* Online Edition available at **www.hrw.com**.

Guided Reading Audio CD

A direct reading of each chapter using instructional visuals as guideposts. For auditory learners, reluctant readers, and Spanish-speaking students. Available in English and Spanish.

SKILLS DEVELOPMENT RESOURCES	SECTION REVIEW AND ASSESSMENT	STANDARDS CORRELATIONS
SE Pre-Reading Activity, p. 2 `GENERAL` **OSP Science Puzzlers, Twisters & Teasers*** `GENERAL`		National Science Education Standards UCP 3; SAI 1
CRF Directed Reading A* ■ `BASIC`**, B*** `SPECIAL NEEDS` **CRF Vocabulary and Section Summary*** ■ `GENERAL` **SE Reading Strategy** Reading Organizer, p. 4 `GENERAL` **TE Inclusion Strategies,** p. 5 ◆ **CRF Reinforcement Worksheet** Stella Star, Ace Reporter* `BASIC`	**SE Reading Checks,** pp. 5, 6 `GENERAL` **TE Reteaching,** p. 6 `BASIC` **TE Quiz,** p. 6 `GENERAL` **TE Alternative Assessment,** p. 6 `GENERAL` **SE Section Review,*** p. 7 ■ `GENERAL` **CRF Section Quiz*** ■ `GENERAL`	UCP 2; SAI 2; SPSP 5; HNS 1, 2, 3; ES 3a, 3b, 3c
CRF Directed Reading A* ■ `BASIC`**, B*** `SPECIAL NEEDS` **CRF Vocabulary and Section Summary*** ■ `GENERAL` **SE Reading Strategy** Mnemonics, p. 8 `GENERAL` **TE Reading Strategy** Prediction Guide, p. 9 `BASIC` **TE Inclusion Strategies,** p. 10	**SE Reading Checks,** pp. 8, 10, 13 `GENERAL` **TE Reteaching,** p. 12 `BASIC` **TE Quiz,** p. 12 `GENERAL` **TE Alternative Assessment,** p. 12 `GENERAL` **SE Section Review,*** p. 13 ■ `GENERAL` **CRF Section Quiz*** ■ `GENERAL`	ST 2; SPSP 5; HNS 1, 3; *Chapter Lab:* SAI 1; ST 1
CRF Directed Reading A* ■ `BASIC`**, B*** `SPECIAL NEEDS` **CRF Vocabulary and Section Summary*** ■ `GENERAL` **SE Reading Strategy** Paired Summarizing, p. 14 `GENERAL` **CRF Critical Thinking** Through the Eyes of a Telescope* `ADVANCED`	**SE Reading Checks,** pp. 15, 17, 18, 20 `GENERAL` **TE Homework,** p. 17 `GENERAL` **TE Reteaching,** p. 20 `BASIC` **TE Quiz,** p. 20 `GENERAL` **TE Alternative Assessment,** p. 20 `GENERAL` **TE Homework,** p. 20 `GENERAL` **SE Section Review,*** p. 21 ■ `GENERAL` **CRF Section Quiz*** ■ `GENERAL`	UCP 3; SAI 1; ST 2; SPSP 5; HNS 1; *LabBook:* UCP 2; SAI 1; HNS 1; ES 3b

One-Stop Planner® CD-ROM

This convenient CD-ROM includes:
- Lab Materials QuickList Software
- Holt Calendar Planner
- Customizable Lesson Plans
- Printable Worksheets
- ExamView® Test Generator

CNN student News™

cnnstudentnews.com

Find the latest news, lesson plans, and activities related to important scientific events.

SCLINKS.
NSTA

www.scilinks.org

Maintained by the **National Science Teachers Association.** See Chapter Enrichment pages for a complete list of topics.

Current Science®

Check out *Current Science* articles and activities by visiting the HRW Web site at **go.hrw.com.** Just type in the keyword **HZ5CS18T.**

Classroom Videos
- **Lab Videos** demonstrate the chapter lab.
- **Brain Food Video Quizzes** help students review the chapter material.

Visual Resources

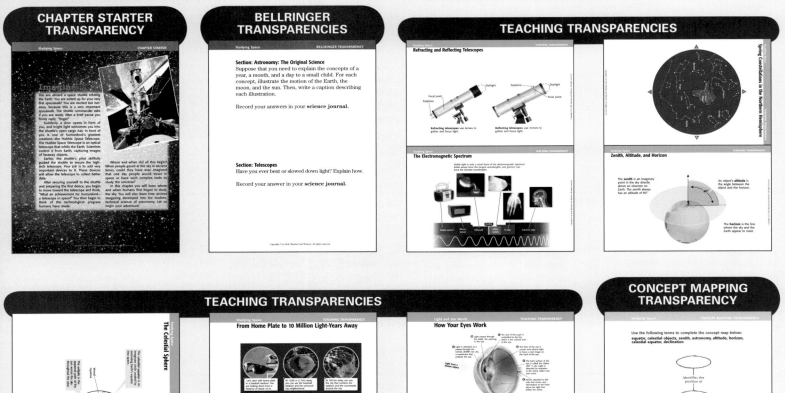

CHAPTER STARTER TRANSPARENCY

BELLRINGER TRANSPARENCIES

TEACHING TRANSPARENCIES

TEACHING TRANSPARENCIES

CONCEPT MAPPING TRANSPARENCY

Planning Resources

LESSON PLANS

Lesson Plan SAMPLE

Section: Waves

Pacing
Regular Schedule: with lab(s):2 days without lab(s):1 days
Block Schedule: with lab(s):1 1/2 days without lab(s):1/2 days

Objectives
1. Relate the seven properties of life to a living organism.
2. Describe seven themes that can help you to organize what you learn about biology.
3. Identify the tiny structures that make up all living organisms.
4. Differentiate between reproduction and heredity and between metabolism and homeostasis.

National Science Education Standards Covered
LSInter6: Cells have particular structures that underlie their functions.
LSMat1: Most cell functions involve chemical reactions.
LSBeh1: Cells store and use information to guide their functions.
UCP1: Cell functions are regulated.
SI1: Cells can differentiate and form complete multicellular organisms.
PS1: Species evolve over time.
ESS1: The great diversity of organisms is the result of more than 3.5 billion years of evolution.
ESS2: Natural selection and its evolutionary consequences provide a scientific explanation for the fossil record of ancient life forms as well as for the striking molecular similarities observed among the diverse species of living organisms.
ST1: The millions of different species of plants, animals, and microorganisms that live on Earth today are related by descent from common ancestors.
ST2: The energy for life primarily comes from the sun.
SPSP1: The complexity and organization of organisms accommodates the need for obtaining, transforming, transporting, releasing, and eliminating the matter and energy used to sustain the organism.
SPSP6: As matter and energy flows through different levels of organization of living systems—cells, organs, communities—and between living systems and the physical environment, chemical elements are recombined in different ways.
HNS1: Organisms have behavioral responses to internal changes and to external stimuli.

PARENT LETTER

Dear Parent, SAMPLE

Your son's or daughter's science class will soon begin exploring the chapter entitled "The World of Physical Science." In this chapter, students will learn about how the scientific method applies to the world of physical science and the role of physical science in the world. By the end of the chapter, students should demonstrate a clear understanding of the chapter's main ideas and be able to discuss the following topics:

1. physical science as the study of energy and matter (Section 1)
2. the role of physical science in the world around them (Section 1)
3. careers that rely on physical science (Section 1)
4. the steps used in the scientific method (Section 1)
5. examples of technology (Section 2)
6. how the scientific method is used to answer questions and solve problems (Section 2)
7. how our knowledge of science changes over time (Section 2)
8. how models represent real objects or systems (Section 3)
9. examples of different ways models are used in science (Section 3)
10. the importance of the International System of Units (Section 4)
11. the appropriate units to use for particular measurements (Section 4)
12. how area and density are derived quantities (Section 4)

Questions to Ask Along the Way

You can help your son or daughter learn about these topics by asking interesting questions such as the following:

- What are some surprising careers that use physical science?
- What is a characteristic of a good hypothesis?
- When is it a good idea to use a model?
- Why do Americans measure things in terms of inches and yards and meters?

ALSO IN SPANISH

TEST ITEM LISTING

TEST ITEM LISTING
The World of Earth Science SAMPLE

MULTIPLE CHOICE

1. A limitation of models is that
   a. they are large enough to see.
   b. they do not exactly like the things that they model.
   c. they are smaller than the things that they model.
   d. they model unfamiliar things.
   Answer: B   Difficulty: 1   Section: 1   Objective: 2

2. The length 10 m is equal to
   a. 100 cm.   c. 10,000 mm.
   b. 1,000 cm.   d. Both (a) and (c)
   Answer: B   Difficulty: 1   Section: 3   Objective: 2

3. To be valid, a hypothesis must be
   a. testable.   c. made into a law.
   b. supported by evidence.   d. Both (a) and (b)
   Answer: D   Difficulty: 1   Section: 3   Objective: 2 1

4. The statement "Sheila has a stain on her shirt" is an example of a(n)
   a. law.   c. observation.
   b. hypothesis.   d. prediction.
   Answer: C   Difficulty: 1   Section: 3   Objective: 2

5. A hypothesis is often developed out of
   a. observations.   c. laws.
   b. experiments.   d. Both (a) and (b)
   Answer: D   Difficulty: 1   Section: 3   Objective: 2

6. How many millimeters are in 3.5 kL?
   a. 3,500 mL.   c. 3,500,000 mL.
   b. 0.0035 mL.   d. 35,000 mL.
   Answer: B   Difficulty: 1   Section: 3   Objective: 2

7. A map of Seattle is an example of a
   a. law.   c. model.
   b. theory.   d. unit.
   Answer: C   Difficulty: 1   Section: 3   Objective: 2

8. A lab has the safety icons shown below. These icons mean that you should wear
   a. only safety goggles.   c. safety goggles and a lab apron.
   b. only a lab apron.   d. safety goggles, a lab apron, and gloves.
   Answer: D   Difficulty: 1   Section: 1   Objective: 2

9. The law of conservation of mass says the tot al mass before a chemical change is
   a. more than the total mass after the change.
   b. less than the total mass after the change.
   c. the same as the total mass after the change.
   d. not the same as the total mass after the change.
   Answer: C   Difficulty: 1   Section: 1   Objective: 2

10. In which of the following areas might you find a geochemist at work?
    a. studying the chemistry of rocks   c. studying fishes
    b. studying forestry   d. studying the atmosphere
    Answer: B   Difficulty: 1   Section: 2   Objective: 2

One-Stop Planner® CD-ROM

This CD-ROM includes all of the resources shown here and the following time-saving tools:

- *Lab Materials QuickList Software*
- *Customizable lesson plans*
- *Holt Calendar Planner*
- *The powerful ExamView® Test Generator*

Meeting Individual Needs

DIRECTED READING A

Skills Worksheet
Directed Reading A — SAMPLE

Section:
THAT'S SCIENCE!
1. How did James Czarnowski get his idea for the penguin... Explain.

BASIC

ALSO IN SPANISH

DIRECTED READING B

Skills Worksheet
Directed Reading B — SAMPLE

Section:
THAT'S SCIENCE!
1. How did James Czarnowski get his idea for the penguin boat, Proteus? Explain.

2. What is unusual about the way that Proteus moves through the water?

SPECIAL NEEDS — PHYSICAL SCIENCE

VOCABULARY ACTIVITY

Activity
Vocabulary Activity — SAMPLE

Getting the Dirt on the Soil

GENERAL

VOCABULARY AND SECTION SUMMARY

Skills Worksheet
Vocabulary & Notes — SAMPLE

Section:
VOCABULARY
In your own words, write a definition of the following term in the space provided.
1. scientific method

2. technology

GENERAL

ALSO IN SPANISH

REINFORCEMENT

Skills Worksheet
Reinforcement — SAMPLE

The Plane Truth

BASIC

CRITICAL THINKING

Skills Worksheet
Critical Thinking — SAMPLE

A Solar Solution

ADVANCED

SCILINKS ACTIVITY

Activity
SciLinks Activity — SAMPLE

MARINE ECOSYSTEMS

GENERAL

SCIENCE PUZZLERS, TWISTERS & TEASERS

CHAPTER
18 SCIENCE PUZZLERS, TWISTERS & TEASERS
Studying Space

Sky Spy
1. Hidden in the puzzle below are six words associated with finding stars in the night sky.

GENERAL

Labs and Activities

LONG-TERM PROJECTS & RESEARCH IDEAS

PROJECT
46 STUDENT WORKSHEET — DESIGN YOUR OWN
Celestial Inspiration

INTERNET KEYWORDS
Egyptian astronomy
Pyramids of Khufu
Egyptian calendar

Ancient Investigation
1. Find out more about ancient Egyptian astronomy.

Other Research Ideas

Long-Term Project Idea

ADVANCED

WHIZ-BANG DEMONSTRATIONS

DEMO
31 TEACHER-LED DEMONSTRATION — DISCOVERY LAB
Refraction Action

Purpose
Students learn why the sun can be seen before it rises above the horizon.

Time Required
10–15 minutes

Lab Ratings

MATERIALS

HELPFUL HINT

What to Do

GENERAL

INQUIRY LABS

LAB
13 TEACHER'S PREPARATORY GUIDE
Constellation Prize

Purpose
Students locate constellations and use an astrolabe and a compass to track the movement of constellations.

Time Required
Two 45-minute class periods and one evening at home to make nighttime observations

Lab Ratings

Advance Preparation

Safety Information
None

Teaching Strategies

Evaluation Strategies

ADVANCED

DATASHEETS FOR QUICKLABS

TEACHER RESOURCE PAGE
Quick Lab
Reaction to Stress — DATASHEET FOR QUICK LAB SAMPLE

Background

DATASHEETS FOR CHAPTER LABS

TEACHER RESOURCE PAGE
Skills Practice Lab
Using Scientific Methods — DATASHEET FOR CHAPTER LAB SAMPLE

Teacher's Notes
TIME REQUIRED
One 45-minute class period.

DATASHEETS FOR LABBOOK

TEACHER RESOURCE PAGE
Skills Practice Lab
Does It All Add Up? — DATASHEET FOR LABBOOK LAB SAMPLE

Teacher's Notes
TIME REQUIRED
One 45-minute class period.

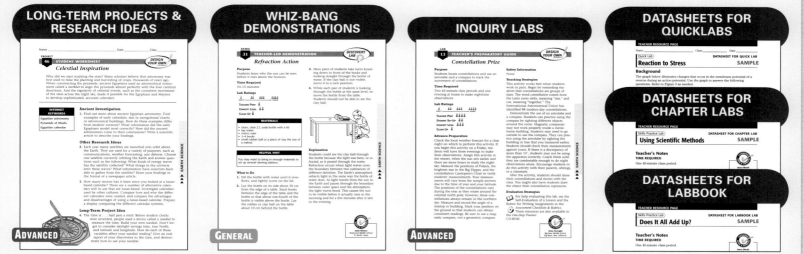

Review and Assessments

SECTION QUIZ

Assessment
Section Quiz — SAMPLE

Section:
In the space provided, write the letter of the description that best matches the term or phrase.
_____ 1. building molecules that can be used as an energy source or breaking down molecules in which energy is stored
_____ 2. the process by which light energy is converted to chemical energy
_____ 3. an organism that uses sunlight or inorganic substances to make organic compounds

GENERAL

ALSO IN SPANISH

SECTION REVIEW

Skills Worksheet
Section Review — SAMPLE

Section:
KEY TERMS
1. What do paleontologist study?

2. How does a trace fossil differ from petrified wood?

GENERAL

ALSO IN SPANISH

CHAPTER REVIEW

Skills Worksheet
Chapter Review — SAMPLE

USING VOCABULARY
1. Define biome in your own words.

2. Describe the characteristics of a savanna and a desert.

GENERAL

ALSO IN SPANISH

CHAPTER TEST A

Assessment
Chapter Test A — SAMPLE

MULTIPLE CHOICE
In the space provided, write the letter of the term or phrase that best completes each statement or best answers each question.
_____ 1. Surface currents are formed by
a. the moon's gravity.   c. wind.
b. the sun's gravity.   d. increased water density.
_____ 2. When waves come near the shore,
a. they speed up.   c. their wavelength increases.
b. they maintain their speed.   d. their wave height increases.

ALSO IN SPANISH

CHAPTER TEST B

Assessment
Chapter Test B — SAMPLE

MULTIPLE CHOICE
In the space provided, write the letter of the term or phrase that best completes each statement or best answers each question.
_____ 1. Surface currents are formed by
a. the moon's gravity.   c. wind.
b. the sun's gravity.   d. increased water density.
When waves come near the shore,
a. they speed up.   c. their wavelength increases.
b. they maintain their speed.   d. their wave height increases.

ADVANCED

CHAPTER TEST C

Assessment
Chapter Test C — SAMPLE

MULTIPLE CHOICE
In the space provided, write the letter of the term or phrase that best completes each statement or best answers each question.
_____ 1. Surface currents are formed by
a. the moon's gravity.   c. wind.
b. the sun's gravity.   d. increased water density.
_____ 2. When waves come near the shore,
a. they speed up.   c. their wavelength increases.
b. they maintain their speed.   d. their wave height increases.

SPECIAL NEEDS

STANDARDIZED TEST PREPARATION

Assessment
Standardized Test Preparation — SAMPLE

READING
Read the passages below. Then, read each question that follows the passage. Decide which is the best answer to each question.

GENERAL

PERFORMANCE-BASED ASSESSMENT

Assessment
Performanced-Based Assessment — SKILL BUILDER SAMPLE

OBJECTIVE
Determine which factors cause some sugar shapes to break down faster than others.

KNOW THE SCORE!

Using Scientific Methods

MATERIALS AND EQUIPMENT

GENERAL

This Chapter Enrichment provides relevant and interesting information to expand and enhance your presentation of the chapter material.

Section 1

Astronomy: The Original Science

Mayan Calendars

● The Maya of Central America used two calendars—a ceremonial calendar of 260 days and an astronomical calendar of 365 days, divided into 18 months of 20 days each. The Maya created an additional 5-day month for religious ceremony. These interlocking calendars enabled the Maya to predict when eclipses would occur and when Venus would rise.

The Herschel Family

● By the time William Herschel was 36, he seemed destined for a career as a musician. He was a gifted organist and conducted an orchestra in Bath, England. But Herschel had a great interest in stargazing, and he began to devote more of his time to astronomy. Finding the available telescopes to be inadequate, he set up his own forge and mirror-grinding shop to make large-mirror telescopes. His telescopes were of extremely high quality and power and surpassed even those used at the Royal Observatory, in Greenwich at that time.

● William Herschel was joined by his sister Caroline, and soon the two were conducting systematic telescopic surveys of the skies. In addition to discovering Uranus, William Herschel developed new observational techniques; made important discoveries about nebulas, star clusters, and double stars; and contributed immeasurably to the cataloging of stars.

Is That a Fact!

◆ Caroline Herschel made many significant discoveries and is considered to be the first modern female astronomer. She discovered eight comets and three nebulas. In 1828, Britain's Astronomical Society awarded her a gold medal for her collaborations with her brother.

Section 2

Telescopes

Linking Radio Telescopes

● To improve image resolution and detect very faint emissions, international teams of astronomers sometimes link telescopes from opposite sides of the world. Recently, scientists have taken this technique a step further with the Very Long Baseline Interferometry (VLBI) Space Observatory Program. In this program, astronomers from around the world link their ground-based radio telescopes with a radio telescope that is orbiting the Earth. Each time two telescopes link, they function as a single telescope that has a width 2.5 times the diameter of the Earth!

Charge Coupled Devices

● Modern professional astronomers who use optical telescopes rarely look through their telescopes. Instead, they view images on computer monitors. Most modern optical telescopes are equipped with a semiconductor device known as a *charge coupled device* (CCD), which converts the individual light particles (photons) from celestial objects into electrons. The electrons are detected, counted, and rendered as an image on a computer screen.

Gamma-Ray Telescopes

- Gamma-ray telescopes are designed to detect gamma rays, which behave more like "bullets of energy" than waves. To detect this type of radiation, a gamma-ray telescope is equipped with a particle detector that collects data resulting from the collision of a gamma-ray photon and an atom. The data can be used to determine both the energy of the ray and the direction of the ray's source. The gamma-ray telescope aboard the *Compton Gamma-ray Observatory* (launched in 1991) has detected objects known as *gamma-ray bursters,* which are brilliant, brief flashes of tremendous energy that last no more than a few minutes and then disappear.

Is That a Fact!

◆ The *Hubble Space Telescope*'s resolution and sensitivity are so acute that the telescope could detect the light from a firefly 16,000 km away!

Section 3

Mapping the Stars

Sky Maps

- Some star maps show what the night sky looks like during a particular season in the Northern or Southern Hemisphere. These star maps are circular, and their edges represent the horizon. They are labeled with cardinal directions, and a "+" represents the zenith. Stars are represented by dots—the larger a dot is, the brighter the star is. To use the map, a stargazer should hold the map overhead and orient it according to the cardinal directions.

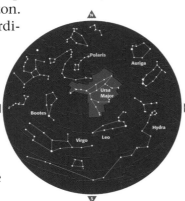

The Messier Catalog

- The *Messier Catalog* (1784) is one of the most well known astronomical catalogs. It was compiled by the French astronomer Charles Messier, whom King Louis XV dubbed the "comet ferret." Through the small telescopes available to Messier, comets looked like indistinct blotches. Many of Messier's blurred blotches weren't comets, however, but star clusters, nebulae, and galaxies. After he realized his mistake, Messier began to compile a catalog of these "noncomets" to spare other comet seekers the frustration that he experienced.

Is That a Fact!

◆ Messier's catalog of more than 100 star clusters, nebulae, and galaxies is widely used today by amateur astronomers with small telescopes. The catalog numbers are still used by professional astronomers. The objects on Messier's list, such as the Crab nebula and the Pleiades, retain their Messier designations, M1 and M45, respectively.

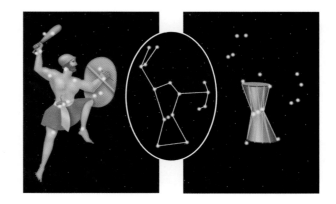

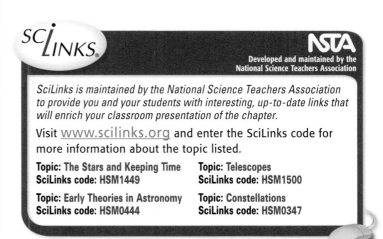

SciLINKS®

NSTA
Developed and maintained by the
National Science Teachers Association

SciLinks is maintained by the National Science Teachers Association to provide you and your students with interesting, up-to-date links that will enrich your classroom presentation of the chapter.

Visit www.scilinks.org and enter the SciLinks code for more information about the topic listed.

Topic: The Stars and Keeping Time
SciLinks code: HSM1449

Topic: Telescopes
SciLinks code: HSM1500

Topic: Early Theories in Astronomy
SciLinks code: HSM0444

Topic: Constellations
SciLinks code: HSM0347

Overview

This chapter introduces some fundamental concepts in astronomy. Students learn about the early history of astronomy and early theories about the structure of the universe. The chapter then introduces optical and non-optical telescopes. The chapter also discusses how the location of stars and other objects in the sky is described. The chapter concludes with a discussion of distance and scale in the universe.

Assessing Prior Knowledge

Students should be familiar with the following topics:

- the solar system is composed of planets and one star
- Newton's law of universal gravitation

Identifying Misconceptions

Students may confuse the apparent movement of stars and other objects in the sky with the actual movement of these objects. As you discuss this chapter, it may be useful to compare apparent and actual movement with phenomena that students are familiar with. For example, compare astronomic observations with observations that students might make from a moving car or a Ferris wheel.

1

Studying Space

About the  PHOTO

This time-exposure photograph was taken at an observatory located high in the mountains of Chile. As the night passed, the photograph recorded the stars as they circled the southern celestial pole. Just as Earth's rotation causes the sun to appear to move across the sky during the day, Earth's rotation also causes the stars to appear to move across the night sky.

PRE-READING ACTIVITY

 FOLDNOTES **Three-Panel Flip Chart**
Before you read the chapter, create the FoldNote entitled "Three-Panel Flip Chart" described in the **Study Skills** section of the Appendix. Label the flaps of the three-panel flip chart with "Astronomy," "Telescopes," and "Mapping the stars." As you read the chapter, write information you learn about each category under the appropriate flap.

Standards Correlations

National Science Education Standards

The following codes indicate the National Science Education Standards that correlate to this chapter. The full text of the standards is at the front of the book.

Chapter Opener
UCP 3; SAI 1

Section 1 Astronomy: The Original Science
UCP 2; SAI 2; SPSP 5; HNS 1, 2, 3; ES 3a, 3b, 3c

Section 2 Telescopes
ST 2; SPSP 5; HNS 1, 3

Section 3 Mapping the Stars
UCP 3; SAI 1; ST 2; SPSP 5; HNS 1; *LabBook:* UCP 2; SAI 1; HNS 1; ES 3b

Chapter Lab
SAI 1; ST 1

Chapter Review
SPSP 5; HNS 1; ES 3b, 3c

Science in Action
SAI 1; SPSP 5; HNS 1

START-UP ACTIVITY

MATERIALS

FOR EACH STUDENT
- paper clip
- protractor
- straw, soda
- thread, 15 cm

Answers

1. Answers may vary. As a student moves closer to an object, the object's altitude should increase.

2. Answers may vary. A similar method could be used to measure the altitude of a star. One advantage of an astrolabe is that it allows an observer to precisely measure the altitude of an object. One disadvantage of an astrolabe is that the measurement depends on the location of the observer and the time that the measurement is taken.

START-UP ACTIVITY

Making an Astrolabe

In this activity, you will make an astronomical device called an *astrolabe* (AS troh LAYB). Ancient astronomers used astrolabes to measure the location of stars in the sky. You will use the astrolabe to measure the angle, or altitude, of an object.

Procedure

1. Tie one end of a **piece of thread** that is 15 cm long to the center of the straight edge of a **protractor.** Attach a **paper clip** to the other end of the string.

2. Tape a **soda straw** lengthwise along the straight edge of the protractor. Your astrolabe is complete!

3. Go outside, and hold the astrolabe in front of you.

4. Look through the straw at a distant object, such as a treetop. The curve of the astrolabe should point toward the ground.

5. Hold the astrolabe still, and carefully pinch the string between your thumb and the protractor. Count the number of degrees between the string and the 90° marker on the protractor. This angle is the altitude of the object.

Analysis

1. What is the altitude of the object? How would the altitude change if you moved closer to the object?

2. Explain how you would use an astrolabe to find the altitude of a star. What are the advantages and disadvantages of this method of measurement?

Chapter Starter Transparency
Use this transparency to help students begin thinking about the significance of space-based observatories.

CHAPTER RESOURCES

Technology

Transparencies
- Chapter Starter Transparency

READING SKILLS

Student Edition on CD-ROM

Guided Reading Audio CD
- English or Spanish

Classroom Videos
- Brain Food Video Quiz

Workbooks

Science Puzzlers, Twisters & Teasers
- Studying Space GENERAL

Overview

This section describes how the units of the calendar are based on the movements of bodies in space. Students will also explore the development of the science of astronomy.

Bellringer

Have students suppose that they need to explain the concepts of a year, a month, and a day to a small child. For each concept, have students illustrate the motion of the Earth and the moon. Students should write a caption describing each illustration.

ACTiViTY ———— GENERAL

Naming the Months The lunar calendar of the Natchez peoples of the Mississippi River Valley reflected the seasonal rhythms of their culture. The names of the months in their calendar—strawberry month, peach month, maize month, turkey month, and chestnut month—reflect the hunter-gatherer nature of their society. Ask students if they can identify the time period that corresponds to each Natchez month. **LS Verbal**

SECTION
1

Astronomy: The Original Science

Imagine that it is 5,000 years ago. Clocks and modern calendars have not been invented. How would you tell the time or know what day it is? One way to tell the time is to study the movement of stars, planets, and the moon.

People in ancient cultures used the seasonal cycles of the stars, planets, and the moon to mark the passage of time. For example, by observing these yearly cycles, early farmers learned the best times of year to plant and harvest various crops. Studying the movement of objects in the sky was so important to ancient people that they built observatories, such as the one shown in **Figure 1.** Over time, the study of the night sky became the science of astronomy. **Astronomy** is the study of the universe. Although ancient cultures did not fully understand how the planets, moons, and stars move in relation to each other, their observations led to the first calendars.

Our Modern Calendar

The years, months, and days of our modern calendar are based on the observation of bodies in our solar system. A **year** is the time required for the Earth to orbit once around the sun. A **month** is roughly the amount of time required for the moon to orbit once around the Earth. (The word *month* comes from the word *moon.*) A **day** is the time required for the Earth to rotate once on its axis.

astronomy the study of the universe

year the time required for the Earth to orbit once around the sun

Figure 1 *This building is located at Chichén Itzá in the Yucatán, Mexico. It is thought to be an ancient Mayan observatory.*

SCIENCE HUMOR

Q: What did Copernicus say about Ptolemy's theory of an Earth-centered universe?

A: Ptolemy another one!

Who's Who of Early Astronomy

Astronomical observations have given us much more than the modern calendar that we use. The careful work of early astronomers helped people understand their place in the universe. The earliest astronomers had no history to learn from. Almost everything they knew about the universe came from what they could discover with their eyes and minds. Not surprisingly, most early astronomers thought that the universe consisted of the sun, the moon, and the planets. They thought that the stars were at the edge of the universe. Claudius Ptolemy (KLAW dee uhs TAHL uh mee) and Nicolaus Copernicus (NIK uh LAY uhs koh PUHR ni kuhs) were two early scientists who influenced the way that people thought about the structure of the universe.

month a division of the year that is based on the orbit of the moon around the Earth

day the time required for Earth to rotate once on its axis

Ptolemy: An Earth-Centered Universe

In 140 CE, Ptolemy, a Greek astronomer, wrote a book that combined all of the ancient knowledge of astronomy that he could find. He expanded ancient theories with careful mathematical calculations in what was called the *Ptolemaic theory.* Ptolemy thought that the Earth was at the center of the universe and that the other planets and the sun revolved around the Earth. Although the Ptolemaic theory, shown in **Figure 2,** was incorrect, it predicted the motions of the planets better than any other theory at the time did. For over 1,500 years in Europe, the Ptolemaic theory was the most popular theory for the structure of the universe.

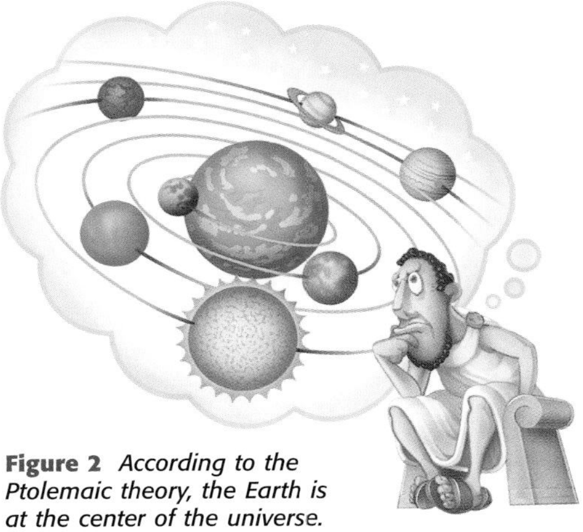

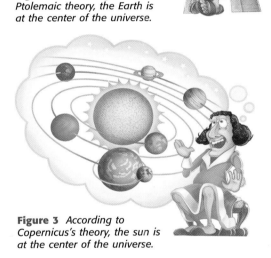

Figure 2 *According to the Ptolemaic theory, the Earth is at the center of the universe.*

Copernicus: A Sun-Centered Universe

In 1543, a Polish astronomer named Copernicus published a new theory that would eventually revolutionize astronomy. According to his theory, which is shown in **Figure 3,** the sun is at the center of the universe, and all of the planets—including the Earth—orbit the sun. Although Copernicus correctly thought that the planets orbit the sun, his theory did not replace the Ptolemaic theory immediately. When Copernicus's theory was accepted, major changes in science and society called the *Copernican revolution* took place.

Figure 3 *According to Copernicus's theory, the sun is at the center of the universe.*

Reading Check What was Copernicus's theory? *(See the Appendix for answers to Reading Checks.)*

Teach

INCLUSION Strategies

- ***Attention Deficit Disorder***
- ***Learning Disabled***
- ***Hearing Impaired***

Have students play a card game that will help them organize the astronomers described in this section. Organize students in groups of three to five students. Give students a deck of cards that has the names of Ptolemy, Copernicus, Brahe, Kepler, Galileo, Newton, and Hubble on separate cards. Give students a second deck of cards that is a different color. Each card in this deck should have five facts about one of the astronomers in the first deck. Ask each group to place the name cards face up in a row. The fact cards should be placed in a pile face down. Then, in turn, each team member should draw a card, read it aloud, and attempt to place it under the correct name. If a student is unable to match the facts with the correct name, his or her team members may help. The game is complete when all fact cards are matched with the correct name cards. At the end of class, hand out a paper copy of the facts and names. **LS Verbal/Interpersonal**

Answer to Reading Check

Copernicus believed in a sun-centered universe.

Group ACTIVITY—ADVANCED

Astronomy Debate Have teams role-play a debate between Ptolemy, Copernicus, Brahe, Kepler, and Newton. Suggest that students prepare for the debate by researching how each scientist formed his theories. For example, Copernicus used mathematics to analyze the observations made by earlier astronomers, and he made few personal observations. Ptolemy, on the other hand, was a keen observer, but he could reconcile his theory with his observations only by developing a complex but inaccurate system to explain the apparent motion of the planets. Encourage students to research the lives of these scientists and base their positions on the scientists' ideas. When students are role-playing, remind them that each scientist based his theories on his own observations and the theories of scientists who came before him. **LS Logical** Co-op Learning

Close

Reteaching — BASIC

Astronomy Review Have student work in groups to choose an astronomer described in this section and spend ten minutes summarizing the contributions of the astronomer chosen. Have groups present their summaries to the class. **LS** Interpersonal

Quiz — GENERAL

1. What was Copernicus's theory about the structure of the universe? (Copernicus argued that the sun is at the center of the universe and that all planets revolve around the sun.)

2. How did Newton's theories explain why planets orbit the sun and why moons orbit planets? (Newton explained that the force of gravity keeps the planets and moons in orbit.)

Alternative Assessment — GENERAL

Timeline Have students make an illustrated timeline that describes 10 events important to the development of modern astronomy. Encourage students to use reference materials to research timeline entries. **LS** Visual

Answer to Reading Check

Newton's law of gravity helped explain why the planets orbit the sun and moons orbit planets.

Figure 4 *Brahe (upper right) used a mural quadrant, which is a large quarter-circle on a wall, to measure the positions of stars and planets.*

Tycho Brahe: A Wealth of Data

In the late-1500s, Danish astronomer Tycho Brahe (TIE koh BRAW uh) used several large tools, including the one shown in **Figure 4,** to make the most detailed astronomical observations that had been recorded so far. Brahe favored a theory of an Earth-centered universe that was different from the Ptolemaic theory. Brahe thought that the sun and the moon revolved around the Earth and that the other planets revolved around the sun. While his theory was not correct, Brahe recorded very precise observations of the planets and stars that helped future astronomers.

Johannes Kepler: Laws of Planetary Motion

After Brahe died, his assistant, Johannes Kepler, continued Brahe's work. Kepler did not agree with Brahe's theory, but he recognized how valuable Brahe's data were. In 1609, after analyzing the data, Kepler announced that all of the planets revolve around the sun in elliptical orbits and that the sun is not in the exact center of the orbits. Kepler also stated three laws of planetary motion. These laws are still used today.

Galileo: Turning a Telescope to the Sky

In 1609, Galileo Galilei became one of the first people to use a telescope to observe objects in space. Galileo discovered craters and mountains on the Earth's moon, four of Jupiter's moons, sunspots on the sun, and the phases of Venus. These discoveries showed that the planets are not "wandering stars" but are physical bodies like the Earth.

Isaac Newton: The Laws of Gravity

In 1687, a scientist named Sir Isaac Newton showed that all objects in the universe attract each other through gravitational force. The force of gravity depends on the mass of the objects and the distance between them. Newton's law of gravity explained why all of the planets orbit the most massive object in the solar system—the sun. Thus, Newton helped explain the observations of the scientists who came before him.

✓ Reading Check How did the work of Isaac Newton help explain the observations of earlier scientists?

SCIENTISTS AT ODDS

Tycho Brahe Tycho Brahe was eccentric and contrary. As a young man, he insulted a fellow student and was challenged to a duel. During the duel, part of his nose was sliced off, and for the rest of his life, he wore a metal nose prosthesis. Brahe was also a notoriously bad landlord. He cheated and abused the peasants who worked for him. It's not surprising that given his bad temperament, Brahe withheld vital information from his assistant, Johannes Kepler. Although Brahe's family fought Kepler for years, Kepler finally gained access to Brahe's observations after Brahe's death. These observations helped prove that the planets revolve around the sun in elliptical orbits.

The invention of the telescope and the discovery of gravity were two milestones in the development of modern astronomy. In the 200 years following Newton's discoveries, scientists made many discoveries about our solar system. But they did not learn that our galaxy has cosmic neighbors until the 1920s.

Edwin Hubble: Beyond the Edge of the Milky Way

Before the 1920s, many astronomers thought that our galaxy, the Milky Way, included every object in space. In 1924, Edwin Hubble proved that other galaxies existed beyond the edge of the Milky Way. His data confirmed the beliefs of many astronomers that the universe is much larger than our galaxy. Today, larger and better telescopes on the Earth and in space, new models of the universe, and spacecraft help astronomers study space. Computers, shown in **Figure 5,** help process data and control the movement of telescopes. These tools have helped answer many questions about the universe. Yet new technology has presented questions that were unthinkable even 10 years ago.

Figure 5 *Computers are used to control telescopes and process large amounts of data.*

SECTION Review

Summary

- Astronomy, the study of the universe, is one of the oldest sciences.
- The units of the modern calendar—days, months, and years—are based on observations of objects in space.
- Ptolemaic theory states that the Earth is at the center of the universe.
- Copernican theory states that the sun is at the center of the universe.
- Modern astronomy has shown that there are billions of galaxies.

Using Key Terms

1. Use each of the following terms in a separate sentence: *year, day, month,* and *astronomy.*

Understanding Key Ideas

2. What happens in 1 year?
   a. The moon completes one orbit around the Earth.
   b. The sun travels once around the Earth.
   c. The Earth revolves once on its axis.
   d. The Earth completes one orbit around the sun.

3. What is the difference between the Ptolemaic and Copernican theories? Who was more accurate: Ptolemy or Copernicus?

4. What contributions did Brahe and Kepler make to astronomy?

5. What contributions did Galileo, Newton, and Hubble make to astronomy?

Math Skills

6. How many times did Earth orbit the sun between 140 CE, when Ptolemy introduced his theories, and 1543, when Copernicus introduced his theories?

Critical Thinking

7. **Analyzing Relationships** What advantage did Galileo have over earlier astronomers?

8. **Making Inferences** Why is astronomy such an old science?

SCILINKS®

Developed and maintained by the National Science Teachers Association

For a variety of links related to this chapter, go to www.scilinks.org
Topic: The Stars and Keeping Time; Early Theories in Astronomy
SciLinks code: HSM1449; HSM0444

Focus

Overview

In this section, students will learn how reflecting and refracting telescopes work. The section discusses the electromagnetic spectrum and the use of non-optical telescopes to detect invisible radiation.

🔔 Bellringer

Ask students to write a brief answer to the following questions: "Have you ever bent or slowed down light? How?" (Students may mention wearing glasses or looking through a microscope.)

Motivate

ACTIVITY ———— GENERAL

Making a Waterdrop Lens Give pairs of students a 6 cm × 6 cm piece of plastic wrap, and have each pair place the plastic wrap over some print on a newspaper. Put a drop of water on each piece of plastic. Have students note what shape the drop is and how the drop magnifies the newsprint. Tell students that the rounded, or *convex*, water-drop is a very simple lens that is similar in principle to the kinds of lenses used in some telescopes. **LS Visual/Kinesthetic**

SECTION

2

READING WARM-UP

Objectives

- Compare refracting telescopes with reflecting telescopes.
- Explain how the atmosphere limits astronomical observations, and explain how astronomers overcome these limitations.
- List the types of electromagnetic radiation that astronomers use to study objects in space.

Terms to Learn

telescope
refracting telescope
reflecting telescope
electromagnetic spectrum

READING STRATEGY

Mnemonics As you read this section, create a mnemonic device to help you remember the characteristics of each type of radiation in the electromagnetic spectrum.

Telescopes

What color are Saturn's rings? What does the surface of the moon look like? To answer these questions, you could use a device called a telescope.

For professional astronomers and amateur stargazers, the telescope is the standard tool for observing the sky. A **telescope** is an instrument that gathers electromagnetic radiation from objects in space and concentrates it for better observation.

Optical Telescopes

Optical telescopes, which are the most common type of telescope, are used to study visible light from objects in the universe. Without using an optical telescope, you can see only about 3,000 stars in the night sky. Using an optical telescope, however, you can see millions of stars and other objects.

An optical telescope collects visible light and focuses it to a focal point for closer observation. A *focal point* is the point where the rays of light that pass through a lens or that reflect from a mirror converge. The simplest optical telescope has two lenses. One lens, called the *objective lens,* collects light and forms an image at the back of the telescope. The bigger the objective lens is, the more light the telescope can gather. The second lens is located in the eyepiece of the telescope. This lens magnifies the image produced by the objective lens. **Figure 1** shows how much you can see by using an optical telescope.

✓ **Reading Check** What are the functions of the two lenses in an optical telescope? (*See the Appendix for answers to Reading Checks.*)

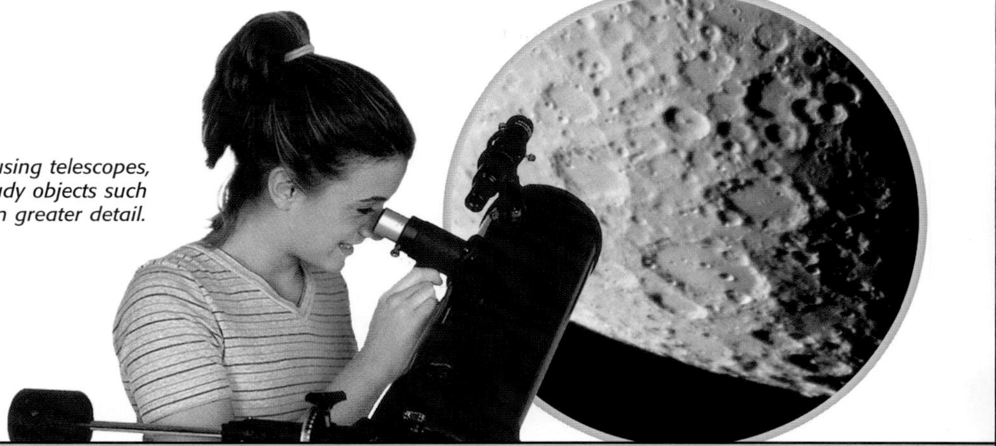

Figure 1 *By using telescopes, people can study objects such as the moon in greater detail.*

Answer to Reading Check

The objective lens collects light and forms an image at the back of the telescope. The eyepiece magnifies the image produced by the objective lens.

Figure 2 Refracting and Reflecting Telescopes

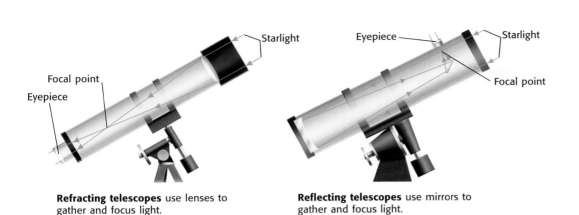

Starlight
Focal point
Eyepiece

Eyepiece
Starlight
Focal point

Refracting telescopes use lenses to gather and focus light.

Reflecting telescopes use mirrors to gather and focus light.

Refracting Telescopes

Telescopes that use lenses to gather and focus light are called **refracting telescopes.** As shown in **Figure 2,** a refracting telescope has an objective lens that bends light that passes through it and focuses the light to be magnified by an eyepiece. Refracting telescopes have two disadvantages. First, lenses focus different colors of light at slightly different distances, so images cannot be perfectly focused. Second, the size of a refracting telescope is also limited by the size of the objective lens. If the lens is too large, the glass sags under its own weight and images are distorted. These limitations are two reasons that most professional astronomers use reflecting telescopes.

Reflecting Telescopes

A telescope that uses a curved mirror to gather and focus light is called a **reflecting telescope.** Light enters the telescope and is reflected from a large, curved mirror to a flat mirror. As shown in **Figure 2,** the flat mirror focuses the image and reflects the light to be magnified by the eyepiece.

One advantage of reflecting telescopes is that the mirrors can be very large. Large mirrors allow reflecting telescopes to gather more light than refracting telescopes do. Another advantage is that curved mirrors are polished on their curved side, which prevents light from entering the glass. Thus, any flaws in the glass do not affect the light. A third advantage is that mirrors can focus all colors of light to the same focal point. Therefore, reflecting telescopes allow all colors of light from an object to be seen in focus at the same time.

telescope an instrument that collects electromagnetic radiation from the sky and concentrates it for better observation

refracting telescope a telescope that uses a set of lenses to gather and focus light from distant objects

reflecting telescope a telescope that uses a curved mirror to gather and focus light from distant objects

CONNECTION to Physical Science — GENERAL

Cones and Rods The human retina contains receptors called *cones* and *rods,* which perceive different wavelengths of light. Cones are found in the central part of the retina and perceive color. Rods, located at the outer part of the retina, perceive only black and white. When little light is present, rods are more sensitive than cones. For this reason, stargazers sometimes look at objects by using their peripheral vision rather than looking at an object straight on. This method takes advantage of the rods' ability to detect faint objects in the sky. Encourage interested students use a telescope, binoculars, or their unaided eyes to test this technique.

Answer to Reading Check

Air pollution, water vapor, and light pollution distort the images produced by optical telescopes.

Figure 3 *The Keck Telescopes are in Hawaii. The 36 hexagonal mirrors in each telescope (shown in the inset) combine to form a light-reflecting surface that is 10 m across.*

Figure 4 *The* Hubble Space Telescope *has produced very clear images of objects in deep space.*

Very Large Reflecting Telescopes

In some very large reflecting telescopes, several mirrors work together to collect light and focus it in the same area. The Keck Telescopes in Hawaii, shown in **Figure 3,** are twin telescopes that each have 36 hexagonal mirrors that work together. Linking several mirrors allows more light to be collected and focused in one spot.

Optical Telescopes and the Atmosphere

The light gathered by telescopes on the Earth is affected by the atmosphere. The Earth's atmosphere causes starlight to shimmer and blur. Also, light pollution from large cities can make the sky look bright. As a result, an observer's ability to view faint objects is limited. Astronomers often place telescopes in dry areas to avoid moisture in the air. Mountaintops are also good locations for telescopes because the air is thinner at higher elevations. In addition, mountaintops generally have less air pollution and light pollution than other areas do, so the visibility of stars is better.

✓ **Reading Check** How does the atmosphere affect the images produced by optical telescopes?

Optical Telescopes in Space

To avoid interference by the atmosphere, scientists have put telescopes in space. Although the mirror in the *Hubble Space Telescope,* shown in **Figure 4,** is only 2.4 m across, this optical telescope can detect very faint objects in space.

Science BlOOpers

Hubble Mirror Flaws When the *Hubble Space Telescope* was deployed in 1990, it became immediately apparent that the telescope was not operating correctly—images transmitted back to Earth were blurred. A minute flaw was discovered in the telescope's main mirror. The mirror had been ground about 0.0002 cm (about one-fiftieth of the width of a human hair) flatter than it should have been. Although much of the image distortion was corrected with computer processing, the telescope was much less powerful than scientists originally hoped it would be. During a 1993 repair mission, space shuttle astronauts placed on the telescope a number of corrective devices that made the telescope fully operational.

The Electromagnetic Spectrum

For thousands of years, humans have used their eyes to observe stars and planets. But scientists eventually discovered that visible light, the light that we can see, is not the only form of radiation. In 1852, James Clerk Maxwell proved that visible light is a part of the electromagnetic spectrum. The **electromagnetic spectrum** is made up of all of the wavelengths of electromagnetic radiation.

Detecting Electromagnetic Radiation

Each color of light is a different wavelength of electromagnetic radiation. Humans can see radiation from red light, which has a long wavelength, to blue light, which has a shorter wavelength. But visible light is only a small part of the electromagnetic spectrum, as shown in **Figure 5.** The rest of the electromagnetic spectrum—radio waves, microwaves, infrared light, ultraviolet light, X rays, and gamma rays—is invisible. The Earth's atmosphere blocks most invisible radiation from objects in space. In this way, the atmosphere functions as a protective shield around the Earth. Radiation that can pass through the atmosphere includes some radio waves, microwaves, infrared light, visible light, and some ultraviolet light.

electromagnetic spectrum all of the frequencies or wavelengths of electromagnetic radiation

Figure 5 *Visible light is only a small band of the electromagnetic spectrum. Radio waves have the longest wavelengths, and gamma rays have the shortest wavelengths.*

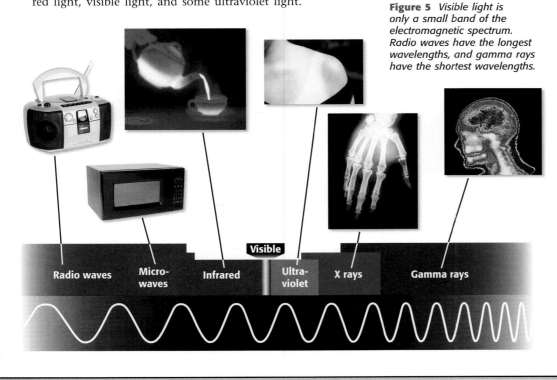

WEIRD SCIENCE

Shortly after Marconi invented the radio in the 1890s, people became interested in listening for messages from intelligent life in the universe. In 1901, a reward of 100,000 francs was offered to the first person to communicate with aliens.

Close

Reteaching — BASIC

Optical Telescope Review
Reproduce **Figure 2** on the board, but omit labels or lines indicating the path of light in the telescopes. Ask students volunteers to help you indicate the path of light in each telescope and to add the labels shown in the figure. **LS** **Visual**

Quiz — GENERAL

1. What limits the size and magnification of a refracting telescope? (If the objective lens is too large, gravity will cause the glass to sag, which distorts the image.)

2. What is the advantage of linking radio telescopes? (When radio telescopes are linked together, they act as a very large unit and are more powerful. Scientists can then make observations that show extremely fine details of distant objects.)

Alternative Assessment — GENERAL

Writing Letters Have students imagine that they are trying to gather support for building the largest radio-telescope array ever. Have them write a persuasive letter describing how radio telescopes differ from optical telescopes and how building such a large array could further our understanding of the universe. Students should include a diagram showing how their array would collect signals from a wide area. **LS** **Intrapersonal**

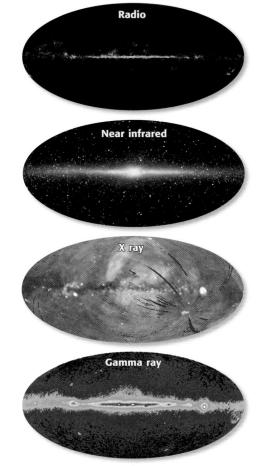

Figure 6 *Each image shows the Milky Way as it would appear if we could see other wavelengths of electromagnetic radiation.*

Nonoptical Telescopes

To study invisible radiation, scientists use nonoptical telescopes. Nonoptical telescopes detect radiation that cannot be seen by the human eye. Astronomers study the entire electromagnetic spectrum because each type of radiation reveals different clues about an object. As **Figure 6** shows, our galaxy looks very different when it is observed at various wavelengths. A different type of telescope was used to produce each image. The "cloud" that goes across the image is the Milky Way galaxy.

Radio Telescopes

Radio telescopes detect radio waves. Radio telescopes have to be much larger than optical telescopes because radio wavelengths are about 1 million times longer than optical wavelengths. Also, very little radio radiation reaches the Earth from space, so radio telescopes must be very sensitive. The surface of radio telescopes does not have to be as flawless as the lenses and mirrors of optical telescopes. In fact, the surface of a radio telescope does not have to be solid.

Linking Radio Telescopes

Astronomers can get more detailed images of the universe by linking radio telescopes together. When radio telescopes are linked together, they work like a single giant telescope. For example, the Very Large Array (VLA) consists of 27 radio telescopes that are spread over 30 km. Working together, the telescopes function as a single telescope that is 30 km across!

CONNECTION TO Physics

Detecting Infrared Radiation In this activity, you will replicate James Maxwell's discovery of invisible infrared radiation. First, paint the bulbs of three thermometers black. Place a sheet of white paper inside a tall cardboard box. Tape the thermometers parallel to each other, and place them inside the box. Cut a small notch in the top of the box, and position a small glass prism so that a spectrum is projected inside the box. Arrange the thermometers so that one is just outside the red end of the spectrum, with no direct light on it. After 10 min, record the temperatures. Which thermometer recorded the highest temperature? Explain why.

ACTIVITY

Group ACTIVITY — GENERAL

Reflecting Telescopes Students can construct a simple reflecting telescope. Instruct students to turn off the lights and place a makeup mirror (a curved, focusing mirror) near a window so that the moon and some stars are reflected in it. One student should hold a hand mirror in front of the makeup mirror so that he or she can see a reflection of the makeup mirror in the hand mirror. Then, the other student should use a magnifying lens to view the reflection in the hand mirror. **LS** **Kinesthetic/Visual**

Answer to Connection to Physics

The thermometer that was placed just outside of the visible spectrum recorded the highest temperature, because infrared radiation is warmer than visible light. (Note: A detailed procedure and diagrams of this experiment are available on the Internet.)

Nonoptical Telescopes in Space

Because most electromagnetic waves are blocked by the Earth's atmosphere, scientists have placed ultraviolet telescopes, infrared telescopes, gamma-ray telescopes, and X-ray telescopes in space. The *Chandra X-Ray Observatory*, a space-based telescope that detects X rays, is illustrated in **Figure 7.** X-ray telescopes in space can be much more sensitive than optical telescopes. For example, NASA has tested an X-ray telescope that can detect an object that is the size of a frisbee on the surface of the sun. If an optical telescope had a similar power, it could detect a hair on the head of an astronaut on the moon!

Reading Check Why are X-ray telescopes placed in space?

Figure 7 *The* Chandra X-Ray Observatory *can detect black holes and some of the most distant objects in the universe.*

SECTION Review

Summary

- Refracting telescopes use lenses to gather and focus light.
- Reflecting telescopes use mirrors to gather and focus light.
- Astronomers study all wavelengths of the electromagnetic spectrum, including radio waves, microwaves, infrared light, visible light, ultraviolet light, X rays, and gamma rays.
- The atmosphere blocks most forms of electromagnetic radiation from reaching the Earth. To overcome this limitation, astronomers place telescopes in space.

Using Key Terms

For each pair of terms, explain how the meanings of the terms differ.

1. *refracting telescope* and *reflecting telescope*
2. *telescope* and *electromagnetic spectrum*

Understanding Key Ideas

3. How does the atmosphere affect astronomical observations?
   a. It focuses visible light.
   b. It blocks most electromagnetic radiation.
   c. It blocks all radio waves.
   d. It does not affect astronomical observations.
4. Describe how reflecting and refracting telescopes work.
5. What limits the size of a refracting telescope? Explain.
6. What advantages do reflecting telescopes have over refracting telescopes?
7. List the types of radiation in the electromagnetic spectrum, from the longest wavelength to the shortest wavelength. Then, describe how astronomers study each type of radiation.

Math Skills

8. A telescope's light-gathering power is proportional to the area of its objective lens or mirror. If the diameter of a lens is 1 m, what is the area of the lens? (Hint: *area* = $3.1416 \times radius^2$)

Critical Thinking

9. **Applying Concepts** Describe three reasons why Hawaii is a good location for a telescope.
10. **Making Inferences** Why doesn't the surface of a radio telescope have to be as flawless as the surface of a mirror in an optical telescope?
11. **Making Inferences** What limitation of a refracting telescope could be overcome by placing the telescope in space?

SCI LINKS.

NSTA

Developed and maintained by the National Science Teachers Association

For a variety of links related to this chapter, go to www.scilinks.org

Topic: Telescopes
SciLinks code: HSM1500

3. b
4. A refracting telescope has an objective lens that gathers and focuses light. Another lens in the eyepiece magnifies the image produced by the objective lens. A reflecting telescope has a large curved mirror that reflects light to a flat mirror. A lens in the eyepiece magnifies the image reflected from the flat mirror for observation.
5. The size of the objective lens limits the size of a refracting telescope. If the lens is too large, it sags and distorts images.
6. Answers may vary. Reflecting telescopes can be more powerful than refracting telescopes because reflecting telescopes are not limited by the size of the objective lens. Also, mirrors reflect all wavelengths of light to the same focal point, and lenses focus different wavelengths at slightly different distances.
7. radio waves, microwaves, infrared, visible light, ultraviolet, X rays, gamma rays; Astronomers use reflecting and refracting telescopes to study visible light, and special nonoptical telescopes to study the other forms of electromagnetic radiation.
8. 3.1416×0.25 m^2 = 0.79 m^2
9. Hawaii has little light pollution. Air pollution is also minimal. In addition, Hawaii has many tall volcanoes, some of which are good sites for observatories.
10. The surface of a radio telescope does not have to be as flawless as the surface of a mirror in an optical telescope because radio waves are much larger than visible light waves.
11. The atmosphere distorts radiation passing through it, and light pollution can limit an astronomer's ability to see faint objects. In addition, Earth's gravity causes the objective lens to sag, which distorts images. Placing a telescope in space can solve these problems.

Answer to Reading Check

because the atmosphere blocks most X-ray radiation from space

Answers to Section Review

1. Sample answer: A refracting telescope uses lenses to gather and focus light. A reflecting telescope uses mirrors to gather and focus light.
2. A telescope is an instrument that is used to observe electromagnetic radiation. The electromagnetic spectrum consists of all of the wavelengths of electromagnetic radiation.

CHAPTER RESOURCES

Chapter Resource File

- Section Quiz GENERAL
- Section Review GENERAL
- Vocabulary and Section Summary GENERAL
- Critical Thinking ADVANCED
- SciLinks Activity GENERAL

Focus

Overview

This section discusses constellations and their significance to ancient and modern astronomers. Students will learn how to use a sky map to find stars in the night sky. The section also explains how the celestial sphere helps astronomers describe the location of stars. The section defines *light-year* and concludes with a discussion of the size and scale of the universe.

Bellringer

Ask students if it is possible to determine the direction of the North Pole by looking at the stars. Have students explain their answers in their **science journal.**

Motivate

Discussion — GENERAL

Naming Constellations Ask students to think of other possible names for the constellation in **Figure 1.** Discuss why people such as farmers, poets, and astronomers have been inspired to name and identify constellations. As an extension, ask students to observe the night sky at home and name their own constellations. Students can write a legend about their constellation and share a sketch of it with the class. **LS Verbal/Visual**

Mapping the Stars

Have you ever seen Orion the Hunter or the Big Dipper in the night sky? Ancient cultures linked stars together to form patterns that represented characters from myths and objects in their lives.

Today, we can see the same star patterns that people in ancient cultures saw. Modern astronomers still use many of the names given to stars centuries ago. But astronomers can now describe a star's location precisely. Advances in astronomy have led to a better understanding of how far away stars are and how big the universe is.

Patterns in the Sky

When people in ancient cultures connected stars in patterns, they named sections of the sky based on the patterns. These patterns are called *constellations*. **Constellations** are sections of the sky that contain recognizable star patterns. Understanding the location and movement of constellations helped people navigate and keep track of time.

Different civilizations had different names for the same constellations. For example, where the Greeks saw a hunter (Orion) in the northern sky, the Japanese saw a drum, as shown in **Figure 1.** Today, different cultures still interpret the sky in different ways, but astronomers have agreed on the names and locations of the constellations.

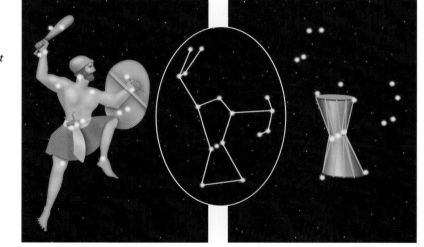

Figure 1 *The ancient Greeks saw Orion as a hunter, but the Japanese saw the same set of stars as a drum.*

CHAPTER RESOURCES

Chapter Resource File

- **Lesson Plan**
- **Directed Reading A** BASIC
- **Directed Reading B** SPECIAL NEEDS

Technology

Transparencies
- Bellringer
- Spring Constellations in the Northern Hemisphere

Is That a Fact!

The International Astronomical Union has standardized the boundaries of the 88 constellations so that the total area of the celestial sphere can be classified according to the constellations. Thus, the constellations fit together like a jigsaw puzzle that surrounds the Earth.

Figure 2 *This sky map shows some of the constellations in the Northern Hemisphere at midnight in the spring. Ursa Major (the Great Bear) is a region of the sky that includes all of the stars that make up that constellation.*

constellation a region of the sky that contains a recognizable star pattern and that is used to describe the location of objects in space

Constellations Help Organize the Sky

When you think of constellations, you probably think of the stick figures made by connecting bright stars with imaginary lines. To an astronomer, however, a constellation is something more. As you can see in **Figure 2,** a constellation is a region of the sky. Each constellation shares a border with neighboring constellations. For example, in the same way that the state of Texas is a region of the United States, Ursa Major is a region of the sky. Every star or galaxy is located within 1 of 88 constellations.

Seasonal Changes

The sky map in **Figure 2** shows what the midnight sky in the Northern Hemisphere looks like in the spring. But as the Earth revolves around the sun, the apparent locations of the constellations change from season to season. In addition, different constellations are visible in the Southern Hemisphere. Thus, a child in Chile can see different constellations than you can. Therefore, this map is not accurate for the other three seasons or for the Southern Hemisphere. Sky maps for summer, fall, and winter in the Northern Hemisphere appear in the Appendix of this book.

✔ **Reading Check** Why are different constellations visible in the Northern and Southern Hemispheres? (*See the Appendix for answers to Reading Checks.*)

Using a Sky Map
1. Hold your **textbook** over your head with the cover facing upward. Turn the book so that the direction at the bottom of the sky map is the same as the direction you are facing.
2. Notice the locations of the constellations in relation to each other.
3. If you look up at the sky at night in the spring, you should see the stars positioned as they are on your map.
4. Why are *E* and *W* on sky maps the reverse of how they appear on land maps?

ACTIVITY ─────── GENERAL

A Compass on Your Wrist Tell students that a watch that has an hour hand can be used to tell direction during the daytime. Have students hold a watch horizontally such that the hour hand (the shorter hand) points directly at the sun. If they halve the distance between the hour hand and the 12 with a toothpick, the toothpick will be pointing south. Ask students how they would adapt this method to work in the Southern Hemisphere. **LS Kinesthetic/Logical**

CONNECTION ACTIVITY
Real World ─────── BASIC

Measuring the Sky with Your Hands Amateur astronomers can measure the sky by using their hands. If you extend your arm and make a fist that begins at the horizon, you can gauge roughly 10°. If you open your hand and align your little finger with the horizon, you have marked off 20°. **LS Kinesthetic**

English Language Learners

Figure 3 *Using an astrolabe, you can determine the altitude of a star by measuring the angle between the horizon and a star. The altitude of any object depends on where you are and when you look.*

zenith the point in the sky directly above an observer on Earth

altitude the angle between an object in the sky and the horizon

horizon the line where the sky and the Earth appear to meet

The **zenith** is an imaginary point in the sky directly above an observer on Earth. The zenith always has an altitude of 90°.

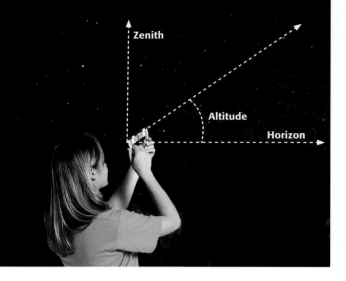

Finding Stars in the Night Sky

Have you ever tried to show someone a star by pointing to it? Did the person miss what you were seeing? If you use an instrument called an *astrolabe,* shown in **Figure 3,** you can describe the location of a star or planet. To use an astrolabe correctly, you need to understand the three points of reference shown in **Figure 4.** This method is useful to describe the location of a star relative to where you are. But if you want to describe a star's location in relation to the Earth, you need to use the celestial sphere, shown in **Figure 5.**

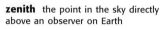

Figure 4   **Zenith, Altitude, and Horizon**

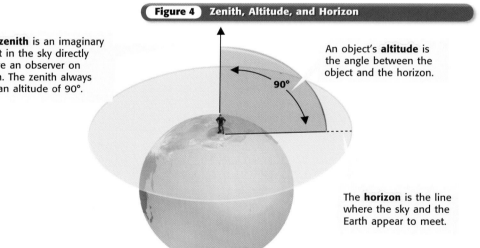

An object's **altitude** is the angle between the object and the horizon.

90°

The **horizon** is the line where the sky and the Earth appear to meet.

MISCONCEPTION ///ALERT\\\

Altitude Students may be confused by the term *altitude* when it is used in astronomy. Point out that altitude does not denote height or elevation but rather the angle between an object, the horizon, and the observer. As the Earth rotates, the altitude of the stars changes. The altitude of an object is also affected by the location of the observer.

Group ACTIVITY ─ GENERAL

Space Science Exploration Engage the class in a space science Internet orientation. There are hundreds of well-maintained Web sites with stunning images from all over the universe. Students will be excited to find these images and share them with the class. **LS Interpersonal**

Figure 5 **The Celestial Sphere**

To talk to each other about the location of a star, astronomers must have a common method of describing a star's location. The method that astronomers have invented is based on a reference system known as the *celestial sphere*. The celestial sphere is an imaginary sphere that surrounds the Earth. Just as we use latitude and longitude to plot positions on Earth, astronomers use right ascension and declination to plot positions in the sky. *Right ascension* is a measure of how far east an object is from the *vernal equinox,* the location of the sun on the first day of spring. *Declination* is a measure of how far north or south an object is from the celestial equator.

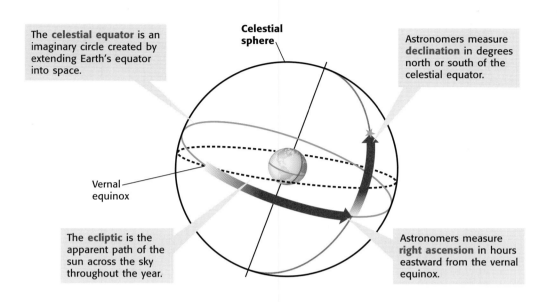

The **celestial equator** is an imaginary circle created by extending Earth's equator into space.

Celestial sphere

Astronomers measure **declination** in degrees north or south of the celestial equator.

Vernal equinox

The **ecliptic** is the apparent path of the sun across the sky throughout the year.

Astronomers measure **right ascension** in hours eastward from the vernal equinox.

The Path of Stars Across the Sky

Just as the sun appears to move across the sky during the day, most stars and planets rise and set throughout the night. This apparent motion is caused by the Earth's rotation. As the Earth spins on its axis, stars and planets appear to move. Near the poles, however, stars are circumpolar. *Circumpolar stars* are stars that can be seen at all times of year and all times of night. These stars never set, and they appear to circle the celestial poles. You also see different stars in the sky depending on the time of year. Why? The reason is that as the Earth travels around the sun, different areas of the universe are visible.

✓ Reading Check How is the apparent movement of the sun similar to the apparent movement of most stars during the night?

CHAPTER RESOURCES

Technology

Transparencies
• Zenith, Altitude, and Horizon
• The Celestial Sphere

Homework ———— GENERAL

Locating Polaris The Big Dipper is an easily recognizable group of stars that can be seen from the mid-northern latitudes all year. If students are unfamiliar with the Big Dipper, draw it on the board or show them a picture of it. Ask students to use a star chart to locate the Big Dipper on a clear night. After students have located the Big Dipper, they can easily find Polaris (the North Star). The two stars that make up the front of the dipper bowl point directly toward Polaris. To find Polaris, students should estimate the distance between the two stars that make up the front of the bowl. Students can pinpoint the North Star by extending from the front two stars of the bowl an imaginary line that is 5 times the length of that distance.

Answer to Reading Check

The apparent movement of the sun and stars is caused by the Earth's rotation on its axis.

CONNECTION ACTIVITY
Math ———— GENERAL

Calculating a Light-Year
The textbook indicates that the distance light travels in a year is 9.46 trillion kilometers. Have students verify this distance by doing the calculation themselves. Ask students to calculate the distance light travels in 1 day. (approximately 26 million kilometers) **LS** Logical

Answer to Reading Check
9.46 trillion kilometers

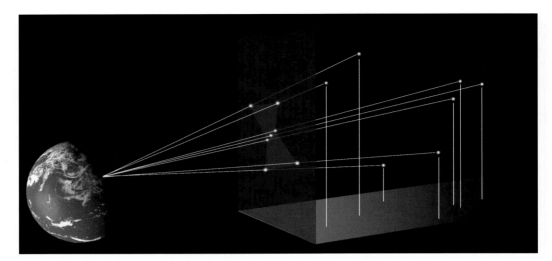

Figure 6 *While the stars in the constellation Orion may appear to be near each other when they are seen from Earth, they are actually very far apart.*

light-year the distance that light travels in one year; about 9.46 trillion kilometers

INTERNET ACTIVITY

For another activity related to this chapter, go to **go.hrw.com** and type in the keyword **HZ5OBSW.**

The Size and Scale of the Universe

Imagine looking out the window of a moving car. Nearby trees appear to move more quickly than farther trees do. Objects that are very far away do not appear to move at all. The same principle applies to stars and planets. In the 1600s, Nicolaus Copernicus noticed that the planets appeared to move relative to each other but that the stars did not. Thus, he thought that the stars must be much farther away than the planets.

Measuring Distance in Space

Today, we know that Copernicus was correct. The stars are much farther away than the planets are. In fact, stars are so distant that a new unit of length—the light-year—was created to measure their distance. A **light-year** is a unit of length equal to the distance that light travels in 1 year. One light-year is equal to about 9.46 trillion kilometers! The farthest objects we can observe are more than 10 billion light-years away. Although the stars may appear to be at similar distances from Earth, their distances vary greatly. For example, **Figure 6** shows how far away the stars that make up part of Orion are.

✓ **Reading Check** How far does light travel in 1 year?

Considering Scale in the Universe

When you think about the universe and all of the objects it contains, it is important to consider scale. For example, stars appear to be very small in the night sky. But we know that most stars are a lot larger than Earth. **Figure 7** will help you understand the scale of objects in the universe.

CONNECTION to
History ———— GENERAL

The North Star in History Polaris has not always been the North Star. Nearly 5,000 years ago, a faint star named Thuban in the constellation Draco held that honor. Because the Earth wobbles on its axis, the location of the north celestial pole changes on a 25,780-year cycle. Some theories argue that the Great Pyramid of Giza was built such that its main passageway aligned with Thuban. Because Thuban did not appear to move in the night sky, it symbolized immortality. In 12,000 years, Vega will replace Polaris as the pole star.

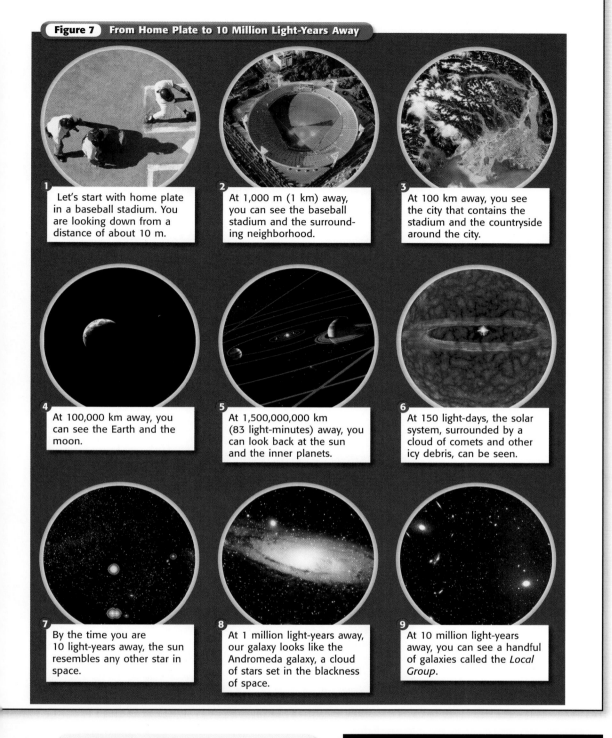

Figure 7 From Home Plate to 10 Million Light-Years Away

1. Let's start with home plate in a baseball stadium. You are looking down from a distance of about 10 m.

2. At 1,000 m (1 km) away, you can see the baseball stadium and the surrounding neighborhood.

3. At 100 km away, you see the city that contains the stadium and the countryside around the city.

4. At 100,000 km away, you can see the Earth and the moon.

5. At 1,500,000,000 km (83 light-minutes) away, you can look back at the sun and the inner planets.

6. At 150 light-days, the solar system, surrounded by a cloud of comets and other icy debris, can be seen.

7. By the time you are 10 light-years away, the sun resembles any other star in space.

8. At 1 million light-years away, our galaxy looks like the Andromeda galaxy, a cloud of stars set in the blackness of space.

9. At 10 million light-years away, you can see a handful of galaxies called the *Local Group*.

Is That a Fact!

The *Hubble Space Telescope* has relayed the first detailed images of distant galaxies. These images have led astronomers to think that there may be 10 times the number of galaxies in the universe than previously thought. Students will enjoy finding images from the *Hubble Space Telescope* on the Internet and sharing them with the class.

CHAPTER RESOURCES

Technology

📦 **Transparencies**
• From Home Plate to 10 Million Light-Years Away

Discussion ——— ADVANCED

Expanding Universe Students may assume that because the universe is expanding, every star they see is moving away from Earth. Point out that the only stars visible with the unaided eye are the ones in our galaxy, which is not expanding. The Milky Way is not expanding for the same reason that our solar system is not expanding—gravity holds moving bodies in orbit. The Milky Way and the Andromeda galaxies are part of the Local Group, a galaxy cluster of 30 galaxies. The galaxies that make up a galaxy cluster are also held together by gravity. The expansion of the universe occurs as galaxy clusters move apart from one another.

If the expansion of the universe were compared to baking a chocolate chip cookie, the chocolate chips would represent galaxy clusters. As the cookie bakes, the chocolate chips do not change in size; however, as the dough expands, the space between the chocolate chips increases. **LS Logical**

CONNECTION ACTIVITY
Math ——— GENERAL

Scientific Notation Have students convert the following numbers to or from scientific notation:

    1,200 (1.2×10^3)

    150,000 (1.5×10^5)

    3.2×10^6 (3,200,000)

    790,000,000 (7.9×10^8)

    5.6×10^{12} (5,600,000,000,000)

LS Logical

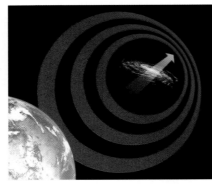

Figure 8 *As an object moves away from an observer at a high speed, the light from the object appears redder. As the object moves toward the observer, the light from the object appears bluer.*

The Doppler Effect

Have you ever noticed that when a driver in an approaching car blows the horn, the horn sounds higher pitched as the car approaches and lower pitched after the car passes? This effect is called the *Doppler effect*. As shown in **Figure 8,** the Doppler effect also occurs with light. If a light source, such as a star or galaxy, is moving quickly away from an observer, the light emitted looks redder than it normally does. This effect is called *redshift*. If a star or galaxy is moving quickly toward an observer, its light appears bluer than it normally does. This effect is known as *blueshift*.

An Expanding Universe

After discovering that the universe is made up of many other galaxies like our own, Edwin Hubble analyzed the light from galaxies and stars to study the general direction that objects in the universe are moving. Hubble soon made another startling discovery—the light from all galaxies except our close neighbors is affected by redshift. This means that galaxies are rapidly moving apart from each other. In other words, because all galaxies except our close neighbors are moving apart, the universe must be expanding. **Figure 9** shows evidence of redshift recorded by the *Hubble Space Telescope* in 2002.

✔ **Reading Check** What logical conclusion could be made if the light from all of the galaxies were affected by blueshift?

Figure 9 *The galaxy that is cut off at the bottom of this image is moving away from us at a much slower speed than the other galaxies are. Distant galaxies are visible as faint disks.*

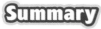

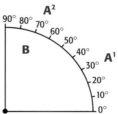

Summary

- Astronomers use constellations to organize the sky.
- Altitude, or the angle between an object and the horizon, can be used to describe the location of an object in the sky.
- The celestial sphere is an imaginary sphere that surrounds the Earth. Using the celestial sphere, astronomers can accurately describe the location of an object without reference to an observer.

- A light-year is the distance that light travels in 1 year.
- The Doppler effect causes the light emitted by objects that are moving away from an observer to appear to shift toward the red end of the spectrum. Objects moving toward an observer are shifted to the blue end of the spectrum.
- Observations of redshift and blueshift indicate that the universe is expanding.

Using Key Terms

The statements below are false. For each statement, replace the underlined term to make a true statement.

1. <u>Zenith</u> is the angle between an object and the horizon.

2. The distance that light travels in 1 year is called a <u>light-meter</u>.

Understanding Key Ideas

3. Stars appear to move across the night sky because of
   a. the rotation of Earth on its axis.
   b. the movement of the Milky Way galaxy.
   c. the movement of stars in the universe.
   d. the revolution of Earth around the sun.

4. How do astronomers use the celestial sphere to plot a star's exact position?

5. How do constellations relate to patterns of stars? How are constellations like states?

6. Why are different sky maps needed for different times of the year?

7. What are redshift and blueshift? Why are these effects useful in the study of the universe?

Critical Thinking

8. **Applying Concepts** Light from the Andromeda galaxy is affected by blueshift. What can you conclude about this galaxy?

9. **Making Comparisons** Explain how Copernicus concluded that stars were farther away than planets. Draw a diagram showing how this principle applies to another example.

Interpreting Graphics

The diagram below shows the altitude of Star A and Star B. Use the diagram below to answer the questions that follow.

10. What is the approximate altitude of star B?

11. In 4 h, star A moved from A^1 to A^2. How many degrees did the star move each hour?

SCI LINKS

NSTA

Developed and maintained by the National Science Teachers Association

For a variety of links related to this chapter, go to www.scilinks.org

Topic: Constellations
SciLinks code: HSM0347

CHAPTER RESOURCES

Chapter Resource File

- Section Quiz GENERAL
- Section Review GENERAL
- Vocabulary and Section Summary GENERAL

Answers to Section Review

1. Altitude
2. light-year
3. a
4. The celestial sphere is a coordinate system that uses the Earth as a reference. Declination indicates how far north or south from the celestial equator an object is. Right ascension indicates how far east or west from the vernal equinox an object is.
5. Constellations are patterns of stars. They are similar to states because both are regions of a much larger area. Like states, constellations share borders with each other.
6. Different sky maps are needed for different times of the year because as the Earth orbits the sun, different stars become visible.
7. Redshift occurs when an object is moving rapidly away from an observer. Light emitted from the object appears to be shifted toward the red end of the spectrum. Blueshift occurs when an object is moving rapidly toward an observer. Light emitted from the object appears to be shifted toward the blue end of the spectrum. Redshift and blueshift indicate whether the universe is expanding.
8. Students should conclude that the Andromeda galaxy is moving toward us.
9. Copernicus concluded that the stars were farther away than the planets because the planets moved more quickly than stars throughout the year. The effect is similar to how nearby objects observed from the window of a moving car appear to move more quickly than distant objects.
10. about 70°
11. 40° ÷ 4 h = 10° per hour (Note that this question is intended to help students sharpen basic math skills and that the actual movement of stars is much more complex.)

Through the Looking Glass

Teacher's Notes

Time Required

One 45-minute class period

Lab Ratings

EASY ——————————— HARD

Teacher Prep 🧪🧪
Student Set-Up 🧪🧪🧪
Concept Level 🧪🧪
Clean Up 🧪

MATERIALS

The materials listed on the student page are enough for a group of 2 or 3 students.

Safety Caution

Remind students to review all safety cautions and icons before beginning this lab activity. Students should never look at the sun through their telescopes. Caution students not to focus sunlight through their telescopes because a fire could start.

Skills Practice Lab

OBJECTIVES

Construct a simple model of a refracting telescope.

Observe distant objects by using your telescope.

MATERIALS

- clay, modeling (1 stick)
- convex lens, 3 cm in diameter (2 of different focal length)
- lamp, desk
- paper, white (1 sheet)
- ruler, metric
- scissors
- tape, masking (1 roll)
- toilet-paper tube, cardboard
- wrapping paper tube, cardboard

SAFETY

Through the Looking Glass

Have you ever looked toward the horizon or up into the sky and wished that you could see farther? Do you think that a telescope might help you see farther? Astronomers use huge telescopes to study the universe. You can build your own telescope to get a glimpse of how these enormous, technologically advanced telescopes help astronomers see distant objects.

Procedure

1. Use modeling clay to form a base that holds one of the lenses upright on your desktop. When the lights are turned off, your teacher will turn on a lamp at the front of the classroom. Rotate your lens so that the light from the lamp passes through the lens.

2. Hold the paper so that the light passing through the lens lands on the paper. To sharpen the image of the light on the paper, slowly move the paper closer to or farther from the lens. Hold the paper in the position in which the image is sharpest.

3. Using the metric ruler, measure the distance between the lens and the paper. Record this distance.

Preparation Notes

One week before the activity, ask students to collect wrapping-paper and toilet-paper cardboard tubes to bring to class. Obtain two double-convex lenses.

You may wish to experiment with the lenses before class to determine their focal length. If you add another lens to the telescope, you can make the image right side up, but light is lost as it passes through the third lens.

CHAPTER RESOURCES

Chapter Resource File

- Datasheet for Chapter Lab
- Lab Notes and Answers

Technology

Classroom Videos
- Lab Video

Lab Book

- The Sun's Yearly Trip Through the Zodiac

4 How far is the paper from the lens? This distance, called the *focal length,* is the distance that the paper has to be from the lens for the image to be in focus.

5 Repeat steps 1–4 using the other lens.

6 Measuring from one end of the long cardboard tube, mark the focal length of the lens that has the longer focal length. Place a mark 2 cm past this line toward the other end of the tube, and label the mark "Cut."

7 Measuring from one end of the short cardboard tube, mark the focal length of the lens that has the shorter focal length. Place a mark 2 cm past this line toward the other end of the tube, and label the mark "Cut."

8 Shorten the tubes by cutting along the marks labeled "Cut." Wear safety goggles when you make these cuts.

9 Tape the lens that has the longer focal length to one end of the longer tube. Tape the other lens to one end of the shorter tube. Slip the empty end of one tube inside the empty end of the other tube. Be sure that there is one lens at each end of this new, longer tube.

10 Congratulations! You have just constructed a telescope. To use your telescope, look through the short tube (the eyepiece) and point the long end at various objects in the room. You can focus the telescope by adjusting its length. Are the images right side up or upside down? Observe birds, insects, trees, or other outside objects. Record the images that you see. **Caution:** NEVER look directly at the sun! Looking directly at the sun could cause permanent blindness.

Analyze the Results

1 **Analyzing Results** Which type of telescope did you just construct: a refracting telescope or a reflecting telescope? What makes your telescope one type and not the other?

2 **Identifying Patterns** What factor determines the focal length of a lens?

Draw Conclusions

3 **Evaluating Results** How would you improve your telescope?

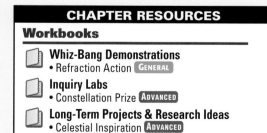

CLASSROOM TESTED & APPROVED

Michael E. Kral
West Hardin Middle School
Cecilia, Kentucky

Background

Begin the activity by discussing the components of a telescope. Simple telescopes, such as the one Galileo made, consist of two lenses and a tube. For the telescope to provide a clear image, the tube should be as long as the total of the focal lengths of the two lenses. Some students may have difficulty knowing when the image is in focus. Explain to these students that they should focus the light bulb's filament on the paper.

Analyze the Results

1. Sample answer: We just constructed a refracting telescope. Refracting telescopes use lenses. Reflecting telescopes use mirrors and lenses.

2. Answers may vary. Students should notice that the size and shape of a lens determine the focal length of the lens.

Draw Conclusions

3. Answers may vary. Accept all reasonable responses.

Chapter Review

Assignment Guide

SECTION	QUESTIONS
1	6, 8–9, 18, 20
2	2, 5, 7, 14, 16, 19
3	3, 10–13, 15, 17, 21–22
1 and 3	4
1, 2, and 3	1

ANSWERS

Using Key Terms

1. Sample answer: A year is the amount of time Earth takes to complete one orbit around the sun. A month is based on the phases of the moon. A day is the amount of time Earth takes to rotate once on its axis. Astronomy is the study of the universe. The electromagnetic spectrum is made up of all forms of electromagnetic radiation. A constellation is a region of the sky that contains a recognizable pattern of stars. Altitude is the angle between an observer, the horizon, and an object in the sky.

2. Sample answer: Refracting telescopes use lenses to magnify and focus an image, and reflecting telescopes use mirrors to magnify and focus an image.

3. Sample answer: The zenith is a point in the sky that is directly above an observer. The horizon is the line where the Earth and the sky appear to meet.

USING KEY TERMS

1 Use each of the following terms in a separate sentence: *year, month, day, astronomy, electromagnetic spectrum, constellation,* and *altitude.*

For each pair of terms, explain how the meanings of the terms differ.

2 *reflecting telescope* and *refracting telescope*

3 *zenith* and *horizon*

4 *year* and *light-year*

UNDERSTANDING KEY IDEAS

Multiple Choice

5 Which of the following answer choices lists types of electromagnetic radiation from longest wavelength to shortest wavelength?

a. radio waves, ultraviolet light, infrared light

b. infrared light, microwaves, X rays

c. X rays, ultraviolet light, gamma rays

d. microwaves, infrared light, visible light

6 The length of a day is based on the amount of time that

a. Earth takes to orbit the sun one time.

b. Earth takes to rotate once on its axis.

c. the moon takes to orbit Earth one time.

d. the moon takes to rotate once on its axis.

7 Which of the following statements about X rays and radio waves from objects in space is true?

a. Both types of radiation can be observed by using the same telescope.

b. Separate telescopes are needed to observe each type of radiation, but both telescopes can be on Earth.

c. Separate telescopes are needed to observe each type of radiation, but both telescopes must be in space.

d. Separate telescopes are needed to observe each type of radiation, but only one of the telescopes must be in space.

8 According to ___, Earth is at the center of the universe.

a. the Ptolemaic theory

b. Copernicus's theory

c. Galileo's theory

d. None of the above

9 Which scientist was one of the first scientists to successfully use a telescope to observe the night sky?

a. Brahe   c. Hubble

b. Galileo   d. Kepler

10 Astronomers divide the sky into

a. galaxies.   c. zeniths.

b. constellations. d. phases.

11 ___ determines which stars you see in the sky.

a. Your latitude

b. The time of year

c. The time of night

d. All of the above

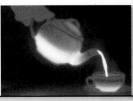

4. Sample answer: A year is the amount of time Earth takes to complete one orbit around the sun. A light-year is the distance that light travels in one year.

Understanding Key Ideas

5. d	**10.** b
6. b	**11.** d
7. d	**12.** a
8. a	**13.** b
9. b	**14.** c

12 The altitude of an object in the sky is the object's angular distance

  a. above the horizon.

  b. from the north celestial pole.

  c. from the zenith.

  d. from the prime meridian.

13 Right ascension is a measure of how far east an object in the sky is from

  a. the observer.

  b. the vernal equinox.

  c. the moon.

  d. Venus.

14 Telescopes that work on Earth's surface include all of the following EXCEPT

  a. radio telescopes.

  b. refracting telescopes.

  c. X-ray telescopes.

  d. reflecting telescopes.

Short Answer

15 Explain how right ascension and declination are similar to latitude and longitude.

16 How does a reflecting telescope work?

CRITICAL THINKING

17 **Concept Mapping** Use the following terms to create a concept map: *right ascension, declination, celestial sphere, degrees, hours, celestial equator,* and *vernal equinox.*

18 **Making Inferences** Why was seeing objects in the sky easier for people in ancient cultures than it is for most people today? What tools help modern people study objects in space in greater detail than was possible in the past?

19 **Making Inferences** Because many forms of radiation from space do not penetrate Earth's atmosphere, astronomers' ability to detect this radiation is limited. But how does the protection of the atmosphere benefit humans?

20 **Analyzing Ideas** Explain why the Ptolemaic theory seems logical based on daily observations of the rising and setting of the sun.

INTERPRETING GRAPHICS

Use the sky map below to answer the questions that follow. (Example: The star Aldebaran is located at about 4 h, 30 min right ascension, 16° declination.)

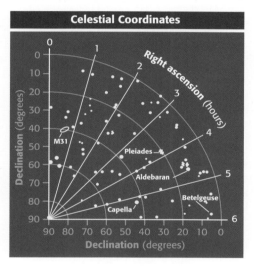

Celestial Coordinates

21 What object is located at 5 h, 55 min right ascension, and 7° declination?

22 What are the celestial coordinates for the Andromeda galaxy (M31)? Round off the right ascension to the nearest half-hour.

15. Sample answer: Right ascension and declination are similar to latitude and longitude because both are coordinate systems that use Earth as a reference point.

16. A reflecting telescope uses a large mirror to gather and focus light. A second mirror, located in front of the focal point, directs light toward an eyepiece for observation.

Critical Thinking

17. 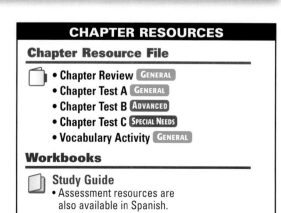 An answer to this exercise can be found at the end of this book.

18. Sample answer: There was not as much light and air pollution in the past as there is now, so fewer stars are visible to the unaided eye now. Tools such as telescopes help modern people study objects in space in greater detail.

19. Sample answer: The Earth's atmosphere protects humans from many forms of harmful radiation. It also traps thermal energy to make Earth warm enough for life.

20. Sample answer: Daily observations of the sun would indicate that the sun is moving. Every day, the sun rises and sets. From this observation, it would be logical but incorrect to conclude that the sun rotates around the Earth.

Interpreting Graphics

21. Betelgeuse

22. 0 h, 30 min right ascension, 40° declination

CHAPTER RESOURCES

Chapter Resource File

- **Chapter Review** `GENERAL`
- **Chapter Test A** `GENERAL`
- **Chapter Test B** `ADVANCED`
- **Chapter Test C** `SPECIAL NEEDS`
- **Vocabulary Activity** `GENERAL`

Workbooks

Study Guide
- Assessment resources are also available in Spanish.

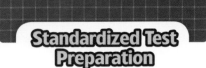
Teacher's Note

To provide practice under more realistic testing conditions, give students 20 minutes to answer all of the questions in this Standardized Test Preparation.

Answers to the standardized test preparation can help you identify student misconceptions and misunderstandings.

READING

Passage 1

1. C

2. H

3. A

TEST DOCTOR

Question 1: Students may choose answer D because 1,800 is divisible by 4, but it is also divisible by 100 and is not divisible by 400.

READING

Read each of the passages below. Then, answer the questions that follow each passage.

Passage 1 In the early Roman calendar, a year had exactly 365 days. The calendar worked well until people realized that the seasons were beginning and ending later each year. To fix this problem, Julius Caesar developed the Julian calendar based on a 365.25-day calendar year. He added 90 days to the year 46 BCE and added an extra day every 4 years. A year in which an extra day is added to the calendar is called a *leap year*. In the mid-1500s, astronomers determined that there are actually 365.2422 days in a year, so Pope Gregory XIII developed the Gregorian calendar. He dropped 10 days from the year 1582 and restricted leap years to years that are divisible by 4 but not by 100 (except for years that are divisible by 400). Today, most countries use the Gregorian calendar.

1. According to the passage, which of the following years is a leap year?

A 46 BCE

B 1582

C 1600

D 1800

2. How long is a year?

F 365 days

G 365.224 days

H 365.2422 days

I 365.25 days

3. Why did Julius Caesar change the early Roman calendar?

A to deal with the fact that the seasons were beginning and ending later each year

B to compete with the Gregorian calendar

C to add an extra day every year

D to shorten the length of a year

Passage 2 The earliest known evidence of astronomical observations is a group of stones near Nabta in southern Egypt that is between 6,000 and 7,000 years old. According to archeoastronomers, some of the stones are positioned such that they would have lined up with the sun during the summer solstice 6,000 years ago. The summer solstice occurs on the longest day of the year. At the Nabta site, the noonday sun is at its zenith (directly overhead) for about three weeks before and after the summer solstice. When the sun is at its zenith, upright objects do not cast shadows. For many civilizations in the Tropics, the zenith sun has had ceremonial significance for thousands of years. The same is probably true for the civilizations that used the Nabta site. Artifacts found at the site near Nabta suggest that the site was created by African cattle herders. These people probably used the site for many purposes, including trade, social bonding, and ritual.

1. In the passage, what does *archeoastronomer* mean?

A an archeologist that studies Egyptian culture

B an astronomer that studies the zenith sun

C an archeologist that studies ancient astronomy

D an astronomer that studies archeologists

2. Why don't upright objects cast a shadow when the sun is at its zenith?

F because the sun is directly overhead

G because the summer solstice is occurring

H because the sun is below the horizon

I because the sun is at its zenith on the longest day of the year

Passage 2

1. C

2. F

TEST DOCTOR

Question 1: Students may choose answer D because the word archeoastronomer combines the words *archeology* and *astronomy*. Explain that because the prefix is *archeo-*, an archeoastronomer is an archeologist.

The diagram below shows a galaxy moving in relation to four observers. The concentric circles illustrate the Doppler effect at each location. Use the diagram below to answer the questions that follow.

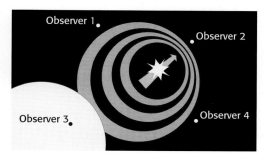

1. Which of the following observers would see the light from the galaxy affected by redshift?

   A observers 1 and 2

   B observer 3

   C observers 3 and 4

   D observers 1 and 4

2. Which of the following observers would see the light from the galaxy affected by blueshift?

   F observer 1

   G observers 2 and 4

   H observers 3 and 4

   I observer 2

3. How would the wavelengths of light detected by observer 4 appear?

   A The wavelengths would appear shorter than they really are.

   B The wavelengths would appear longer than they really are.

   C The wavelengths would appear unchanged.

   D The wavelengths would alternate between blue and red.

Read each question below, and choose the best answer.

1. If light travels 300,000 km/s, how long does light reflected from Mars take to reach Earth when Mars is 65,000,000 km away ?

   A 22 s

   B 217 s

   C 2,170 s

   D 2,200 s

2. Star A is 8 million kilometers from star B. What is this distance expressed in meters?

   F 0.8 m

   G 8,000 m

   H 8×10^6 m

   I 8×10^9 m

3. If each hexagonal mirror in the Keck Telescopes is 1.8 m across, how many mirrors would be needed to create a light-reflecting surface that is 10.8 m across?

   A 3.2

   B 5

   C 6

   D 6.2

4. If the altitude of a star is 37°, what is the angle between the star and the zenith?

   F 143°

   G 90°

   H 53°

   I 37°

5. You are studying an image made by the *Hubble Space Telescope*. If you observe 90 stars in an area that is 1 cm², which of the following estimates is the best estimate for the number of stars in 15 cm²?

   A 700

   B 900

   C 1,200

   D 1,350

Standardized Test Preparation

✚ TEST DOCTOR

Question 2: Remind students that blueshift occurs when an object is moving toward an observer at a high speed. Based on this information, observer 2 is the only reasonable choice.

MATH

1. B

2. I

3. C

4. H

5. D

✚ TEST DOCTOR

Question 2: Remind students that 1,000 m = 1 km. If students chose answer H, they probably converted 8 million kilometers to scientific notation but may have forgotten that the answer should be expressed in meters. Remind students to read every question in a standardized test at least twice.

CHAPTER RESOURCES

Chapter Resource File

• Standardized Test Preparation **GENERAL**

State Resources

For specific resources for your state, visit **go.hrw.com** and type in the keyword **HSMSTR**.

Science Fiction

Background

Lawrence Watt-Evans's memorable characters and stories have earned him high honors in the fields of science fiction and fantasy writing. Readers of *Asimov's Science Fiction* magazine nominated "Why I Left Harry's All-Night Hamburgers" for the best short story of 1987. The next year, that story earned Watt-Evans the Hugo Award and was nominated for the Nebula Award. Two years later, another story, "Windwagon Smith and the Martians," captured an Asimov's Reader's Choice Award. Check out that story in *Crosstime Traffic*, a collection of short stories that Watt-Evans published in 1992.

Science, Technology, and Society

Background

Light pollution can be reduced by fitting lights with "full cut-off fixtures," which are reflector shields that direct light downward. Another simple solution to reduce light pollution is to aim spotlights down rather than up. People can also reduce light pollution and save energy by using motion sensor lights. Timer-controlled lighting is another effective remedy to reduce light pollution.

Science in Action

Science Fiction

"Why I Left Harry's All-Night Hamburgers" by Lawrence Watt-Evans

The main character was 16, and he needed to find a job. So, he began working at Harry's All-Night Hamburgers. His shift was from midnight to 7:30 A.M. so that he could still go to school. Harry's All-Night Hamburgers was pretty quiet most nights, but once in a while some unusual characters came by. For example, one guy came in dressed for Arctic weather even though it was April. Then there were the folks who parked a very strange vehicle in the parking lot for anyone to see. The main character starts questioning the visitors, and what he learns startles and fascinates him. Soon, he's thinking about leaving Harry's. Find out why when you read "Why I Left Harry's All-Night Hamburgers," in the *Holt Anthology of Science Fiction*.

Social Studies ACTIVITY

WRITING SKILL The main character in the story learns that Earth is a pretty strange place. Find out about some of the places mentioned in the story, and create an illustrated travel guide that describes some of the foreign places that interest you.

Science, Technology, and Society

Light Pollution

When your parents were your age, they could look up at the night sky and see many more stars than you can now. In a large city, seeing more than 50 stars or planets in the night sky can be difficult. Light pollution is a growing—or you could say "glowing"—problem. If you have ever seen a white glow over the horizon in the night sky, you have seen the effects of light pollution. Most light pollution comes from outdoor lights that are excessively bright or misdirected. Light pollution not only limits the number of stars that the average person can see but also limits what astronomers can detect. Light pollution affects migrating animals, too. Luckily, there are ways to reduce light pollution. The International Dark Sky Association is working to reduce light pollution around the world. Find out how you can reduce light pollution in your community or home.

Math ACTIVITY

A Virginia high school student named Jennifer Barlow started "National Dark Sky Week." If light pollution is reduced for 1 week each year, for what percentage of the year would light pollution be reduced?

Answer to Social Studies Activity
Answers may vary.

Answer to Math Activity
7 days ÷ 365 days × 100 = 1.9%

People in Science

Neil deGrasse Tyson

Star Writer When Neil deGrasse Tyson was nine years old, he visited a planetarium for the first time. Tyson was so affected by the experience he decided at that moment to dedicate his life to studying the universe. Tyson began studying the stars through a telescope on the roof of his apartment building. This interest led Tyson to attend the Bronx High School of Science, where he studied astronomy and physics. Tyson's passion for astronomy continued when he was a student at Harvard. However, Tyson soon realized that he wanted to share his love of astronomy with the public. So, today Tyson is America's best-known astrophysicist. When something really exciting happens in the universe, such as the discovery of evidence of water on Mars, Tyson is often asked to explain the discovery to the public. He has been interviewed hundreds of times on TV programs and has written several books. Tyson also writes a monthly column in the magazine *Natural History*. But writing and appearing on TV isn't even his day job! Tyson is the director of the Hayden Planetarium in New York—the same planetarium that ignited his interest in astronomy when he was nine years old!

Language Arts ACTIVITY

WRITING SKILL Be a star writer! Visit a planetarium or find a Web site that offers a virtual tour of the universe. Write a magazine-style article about the experience.

To learn more about these Science in Action topics, visit **go.hrw.com** and type in the keyword **HZ5OBSF.**

Check out Current Science® articles related to this chapter by visiting go.hrw.com. Just type in the keyword **HZ5CS18.**

Stars, Galaxies, and the Universe
Chapter Planning Guide

Compression guide:
To shorten instruction because of time limitations, omit the Chapter Lab.

OBJECTIVES	LABS, DEMONSTRATIONS, AND ACTIVITIES	TECHNOLOGY RESOURCES
PACING • 90 min pp. 30–39 **Chapter Opener**	**SE** Start-up Activity, p. 31 ◆ GENERAL	**OSP** Parent Letter ■ GENERAL **CD** Student Edition on CD-ROM **CD** Guided Reading Audio CD ■ **TR** Chapter Starter Transparency* **VID** Brain Food Video Quiz
Section 1 Stars • Describe how color indicates the temperature of a star. • Explain how a scientist can identify a of star's composition. • Describe how scientists classify stars. • Compare absolute magnitude with apparent magnitude. • Identify how astronomers measure distances from Earth to stars. • Describe the difference between the apparent motion and the actual motion of stars.	**TE** Demonstration Light Pollution, p. 32 ◆ BASIC **SE** Connection to Physics Fingerprinting Cars, p. 33 GENERAL **TE** Demonstration Color Indicates Temperature, p. 33 GENERAL **SE** Connection to Biology Rods, Cones, and Stars, p. 34 GENERAL **SE** School-to-Home Activity Stargazing, p. 35 GENERAL **TE** Connection Activity Math, p. 36 GENERAL **SE** Quick Lab Not All Thumbs!, p. 37 GENERAL **SE** Skills Practice Lab Red Hot, or Not?, p. 54 ◆ GENERAL **CRF** Datasheet for Chapter Lab* **SE** Skills Practice Lab I See the Light!, p. 112 GENERAL **CRF** Datasheet for LabBook* **LB** Whiz-Bang Demonstrations Where Do the Stars Go?* BASIC	**CRF** Lesson Plans* **TR** Bellringer Transparency* **TR** Finding the Distance to Stars with Parallax* **CRF** SciLinks Activity* GENERAL **VID** Lab Videos for Earth Science
PACING • 45 min pp. 40–45 **Section 2 The Life Cycle of Stars** • Describe different types of stars. • Describe the quantities that are plotted in the H-R diagram. • Explain how stars at different stages in their life cycle appear on the H-R diagram.	**SE** Connection to Astronomy Long Live the Sun, p. 41 GENERAL **TE** Activity Researching Red Giants, p. 41 ADVANCED **TE** Group Activity Star Cycle, p. 42 GENERAL **TE** Activity Using the H-R Diagram, p. 43 BASIC **TE** Connection Activity Language Arts, p. 43 BASIC	**CRF** Lesson Plans* **TR** Bellringer Transparency* **TR** The H-R Diagram A* **TR** The H-R Diagram B* **TE** Internet Activity, p. 43 GENERAL
PACING • 45 min pp. 46–49 **Section 3 Galaxies** • Identify three types of galaxies. • Describe the contents and characteristics of galaxies. • Explain why looking at distant galaxies reveals what young galaxies looked like.	**SE** Connection to Language Arts Alien Observer, p. 47 GENERAL **SE** Science in Action Math, Social Studies, and Language Arts Activities, pp. 60–61 GENERAL	**CRF** Lesson Plans* **TR** Bellringer Transparency*
PACING • 45 min pp. 50–53 **Section 4 Formation of the Universe** • Describe the big bang theory. • Explain evidence used to support the big bang theory. • Describe the structure of the universe. • Describe two ways scientists calculate the age of the universe. • Explain what will happen if the universe expands forever.	**TE** Activity The Expanding Universe, p. 50 ◆ GENERAL **LB** Long-Term Projects & Research Ideas Contacting the Aliens* ADVANCED	**CRF** Lesson Plans* **TR** Bellringer Transparency* **TR** *LINK TO PHYSICAL SCIENCE* The Doppler Effect*

PACING • 90 min

CHAPTER REVIEW, ASSESSMENT, AND STANDARDIZED TEST PREPARATION

CRF Vocabulary Activity* GENERAL
SE Chapter Review, pp. 56–57 GENERAL
CRF Chapter Review* ■ GENERAL
CRF Chapter Tests A* ■ GENERAL, B* ADVANCED, C* SPECIAL NEEDS
SE Standardized Test Preparation, pp. 58–59 GENERAL
CRF Standardized Test Preparation* GENERAL
CRF Performance-Based Assessment* GENERAL
OSP Test Generator GENERAL
CRF Test Item Listing* GENERAL

Online and Technology Resources

go.hrw.com

Visit go.hrw.com for a variety of free resources related to this textbook. Enter the keyword **HZ5UNV**.

Holt Online Learning

Students can access interactive problem-solving help and active visual concept development with the *Holt Science and Technology* Online Edition available at **www.hrw.com**.

Guided Reading Audio CD

A direct reading of each chapter using instructional visuals as guideposts. For auditory learners, reluctant readers, and Spanish-speaking students. Available in English and Spanish.

SKILLS DEVELOPMENT RESOURCES	SECTION REVIEW AND ASSESSMENT	STANDARDS CORRELATIONS
SE Pre-Reading Activity, p. 30 `GENERAL` **OSP** Science Puzzlers, Twisters & Teasers `GENERAL`		National Science Education Standards UCP 1, 2
CRF Directed Reading A* ■ `BASIC`, B* `SPECIAL NEEDS` **CRF** Vocabulary and Section Summary* ■ `GENERAL` **SE** Reading Strategy Prediction Guide, p. 32 `GENERAL` **TE** Reading Strategy Mnemonics, p. 33 `GENERAL` **TE** Inclusion Strategies, p. 35 ♦ **SE** Math Practice Starlight, Star Bright, p. 36 `GENERAL` **MS** Math Skills for Science Arithmetic with Positive and Negative Numbers* `GENERAL` **MS** Math Skills for Science Distances in Space* `GENERAL`	**SE** Reading Checks, pp. 32, 34, 36, 37, 38 `GENERAL` **TE** Reteaching, p. 38 `BASIC` **TE** Quiz, p. 38 `GENERAL` **TE** Alternative Assessment, p. 38 `GENERAL` **SE** Section Review,* p. 39 `GENERAL` **CRF** Section Quiz* ■ `GENERAL`	UCP 1, 3; SAI 1, 2; SPSP 5; *Chapter Lab:* UCP 1, 3; SAI 1, 2; ST 1; *LabBook:* UCP 2, 3; SAI 1, 2
CRF Directed Reading A* ■ `BASIC`, B* `SPECIAL NEEDS` **CRF** Vocabulary and Section Summary* ■ `GENERAL` **SE** Reading Strategy Paired Summarizing, p. 40 `GENERAL` **CRF** Reinforcement Worksheet Diagramming the Stars* `BASIC`	**SE** Reading Checks, pp. 41, 45 `GENERAL` **TE** Reteaching, p. 44 `BASIC` **TE** Quiz, p. 44 `GENERAL` **TE** Alternative Assessment, p. 44 `GENERAL` **SE** Section Review,* p. 45 ■ `GENERAL` **CRF** Section Quiz* ■ `GENERAL`	UCP 1, 2, 3, 5; SAI 1; HNS 1, 2
CRF Directed Reading A* ■ `BASIC`, B* `SPECIAL NEEDS` **CRF** Vocabulary and Section Summary* ■ `GENERAL` **SE** Reading Strategy Reading Organizer, p. 46 `GENERAL` **TE** Inclusion Strategies, p. 47	**SE** Reading Checks, pp. 46, 48, 49 `GENERAL` **TE** Reteaching, p. 48 `BASIC` **TE** Quiz, p. 48 `GENERAL` **TE** Alternative Assessment, p. 48 `ADVANCED` **SE** Section Review,* p. 49 ■ `GENERAL` **CRF** Section Quiz* ■ `GENERAL`	UCP 1, 5; SAI 1; ST 1, 2; HNS 1, 2; SPSP 5
CRF Directed Reading A* ■ `BASIC`, B* `SPECIAL NEEDS` **CRF** Vocabulary and Section Summary* ■ `GENERAL` **SE** Reading Strategy Prediction Guide, p. 50 `GENERAL` **CRF** Critical Thinking Fleabert and the Amazing Watermelon Seed* `ADVANCED`	**SE** Reading Checks, pp. 51, 52, 53 `GENERAL` **TE** Reteaching, p. 52 `BASIC` **TE** Quiz, p. 52 `GENERAL` **TE** Alternative Assessment, p. 52 `GENERAL` **SE** Section Review,* p. 53 ■ `GENERAL` **TE** Homework, p. 53 `ADVANCED` **CRF** Section Quiz* ■ `GENERAL`	UCP 1, 2, 3, 5; SAI 1

One-Stop Planner® CD-ROM

This convenient CD-ROM includes:
- Lab Materials QuickList Software
- Holt Calendar Planner
- Customizable Lesson Plans
- Printable Worksheets
- ExamView® Test Generator

cnnstudentnews.com

Find the latest news, lesson plans, and activities related to important scientific events.

SCILINKS. NSTA

www.scilinks.org

Maintained by the **National Science Teachers Association.** See Chapter Enrichment pages for a complete list of topics.

Current Science®

Check out *Current Science* articles and activities by visiting the HRW Web site at **go.hrw.com.** Just type in the keyword **HZ5CS19T.**

Classroom Videos
- **Lab Videos** demonstrate the chapter lab.
- **Brain Food Video Quizzes** help students review the chapter material.

Visual Resources

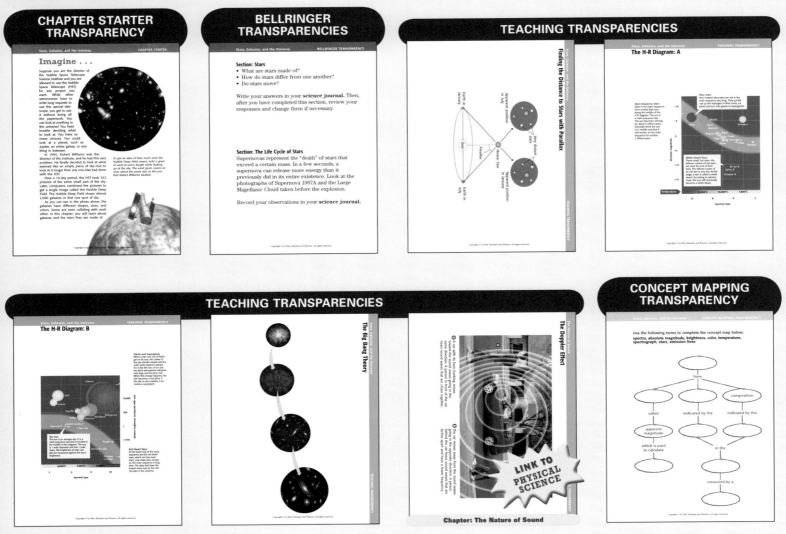

CHAPTER STARTER TRANSPARENCY

Imagine . . .

Suppose you are the director of the Hubble Space Telescope Science Institute and you are allowed to use the Hubble Space Telescope (HST) for any project you want. While other astronomers have to write long requests to use this special telescope, you get to use it without doing all the paperwork. You can look at anything in the universe! You have trouble deciding what to look at. You have so many choices. You could look at a planet, such as Jupiter, an entire galaxy, or anything in between.

In 1995, Robert Williams was the director of the institute, and he had this very problem. He finally decided to look at what seemed like an empty piece of sky—but to look at it longer than any one else had done with the HST.

Over a 10-day period, the HST took 342 pictures of the same small part of the sky. Later, computers combined the pictures to get a single image called the Hubble Deep Field. The Hubble Deep Field shows almost 2,000 galaxies in that one spot of sky.

As you can see in the photo above, the galaxies have different shapes, sizes, and colors. Some are even colliding with each other. In this chapter, you will learn about galaxies and the stars they are made of.

To get an idea of how much area the Hubble Deep Field covers, hold a grain of sand at arm's length while looking up at the sky. The sand grain covers an area about the same size as the one Robert Williams studied.

BELLRINGER TRANSPARENCIES

Section: Stars
- What are stars made of?
- How do stars differ from one another?
- Do stars move?

Write your answers in your **science journal.** Then, after you have completed this section, review your responses and change them if necessary.

Section: The Life Cycle of Stars
Supernovas represent the "death" of stars that exceed a certain mass. In a few seconds, a supernova can release more energy than it previously did in its entire existence. Look at the photographs of Supernova 1987A and the Large Magellanic Cloud taken before the explosion.

Record your observations in your **science journal.**

TEACHING TRANSPARENCIES

Finding the Distance to Stars with Parallax

The H-R Diagram: A

The H-R Diagram: B

The Big Bang Theory

The Doppler Effect

Chapter: The Nature of Sound

CONCEPT MAPPING TRANSPARENCY

Use the following terms to complete the concept map below: spectra, absolute magnitude, brightness, color, temperature, spectrograph, stars, emission lines

Planning Resources

LESSON PLANS

Lesson Plan   SAMPLE

Section: Waves

Pacing
Regular Schedule: with lab(s):2 days   without lab(s):2 days
Block Schedule: with lab(s): 1 1/2 days without lab(s):1 day

Objectives
1. Relate the seven properties of life to a living organism.
2. Describe seven themes that can help you to organize what you learn about biology.
3. Identify the tiny structures that make up all living organisms.
4. Differentiate between reproduction and heredity and between metabolism and homeostasis.

National Science Education Standards Covered
LSInter6:Cells have particular structures that underlie their functions.
LSMat1:Most functions involve chemical reactions.
LSBeh1:Cells store and use information to guide their functions.
UCP1:Cell functions are regulated.
SI1: Cells can differentiate and form complete multicellular organisms.
PS1: Species evolve over time.
ESS1: The great diversity of organisms is the result of more than 3.5 billion years of evolution.
ESS2: Natural selection and its evolutionary consequences provide a scientific explanation for the fossil record of ancient life forms as well as for the striking molecular similarities observed among the diverse species of living organisms.
ST1: The millions of different species of plants, animals, and microorganisms that live on Earth today are related by descent from common ancestors.
ST2: The energy for life primarily comes from the sun.
SPSP1: The complexity and organization of organisms accommodates the need for obtaining, transforming, transporting, releasing, and eliminating the matter and energy used to sustain the organism.
SPSP6: As matter and energy flows through different levels of organization of living systems—cells, organs, communities—and between living systems and the physical environment, chemical elements are recombined in different ways.
HNS1: Organisms have behavioral responses to internal changes and to external stimuli.

PARENT LETTER   SAMPLE

Dear Parent,

Your son's or daughter's science class will soon begin exploring the chapter entitled "The World of Physical Science." In this chapter, students will learn about how the scientific method applies to the world of physical science and the role of physical science in the world. By the end of the chapter, students should demonstrate a clear understanding of the chapter's main ideas and be able to discuss the following topics:

1. physical science is the study of energy and matter (Section 1)
2. the role of physical science in the world around them (Section 1)
3. careers that rely on physical science (Section 1)
4. the steps used in the scientific method (Section 2)
5. examples of technology (Section 2)
6. how the scientific method is used to answer questions and solve problems (Section 2)
7. how our knowledge of science changes over time (Section 2)
8. how models represent real objects or systems (Section 3)
9. examples of different ways models are used in science (Section 3)
10. the importance of the International System of Units (Section 4)
11. the appropriate units to use for particular measurements (Section 4)
12. how area and density are derived quantities (Section 4)

Questions to Ask Along the Way

You can help your son or daughter learn about these topics by asking interesting questions such as the following:

- What are some surprising careers that use physical science?
- What is a characteristic of a good hypothesis?
- When is it a good idea to use a model?
- Why do Americans measure things in terms of inches and yards and meters?

ALSO IN SPANISH

TEST ITEM LISTING

TEST ITEM LISTING
The World of Earth Science   SAMPLE

MULTIPLE CHOICE

1. A limitation of models is that
   a. they are large enough to see.
   b. they do not act exactly like the things that they model.
   c. they are smaller than the things that they model.
   d. they model unfamiliar things.
   Answer: B   Difficulty: 1   Section: 3   Objective: 2

2. The length 10 m is equal to
   a. 100 cm.   c. 10,000 mm.
   b. 1,000 cm.   d. Both (b) and (c).
   Answer: B   Difficulty: 1   Section: 1   Objective: 2

3. To be valid, a hypothesis must be
   a. testable.   c. made into a law.
   b. supported by evidence.   d. Both (a) and (b)
   Answer: D   Difficulty: 1   Section: 1   Objective: 2 1

4. The statement "Sheila has a stain on her shirt" is an example of a(n)
   a. law.   c. observation.
   b. hypothesis.   d. prediction.
   Answer: B   Difficulty: 1   Section: 2   Objective: 2

5. A hypothesis is often developed out of
   a. observations.   c. laws.
   b. experiments.   d. Both (a) and (b)
   Answer: B   Difficulty: 1   Section: 2   Objective: 2

6. How many milliliters are in 3.5 kL?
   a. 3,500 mL   c. 3,500,000 mL
   b. 0.0035 mL.   d. 35,000 mL.
   Answer: B   Difficulty: 1   Section: 3   Objective: 2

7. A map of Seattle is an example of a
   a. law.   c. model.
   b. theory.   d. unit.
   Answer: B   Difficulty: 1   Section: 3   Objective: 2

8. The law of conservation of mass says the total all mass before a chemical change is
   a. more than the total mass after the change.
   b. less than the total mass after the change.
   c. the same as the total mass after the change.
   d. not the same as the total mass after the change.
   Answer: B   Difficulty: 1   Section: 3   Objective: 2

9. A lab has the safety icons shown below. These icons mean that you should wear
   a. only safety goggles.   c. safety goggles and a lab apron.
   b. only a lab apron.   d. safety goggles, a lab apron, and gloves.
   Answer: B   Difficulty: 1   Section: 3   Objective: 2

10. In which of the following areas might you find a geochemist at work?
    a. studying the chemistry of rocks   c. studying fishes
    b. studying forestry   d. studying the atmosphere
    Answer: B   Difficulty: 1   Section: 3

One-Stop Planner® CD-ROM

This CD-ROM includes all of the resources shown here and the following time-saving tools:

- *Lab Materials QuickList Software*
- *Customizable lesson plans*
- *Holt Calendar Planner*
- *The powerful ExamView® Test Generator*

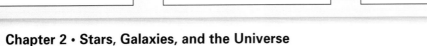

Meeting Individual Needs

DIRECTED READING A

Skills Worksheet
Directed Reading A   SAMPLE

Section:
THAT'S SCIENCE!
1. How did James Czarnowski get his idea for the penguin
Explain.

BASIC

ALSO IN SPANISH

What is unusual about the way that Proteus moves through

DIRECTED READING B

Skills Worksheet
Directed Reading B   SAMPLE

Section:
THAT'S SCIENCE!
1. How did James Czarnowski get his idea for the penguin boat, Proteus?
Explain.

2. What is unusual about the way that Proteus moves through the water?

SPECIAL NEEDS   PHYSICAL SCIENCE
and a cheetah have in common?

VOCABULARY ACTIVITY

Activity
Vocabulary Activity   SAMPLE

Getting the Dirt on the Soil
After you finish reading Chapter: [Unique Title], try this puzzle! Use the clues below to unscramble the vocabulary words. Write your answer in the space provided.

breakdown of rock into   9. the chemical breakdown of rocks
and smaller pieces: and minerals into new
GNETH   substances: CAMILCHE
                                    THEARGWEN
lieces of rock being beneath soil

GENERAL

VOCABULARY AND SECTION SUMMARY

Skills Worksheet
Vocabulary & Notes   SAMPLE

Section:
VOCABULARY
In your own words, write a definition of the following term in the space provided.

1. scientific method

2. technology

ALSO IN SPANISH

GENERAL

REINFORCEMENT

Skills Worksheet
Reinforcement   SAMPLE

The Plane Truth
Complete this worksheet after you finish reading the Section: [Unique Section Title]

You plan to enter a paper airplane contest sponsored by Talkin' Physical Science magazine. The person whose airplane flies the farthest wins a lifetime subscription to the magazine! The week before the contest you watch an airplane landing at a nearby airport. You notice that the wings of the airplane have flaps, as shown in the illustration at right. The paper airplanes you've been testing do not have wing flaps. What question would you ask yourself based on these observations? Write your

Flaps

BASIC

CRITICAL THINKING

Skills Worksheet
Critical Thinking   SAMPLE

A Solar Solution

Dear Mr. Burns,
I've got two great ideas for a new product called the Glins Heater. It's a portable, solar-powered space heater. The heater's design includes these features:
• a heater will be so tiny
• it is an adult's arms and sit while at a sitting bar
• to nestle will have a
• 2 glass top set at an angle to catch the sun's rays
• be inside of the heater it will be dark colored to absorb heat best

Joseph D. Burns
Inventors' Advisory Consultants
Portland, OR 97201

Thank my idea will work. I will make the Glins if using solar setting have and wrong on tests using models. Please write back here with your

ADVANCED

SCILINKS ACTIVITY

Activity
SciLinks Activity   SAMPLE

MARINE ECOSYSTEMS
Go to www.scilinks.com. To find links related to marine ecosystems, type in the keyword HL5600. Then, use the links to answer the questions about marine ecosys-

SciLINKS
NSTA
Go to: www.scilinks.org
Topic: Reproductive System
Irregularities
SciLinks code: HL5600

percentage of the Earth's surface is covered by water?

GENERAL

SCIENCE PUZZLERS, TWISTERS & TEASERS

Name _____ Class _____
21 SCIENCE PUZZLERS, TWISTERS & TEASERS
CHAPTER
Stars, Galaxies, and the Universe

Fractured Frames
1. Each frame represents a word from the chapter or the name of a star, if you read it in just the right way. What word or phrase does the puzzle represent?

LACK LACK   Procy
                     B

b.

tauALPHAALPHAtau   prot
                          star

GENERAL

Labs and Activities

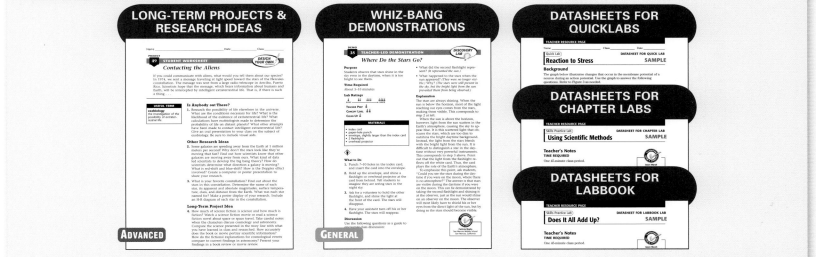

LONG-TERM PROJECTS & RESEARCH IDEAS

Name _____ Date _____ Class _____
PROJECT
49 STUDENT WORKSHEET   *DESIGN YOUR OWN*
Contacting the Aliens

If you could communicate with aliens, what would you tell them about our species? In 1974, we sent a message traveling at light speed toward the stars of the Hercules constellation. The message was sent from a large radio telescope in Arecibo, Puerto Rico. Scientists hope that the message, which bears information about humans and Earth, will be intercepted by intelligent extraterrestrial life. That is, if there is such a thing. . . .

USEFUL TERM
exobiology
the investigation of the possibility of extraterrestrial life

Is Anybody out There?
1. Research the possibility of life elsewhere in the universe. What are the conditions necessary for life? What is the likelihood of the existence of extraterrestrial life? What calculations have exobiologists made to determine the probability of life on distant planets? What other attempts have been made to contact intelligent extraterrestrial life? Give an oral presentation to your class on the subject of exobiology. Be sure to include visual aids.

Other Research Ideas
2. Some galaxies are speeding away from the Earth at 1 million meters per second! Why don't the stars look like they're moving that fast? Find out how scientists know that other galaxies are moving away from us. How did data led scientists to develop the big bang theory? How do scientists determine what direction a galaxy is moving? What is red-shift and blue-shift? How is the Doppler effect involved? Create a computer or poster presentation to share your research.

3. What is your favorite constellation? Find out about the stars in this constellation. Determine the name of each star, its apparent and absolute magnitudes, surface temperature, class, and distance from the Earth. What was each star named for? Make a poster display of your research. Include an H-R diagram of each star in the constellation.

Long-Term Project Idea
4. How much of science fiction is science and how much is fiction? Watch a science fiction movie or read a science fiction novel about space or space travel. Take careful notes when the characters discuss cosmology and astronomy. Compare the science presented in the story line with what you have learned in class and researched. How accurately does the book or movie portray scientific information? How do the fictional explanations for cosmological events compare to current findings in astronomy? Present your findings in a book review or movie review.

ADVANCED

WHIZ-BANG DEMONSTRATIONS

DEMO
35 TEACHER-LED DEMONSTRATION   *DISCOVERY LAB*
Where Do the Stars Go?

Purpose
Students observe that stars shine in the sky even in the daytime, when it is too bright to see them.

Time Required
About 5–10 minutes

Lab Ratings
TEACHER PREP
CONCEPT LEVEL
CLEAN UP

MATERIALS
• index card
• paper-hole punch
• envelope, slightly larger than the index card
• 2 flashlights
• overhead projector

What to Do
1. Punch 7–10 holes in the index card, and insert the card into the envelope.

2. Hold up the envelope, and shine a flashlight or overhead projector at the card from behind. Tell students to imagine they are seeing stars in the night sky.

3. Ask for a volunteer to hold the other flashlight, and shine the light at the front of the card. The stars will disappear.

4. Have your assistant turn off his or her flashlight. The stars will reappear.

Discussion
Use the following questions as a guide to encourage class discussion:

• What did the second flashlight represent? *(It represented the sun.)*

• What happened to the stars when the sun appeared? *(They were no longer visible.) Why? (The stars were still present in the sky, but the bright light from the sun prevented them from being observed.)*

Explanation
The stars are always shining. When the sun is below the horizon, most of the light reaching our eyes comes from the stars, making them visible. This corresponds to step 2 at left.

When the sun is above the horizon, however, light from the sun scatters in the Earth's atmosphere, causing the sky to appear blue. It is this scattered light that obscures the stars, which are too dim to outshine the bright daytime background. Instead, the light from the stars blends with the bright light from the sun. It is difficult to distinguish a star in the daytime without very powerful instruments. This corresponds to step 3 above. Point out that the light from the flashlight reflects off the white card. Thus, the card plays the role of the Earth's atmosphere.

To emphasize this point, ask students, "Could you see the stars during the daytime if you were on the moon, where there is no atmosphere?" The answer is that stars are visible during the daytime if you were on the moon. This can be demonstrated by taking the second flashlight and shining it at the observer, just as the sun would shine on an observer on the moon. The observer will most likely have to shield his or her eyes from the direct light of the sun, but by doing so the stars should become visible.

Patricia Marsh
San Marcos Middle School
San Marcos, California

GENERAL

DATASHEETS FOR QUICKLABS

TEACHER RESOURCE PAGE
Name _____ Class _____ Date _____
Quick Lab
Reaction to Stress   DATASHEET FOR QUICK LAB   SAMPLE

Background
The graph below illustrates changes that occur in the membrane potential of a neuron during an action potential. Use the graph to answer the following questions. Refer to Figure 3 as needed.

DATASHEETS FOR CHAPTER LABS

TEACHER RESOURCE PAGE
Skills Practice Lab
Using Scientific Methods   DATASHEET FOR CHAPTER LAB   SAMPLE

Teacher's Notes
TIME REQUIRED
One 45-minute class period.

DATASHEETS FOR LABBOOK

TEACHER RESOURCE PAGE
Skills Practice Lab
Does It All Add Up?   DATASHEET FOR LABBOOK LAB   SAMPLE

Teacher's Notes
TIME REQUIRED
One 45-minute class period.

Review and Assessments

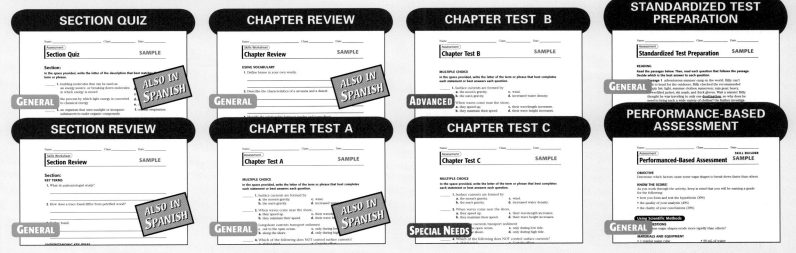

SECTION QUIZ

Assessment
Section Quiz   SAMPLE

Section:
In the space provided, write the letter of the description that best matches the term or phrase.

_____ 1. building molecules that can be used as an energy source, or breaking down molecules in which energy is stored
_____ 2. the process by which light energy is converted to chemical energy
_____ 3. an organism that uses sunlight or inorganic substances to make organic compounds

ALSO IN SPANISH

GENERAL

SECTION REVIEW

Skills Worksheet
Section Review   SAMPLE

Section:
KEY TERMS
1. What do paleontologist study?

2. How does a trace fossil differ from petrified wood?

3. _____ fossil.

UNDERSTANDING KEY IDEAS

GENERAL

ALSO IN SPANISH

CHAPTER REVIEW

Skills Worksheet
Chapter Review   SAMPLE

USING VOCABULARY
1. Define biome in your own words.

2. Describe the characteristics of a savanna and a desert.

3. Identify the relationship between tundra and permafrost

ALSO IN SPANISH

GENERAL

CHAPTER TEST A

Assessment
Chapter Test A   SAMPLE

MULTIPLE CHOICE
In the space provided, write the letter of the term or phrase that best completes each statement or best answers each question.

_____ 1. Surface currents are formed by
a. the moon's gravity.   c. wind.
b. the sun's gravity.   d. increased

_____ 2. When waves come near the shore,
a. they speed up.   c. their wavele
b. they maintain their speed.   d. their wave h

_____ 3. Longshore currents transport sediment
a. out to the open ocean.   c. only during low
b. along the shore.   d. only during hig

_____ 4. Which of the following does NOT control surface currents?

ALSO IN SPANISH

GENERAL

CHAPTER TEST B

Assessment
Chapter Test B   SAMPLE

MULTIPLE CHOICE
In the space provided, write the letter of the term or phrase that best completes each statement or best answers each question.

_____ 1. Surface currents are formed by
a. the moon's gravity.   c. wind.
b. the sun's gravity.   d. increased water density.

_____ 2. When waves come near the shore,
a. they speed up.   c. their wavelength increases.
b. they maintain their speed.   d. their wave height increases.

ADVANCED

CHAPTER TEST C

Assessment
Chapter Test C   SAMPLE

MULTIPLE CHOICE
In the space provided, write the letter of the term or phrase that best completes each statement or best answers each question.

_____ 1. Surface currents are formed by
a. the moon's gravity.   c. wind.
b. the sun's gravity.   d. increased water density.

_____ 2. When waves come near the shore,
a. they speed up.   c. their wavelength increases.
b. they maintain their speed.   d. their wave height increases.

_____ 3. Longshore currents transport sediment
a. out to the open ocean.   c. only during low tide.
b. along the shore.   d. only during high tide.

_____ 4. Which of the following does NOT control surface currents?

SPECIAL NEEDS

STANDARDIZED TEST PREPARATION

Name _____ Class _____ Date _____
Assessment
Standardized Test Preparation   SAMPLE

READING
Read the passages below. Then, read each question that follows the passage. Decide which is the best answer to each question.

Passage 1 adventurous summer camp in the world. Billy can't wait to head for the outdoors. Billy checked the recommended supply list: light, summer clothes; sunscreen; rain gear; heavy, waterproof-lined jacket; ski mask; and thick gloves. Wait a minute! Billy thought he was traveling to only one destination, so why does he need to bring such a wide variety of clothes? On further investiga-

GENERAL

PERFORMANCE-BASED ASSESSMENT

Name _____ Class _____ Date _____
Assessment
Performanced-Based Assessment   SKILL BUILDER   SAMPLE

OBJECTIVE
Determine which factors cause some sugar shapes to break down faster than others.

KNOW THE SCORE
As you work through the activity, keep in mind that you will be earning a grade for the following:
• how you form and test the hypothesis (30%)
• the quality of your analysis (40%)
• the clarity of your conclusions (30%)

Using Scientific Methods
QUESTIONS
me sugar shapes erode more rapidly than others?

MATERIALS AND EQUIPMENT
• 1 regular sugar cube   • 90 mL of water

GENERAL

This Chapter Enrichment provides relevant and interesting information to expand and enhance your presentation of the chapter material.

Section 1

Stars

Space Distances

- After the sun, the star closest to Earth is Proxima Centauri, located more than 4 light-years away. To walk an equivalent distance, a person would have to walk around the Earth more than 944 million times!

- Four space probes—*Voyagers 1* and *2* and *Pioneers 10* and *11*—are en route to interstellar space. They are traveling at a rate of approximately 40,000 km/h. Even at this astounding speed, it would take 150,000 years for the probes to reach Proxima Centauri.

Is That a Fact!

◆ The Big Dipper is not a constellation; it is an asterism, a familiar pattern of stars that may or may not be part a constellation. The Big Dipper is part of the constellation known as *Ursa Major,* or Big Bear.

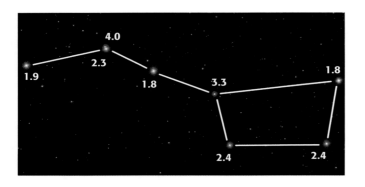

Section 2

The Life Cycle of Stars

The Birth of a Star

- Like humans, stars undergo a life cycle that consists of birth, infancy, maturity, old age, and death. In space, clouds of gas and dust abound; drawn together by gravity, they eventually form a protostar. This fledgling star gives off no visible light and must undergo many changes before it is recognizable as a star. In a process that takes millions of years, the protostar contracts. This shrinkage causes an enormous buildup of pressure and heat. When its temperature reaches about 10 million degrees Celsius, the protostar stops contracting and the process of nuclear fusion begins. Once this hydrogen-fusing process is initiated, the star is born!

Is That a Fact!

◆ From the time our sun emerged as a protostar, it took about 10 million years to become a main-sequence star. A star with one-tenth the mass of the sun would mature in 100 million years, and a star with 3 times the mass of the sun would mature in 1 million years. Stars that have large mass are hotter and take less time to mature than stars that have less mass.

Section 3

Galaxies

Observing Spiral Galaxies

- Most spiral galaxies appear very thin when seen edge-on. The thickness of the spiral disk is only about one-fifth to one-twentieth the width of the disk. When seen edge-on, spiral galaxies resemble a fried egg. The relative size and shape of the bulge are important clues to determining the type of galaxy.

How Many Stars Are in a Galaxy?

- To estimate the number of stars in a galaxy, astronomers consider the sun as one unit of mass. Large spiral galaxies, for example, have a mass of 1 billion to 1 trillion solar masses. Dwarf elliptical galaxies have only a few million solar masses, about one-thousandth the mass of a spiral galaxy. Giant elliptical galaxies have more mass than large spiral galaxies do.

Is That a Fact!

- ◆ The word *galaxy* comes from the Greek word *gala,* meaning "milk." The visible portion of our galaxy looks like a milky cloud in the night sky.

Section 4

Formation of the Universe

Top-Down or Bottom-Up?

- No one is certain how large-scale structures in the universe emerged. Some scientists support the top-down theory. This theory explains that areas of the universe that contained large-scale objects (the size of clusters and super-clusters) were the first to collapse into gaseous, pancake-like shapes. Galaxies condensed from these structures.

- Other scientists support the bottom-up theory. This theory argues that areas of the universe that had small-scale objects (the size of galaxies or smaller) were the first to form. Because of the gravitational forces, these areas aggregated into clusters and superclusters.

Life on Other Planets?

- Many scientists believe that our galaxy alone contains hundreds of millions of planets similar to Earth. These planets may be able to support carbon-based life.

Is That a Fact!

- ◆ The William Herschel telescope on La Palma in the Canary Islands is one of the world's biggest optical telescopes. With its 4.2 m mirror, it could detect a single candle burning 160,000 km away.

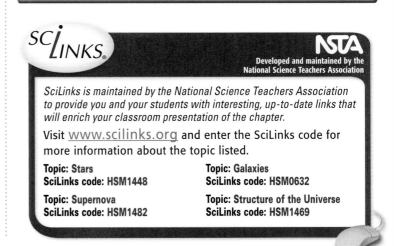

SciLinks is maintained by the National Science Teachers Association to provide you and your students with interesting, up-to-date links that will enrich your classroom presentation of the chapter.

Visit www.scilinks.org and enter the SciLinks code for more information about the topic listed.

Topic: Stars
SciLinks code: HSM1448

Topic: Galaxies
SciLinks code: HSM0632

Topic: Supernova
SciLinks code: HSM1482

Topic: Structure of the Universe
SciLinks code: HSM1469

Overview

Tell students that this chapter will help them learn about the characteristics and life cycles of stars. Students will also learn about types of galaxies, and the structure of the universe.

Assessing Prior Knowledge

Students should be familiar with the following topics:

• the history of astronomy

• the size and scale of the universe

Identifying Misconceptions

As students learn about the expansion of the universe, some of them may think that the expansion of the universe is caused by the expansion of the particles rather than an increase in the space between the particles. Students are also often confused by the terms *mass* and *volume* and often define both as "the amount of matter." It may be helpful to review the definitions of *matter, mass,* and *volume* before teaching the material about universal expansion.

Stars, Galaxies, and the Universe

About the

This image was taken by the *Hubble Space Telescope* and shows the IC 2163 galaxy (right) swinging past the NGC 2207 galaxy (left). Strong forces from NGC 2207 have caused stars and gas to fling out of IC 2163 into long streamers.

PRE-READING ACTIVITY

FOLDNOTES **Three-Panel Flip Chart**
Before you read the chapter, create the FoldNote entitled "Three-Panel Flip Chart" described in the **Study Skills** section of the Appendix. Label the flaps of the three-panel flip chart with "Stars," "Galaxies," and "The universe." As you read the chapter, write information you learn about each category under the appropriate flap.

Standards Correlations

National Science Education Standards

The following codes indicate the National Science Education Standards that correlate to this chapter. The full text of the standards is at the front of the book.

Chapter Opener
UCP 1, 2

Section 1 Stars
UCP 1, 3; SAI 1; SPSP 5; *LabBook:* UCP 2, 3; SAI 1

Section 2 The Life Cycle of Stars
UCP 1, 2, 3, 5; SAI 1; HNS 1, 2

Section 3 Galaxies
UCP 1, 5; SAI 1; ST 1, 2; HNS 1, 2; SPSP 5

Section 4 Formation of the Universe
UCP 1, 2, 3, 5; SAI 1

Chapter Lab
UCP 1, 3; SAI 1, 2; ST 1

Chapter Review
UCP 1, 2, 3, 5; SAI 1; ST 2; HNS 1, 2

Science in Action
UCP 1, 2, 3; HNS 1, 2, 3

START-UP ACTIVITY

MATERIALS

FOR EACH GROUP
- glitter (any color)
- glass jar, 1 qt
- spoon, wooden
- water

Teacher's Notes: Use glass jars that are clear on all sides so that it is easy for students to view the swirling galaxies.

Answers

1. The water should make a swirling, spiral motion after being stirred.
2. The motion of the glitter should appear to be similar to the galaxies in the photo.
3. Answers may vary.

START-UP ACTIVITY

Exploring the Movement of Galaxies in the Universe

Not all galaxies are the same. Galaxies can differ by size, shape, and how they move in space. In this activity, you will explore how the galaxies in the photo move in space.

Procedure

1. Fill a **one-quart glass jar** three-fourths of the way with **water.**
2. Take a pinch of **glitter,** and sprinkle it on the surface of the water.
3. Quickly stir the water with a **wooden spoon.** Be sure to stir the water in a circular pattern.
4. After you stop stirring, look at the water from the sides of the jar and from the top of the jar.

Analysis

1. What kind of motion did the water make after you stopped stirring the water?
2. How is the motion similar to the galaxies in the photo?
3. Make up a name that describes the galaxies in the photo.

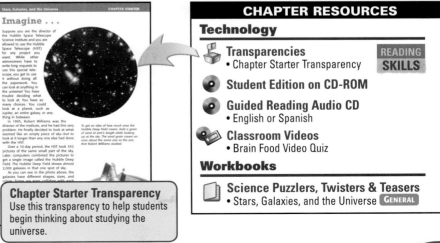

Chapter Starter Transparency
Use this transparency to help students begin thinking about studying the universe.

CHAPTER RESOURCES

Technology

Transparencies
- Chapter Starter Transparency

READING SKILLS

Student Edition on CD-ROM

Guided Reading Audio CD
- English or Spanish

Classroom Videos
- Brain Food Video Quiz

Workbooks

Science Puzzlers, Twisters & Teasers
- Stars, Galaxies, and the Universe **GENERAL**

SECTION
1

Focus

Overview

This section discusses the classification of stars by their color, composition, and brightness. This section also explores how apparent magnitude differs from absolute magnitude and how apparent motion of the stars differs from actual motion of stars.

 **Bellringer**

On the board, write the following questions: "What are stars made of? How do stars differ from one another? Do stars move?" Have students review their responses after completing this section.

Motivate

Demonstration — GENERAL

Light Pollution Demonstrate how ambient light affects the number of visible stars by using a slide projector, a piece of aluminum foil, and a flashlight. Poke small holes in the foil, and in a dark room, project light through the foil. Ask students to count the stars. Then, shine the flashlight on the screen, and ask students to count the stars again. Discuss with students some artificial sources of light pollution. **LS** Visual

---

READING WARM-UP

Objectives

- Describe how color indicates the temperature of a star.
- Explain how a scientist can identify a of star's composition.
- Describe how scientists classify stars.
- Compare absolute magnitude with apparent magnitude.
- Identify how astronomers measure distances from Earth to stars.
- Describe the difference between the apparent motion and the actual motion of stars.

Terms to Learn

spectrum
apparent magnitude
absolute magnitude
light-year
parallax

READING STRATEGY

Prediction Guide Before reading this section, write the title of each heading in this section. Next, under each heading, write what you think you will learn.

Stars

Do you remember the children's song "Twinkle, Twinkle Little Star"? In the song, you sing "How I wonder what you are!" Well, what are stars? And what are they made of?

Most stars look like faint dots of light in the night sky. But stars are actually huge, hot, bright balls of gas that are trillions of kilometers away from Earth. How do astronomers learn about stars when the stars are too far away to visit? Astronomers study starlight!

Color of Stars

Look at the flames on the candle and the Bunsen burner shown in **Figure 1.** Which flame is hottest? How can you tell? Although red and yellow may be thought of as "warm" colors and blue may be thought of as a "cool" color, scientists consider red and yellow to be cool colors and blue to be a warm color. For example, the blue flame of the Bunsen burner is much hotter than the yellow flame of the candle.

If you look carefully at the night sky, you might notice the different colors of some stars. Betelgeuse (BET uhl JOOZ), which is red, and Rigel (RIE juhl), which is blue, are the stars that form two corners of the constellation Orion, shown in **Figure 1.** Because these two stars are different colors, we can conclude that they have different temperatures.

✓ *Reading Check* Which star is hotter, Betelgeuse or Rigel? **Explain your answer.** (*See the Appendix for answers to Reading Checks.*)

Figure 1 In the same way that we know the blue flame of the Bunsen burner is hotter than the yellow flame of the candle, astronomers know that Rigel is hotter than Betelgeuse.

Betelgeuse

Rigel

CHAPTER RESOURCES

Chapter Resource File

- **Lesson Plan**
- **Directed Reading A** BASIC
- **Directed Reading B** SPECIAL NEEDS

Technology

Transparencies
- Bellringer

Answer to Reading Check

Rigel is hotter than Betelgeuse because blue stars are hotter than red stars.

Composition of Stars

A star is made up of different elements in the form of gases. The inner layers of a star are very dense and hot. But the outer layers of a star, or a star's atmosphere, are made up of cool gases. Elements in a star's atmosphere absorb some of the light that radiates from the star. Because different elements absorb different wavelengths of light, astronomers can tell what elements a star is made of from the light they observe from the star.

The Colors of Light

When you look at white light through a glass prism, you see a rainbow of colors called a **spectrum.** The spectrum consists of millions of colors, including red, orange, yellow, green, blue, indigo, and violet. A hot, solid object, such as the glowing wire inside a light bulb, gives off a *continuous spectrum*—a spectrum that shows all the colors. However, the spectrum of a star is different. Astronomers use an instrument called a *spectrograph* to break a star's light into a spectrum. The spectrum gives astronomers information about the composition and temperature of a star. To understand how to read a star's spectrum, think about something more familiar—a neon sign.

Making an ID

Many restaurants use neon signs to attract customers. The gas in a neon sign glows when an electric current flows through the gas. If you were to look at the sign with a spectrograph, you would not see a continuous spectrum. Instead, you would see *emission lines*. Emission lines are lines that are made when certain wavelengths of light, or colors, are given off by hot gases. When an element emits light, only some colors in the spectrum show up, while all the other colors are missing. Each element has a unique set of bright emission lines. Emission lines are like fingerprints for the elements. You can see emission lines for four elements in **Figure 2.**

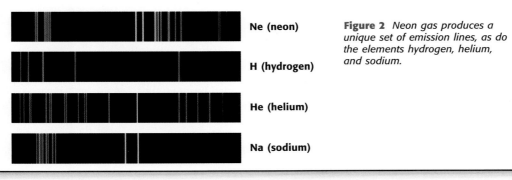

Ne (neon)

H (hydrogen)

He (helium)

Na (sodium)

Figure 2 *Neon gas produces a unique set of emission lines, as do the elements hydrogen, helium, and sodium.*

spectrum the band of color produced when white light passes through a prism

MISCONCEPTION ALERT

Planets and Stars Sometimes, differentiating between stars and planets in the night sky is difficult. If an object twinkles, it's probably a star. Because of their proximity to Earth, planets appear as tiny disks shining with a steady light. Venus is an easy planet to spot—it can often be seen shining brightly in the west, immediately after the sun sets. Venus can also be seen in the morning, accompanying the rising sun. Venus is then called "the morning star." Although stars appear to twinkle in the sky, they actually shine with a steady light. They appear to twinkle because their light is distorted when it passes through Earth's atmosphere. If you were standing on the moon, where there is not an atmosphere, the stars would appear to shine steadily.

Learning About Spectra

When electrons become excited or absorb enough energy, they are boosted to a higher energy level. When the electrons return to their normal energy level, they release energy at specific wavelengths. The specific wavelength emitted depends on the amount of energy that the electron releases when it returns to its normal energy level. This wavelength is unique to each element. By studying the wavelengths emitted from a substance, scientists can determine the elements that are present in that substance. Have students research the elements and compounds that scientists studying starlight have found, and have students use colored pencils to reproduce some of the spectra in their **science journal**.

LS Interpersonal

Answer to Connection to Biology

Rods are good at distinguishing shades of light and dark and at distinguishing shape and movement. Cones are good for distinguishing colors. However, cones do not work well in low light. For this reason, distinguishing star colors is difficult.

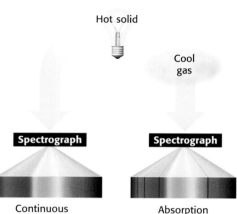

Hot solid

Cool gas

Spectrograph

Spectrograph

Continuous spectrum

Absorption spectrum

Figure 3 *A continuous spectrum (left) shows all colors while an absorption spectrum (right) absorbs some colors. Black lines appear in the spectrum where colors are absorbed.*

CONNECTION TO Biology

Rods, Cones, and Stars

WRITING SKILL Have you ever wondered why it's hard to see the different colors of stars? Our eyes are not sensitive to colors when light levels are low. There are two types of light-sensitive cells in the eye: rods and cones. Research the functions of rods and cones. In your **science journal,** write a paragraph that explains why we can't see colors well in low light.

Trapping the Light–Cosmic Detective Work

Like an element that is charged by an electric current, a star also produces a spectrum. However, while the spectrum of an electrically charged element is made of bright emission lines, a star's spectrum is made of dark emission lines. A star's atmosphere absorbs certain colors of light in the spectrum, which causes black lines to appear.

Identifying Elements Using Dark Lines

Because a star's atmosphere absorbs colors of light instead of emitting them, the spectrum of a star is called an *absorption spectrum*. An absorption spectrum is produced when light from a hot solid or dense gas passes through a cooler gas. Therefore, a star gives off an absorption spectrum because a star's atmosphere is cooler than the inner layers of the star. The black lines of a star's spectrum represent places where less light gets through. **Figure 3** compares a continuous spectrum and an absorption spectrum. What do you notice about the absorption spectrum that is different?

The pattern of lines in a star's absorption spectrum shows some of the elements that are in the star's atmosphere. If a star were made of one element, we could easily identify the element from the star's absorption spectrum. But a star is a mixture of elements and all the different sets of lines for a star's elements appear together in its spectrum. Sorting the patterns is often a puzzle.

✓ **Reading Check** What does a star's absorption spectrum show?

Classifying Stars

In the 1800s, astronomers started to collect and classify the spectra of many stars. At first, letters were assigned to each type of spectra. Stars were classified according to the elements of which they were made. Later, scientists realized that the stars were classified in the wrong order.

Differences in Temperature

Stars are now classified by how hot they are. Temperature differences between stars result in color differences that you can see. For example, the original class O stars are blue—the hottest stars. Look at **Table 1.** Notice that the stars are arranged in order from highest temperature to lowest temperature.

Science Bloopers

The composition of the sun has been the subject of much speculation. In the 19th century, some scientists thought that the sun was made of pure anthracite, because coal was one of the best heat-generating fuels of the time. But given the energy output of the sun, coal would have lasted only about 10,000 years. It took the discovery of radiation and nuclear energy for scientists to develop the current model of the sun's composition and structure.

Answer to Reading Check

A star's absorption spectrum indicates some of the elements that are in the star's atmosphere.

Table 1 Types of Stars				
Class	Color	Surface temperature (°C)	Elements detected	Examples of stars
O	blue	above 30,000	helium	10 Lacertae
B	blue-white	10,000–30,000	helium and hydrogen	Rigel, Spica
A	blue-white	7,500–10,000	hydrogen	Vega, Sirius
F	yellow-white	6,000–7,500	hydrogen and heavier elements	Canopus, Procyon
G	yellow	5,000–6,000	calcium and other metals	the sun, Capella
K	orange	3,500–5,000	calcium and molecules	Arcturus, Aldebaran
M	red	less than 3,500	molecules	Betelgeuse, Antares

Differences in Brightness

With only their eyes to aid them, early astronomers created a system to classify stars based on their brightness. They called the brightest stars in the sky *first-magnitude* stars and the dimmest stars *sixth-magnitude* stars. But when they began to use telescopes, astronomers were able to see many stars that had been too dim to see before. Rather than replace the old system of magnitudes, they added to it. Positive numbers represent dimmer stars, and negative numbers represent brighter stars. For example, by using large telescopes, astronomers can see stars as dim as 29th magnitude. And the brightest star in the night sky, Sirius, has a magnitude of -1.4. The Big Dipper, shown in **Figure 4,** contains both bright stars and dim stars.

Figure 4 *The Big Dipper contains both bright stars and dim stars. What is the magnitude of the brightest star in the Big Dipper?*

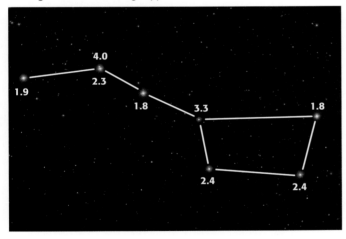

SCHOOL to HOME

Stargazing

WRITING SKILL Someone looking at the night sky in a city would not see as many stars as someone looking at the sky in the country. With a parent, research why this is true. Try to find a place near your home that would be ideal for stargazing. If you find one, schedule a night to stargaze. Write down what you see in the night sky.

ACTIVITY

Using the Table — BASIC

Star Types Draw students' attention to **Table 1.** Have students explain how the stars are arranged (from the hottest to the coolest), and have them identify the hottest and the coolest stars. (The hottest is 10 Lacertae; the coolest are Betelgeuse and Antares.)

Have students locate our sun on the table and describe its temperature relative to the temperatures of other stars. (The sun is a class G star and has a surface temperature between 5,000°C and 6,000°C.)

Be sure that students notice that the temperature of a star indicates the elements detected in star's spectrum. **LS** Visual

CONNECTION to History — GENERAL

Early Systems of Star Classification Hipparchus, a second-century Greek astronomer, developed the first system for star classification. His system divided the stars into six categories based on their apparent brightness. He called the stars that appeared brightest *first magnitude* and called the faintest stars *sixth magnitude*. When telescopes were invented, people learned that many stars were brighter than first-magnitude stars. By the 18th century, scientists decided that a star of a specific magnitude would be about 2.5 times brighter than a star of the previous magnitude. The brightest stars were reclassified to have negative magnitudes.

Is That a Fact!

The brightness of astronomical objects that are not stars is also measured in star magnitudes. For example, Venus shines with an apparent magnitude of −4.6, while the full moon shines with an apparent magnitude of −12.5. With practice, the human eye can discern differences in brightness to one-tenth of a magnitude!

INCLUSION Strategies

- *Learning Disabled*   • *Attention Deficit Disorder*
- *Gifted and Talented*

Organize students into small teams that will create a mnemonic device to help them remember star classification. Next, hand out one deck of star cards comprising star classes and a second deck of star cards comprising the colors, surface temperatures, elements detected, and examples of stars. Ask each team to match each star class with the appropriate attributes. **LS** Interpersonal

Discussion ———— BASIC

Magnitude Scale You may wish to draw a number line on the board to show students how the magnitude scale works. Explain how stars were originally classified on a scale of 1 to 6. Later, as advances in technology enabled the discovery of brighter and fainter stars, the scale was expanded. **LS** Visual

CONNECTION ACTIVITY
Math ———————— GENERAL

Apparent Magnitude Encourage students to compare the apparent magnitude of stars. Point out that if two stars differ by a magnitude of 1, the brighter star is 2.5 times brighter than the dimmer star. Tell students that the star Rigel has an apparent magnitude of 0.18, while Pollux has an apparent magnitude of 1.16. Have students calculate how much brighter Rigel appears than Pollux. (1.16 − 0.18 = 0.98); (0.98 × 2.5 = approximately 2.5; Students should find that Rigel appears about 2.5 times brighter than Pollux.) **LS** Logical

Answer to Reading Check

Apparent magnitude is the brightness of a light or star.

Figure 5 *You can estimate how far away each street light is by looking at its apparent brightness. Does this process work when estimating the distance of stars from Earth?*

apparent magnitude the brightness of a star as seen from the Earth

absolute magnitude the brightness that a star would have at a distance of 32.6 light-years from Earth

MATH PRACTICE

Starlight, Star Bright

Magnitude is used to show how bright one object is compared with another object. Every five magnitudes is equal to a factor of 100 times in brightness. The brightest blue stars, for example, have an absolute magnitude of −10. The sun has an absolute magnitude of about +5. How much brighter is a blue star than the sun? Because each five magnitudes is a factor of 100 and the blue star is 15 magnitudes greater than the sun, the blue star must be 100 × 100 × 100, or 1,000,000 (1 million), times brighter than the sun!

How Bright Is That Star?

If you look at a row of street lights, such as those shown in **Figure 5,** do they all look the same? Of course not! The nearest ones look bright, and the farthest ones look dim.

Apparent Magnitude

The brightness of a light or star is called **apparent magnitude.** If you measure the brightness of a street light with a light meter, you will find that the light's brightness depends on the square of the distance between the light and the light meter. For example, a light that is 10 m away from you will appear 4 (2×2, or 2^2) times brighter than a light that is 20 m away from you. The same light will appear 9 (3×3, or 3^2) times brighter than a light that is 30 m away. But unlike street lights, some stars are brighter than other stars because of their size or energy output, not because of their distance from Earth. So, how can you tell how bright a star is and why?

✓ Reading Check What is apparent magnitude?

Absolute Magnitude

Astronomers use a star's apparent magnitude and its distance from Earth to calculate its absolute magnitude. **Absolute magnitude** is the actual brightness of a star. If all stars were the same distance away, their absolute magnitudes would be the same as their apparent magnitudes. The sun, for example, has an absolute magnitude of +4.8, which is ordinary for a star. But because the sun is so close to Earth, the sun's apparent magnitude is −26.8, which makes it the brightest object in the sky.

MISCONCEPTION
///ALERT

Negative Numbers Are Brighter
Students may think that stars that have negative absolute magnitude values are fainter than those stars that have positive values. Point out that *decreasing* values indicate *increasing* brightness.

CHAPTER RESOURCES

Technology

Transparencies
• Finding the Distance to Stars with Parallax

Workbooks

Math Skills for Science
• Arithmetic with Positive and Negative Numbers **GENERAL**

Figure 6 Measuring a Star's Parallax

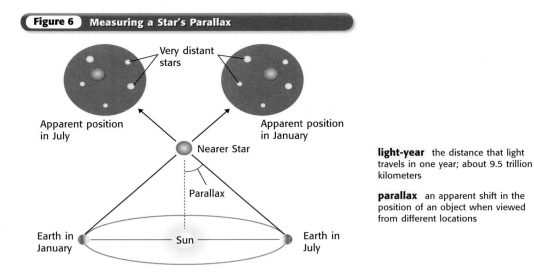

Very distant stars

Apparent position in July

Apparent position in January

Nearer Star

Parallax

Earth in January

Sun

Earth in July

light-year the distance that light travels in one year; about 9.5 trillion kilometers

parallax an apparent shift in the position of an object when viewed from different locations

Distance to the Stars

Because stars are so far away, astronomers use light-years to measure the distances from Earth to the stars. A **light-year** is the distance that light travels in one year. Obviously, it would be easier to give the distance to the North Star as 431 light-years than as 4,080,000,000,000,000 km. But how do astronomers measure a star's distance from Earth?

Stars near the Earth seem to move, while more-distant stars seem to stay in one place as Earth revolves around the sun, as shown in **Figure 6.** A star's apparent shift in position is called **parallax.** Notice that the location of the nearer star in **Figure 6** seems to shift in relation to the pattern of more-distant stars. This shift can be seen only through telescopes. Astronomers use parallax and simple trigonometry (a type of math) to find the actual distance to stars that are close to Earth.

✓ Reading Check What is a light-year?

Motions of Stars

As you know, daytime and nighttime are caused by the Earth's rotation. The Earth's tilt and revolution around the sun cause the seasons. During each season, the Earth faces a different part of the sky at night. Look again at **Figure 6.** In January, the Earth's night side faces a different part of the sky than it faces in July. This is why you see a different set of constellations at different times of the year.

Not All Thumbs!

1. Hold your thumb in front of your face at arm's length.
2. Close one eye, and focus on an **object** some distance behind your thumb.
3. Slowly turn your head side to side a small amount. Notice how your thumb seems to be moving compared with the background you are looking at.
4. Now, move your thumb in close to your face, and move your head the same amount. Does your thumb seem to move more?

Answer to Reading Check
A light-year is the distance that light travels in 1 year.

Reteaching — BASIC

Classifying the Stars Ask students to describe the different ways in which scientists classify stars. **LS** Verbal

Quiz — GENERAL

1. How is the distance from Earth to a star measured? (Scientists determine the distance to a star by using parallax and trigonometry.)

2. How is the apparent movement of the stars in the night sky different from the movement of the stars within a constellation? (The stars in the night sky rise and set as Earth rotates. All of the stars in a constellation are moving relative to one another. It takes thousands of years to observe their movement.)

Alternative Assessment — GENERAL

PORTFOLIO **Illustrating Parallax** Have students make an illustration of the phenomenon of parallax in their **science journal**. Then, have students explain their illustration to a partner.
LS Intrapersonal/Interpersonal

Figure 7 *As Earth rotates on its axis, the stars appear to rotate around Polaris.*

The Apparent Motion of Stars

Because of Earth's rotation, the sun appears to move across the sky. Likewise, if you look at the night sky long enough, the stars also appear to move. In fact, at night you can observe that the whole sky is rotating above us. Look at **Figure 7.** All the stars you see appear to rotate around Polaris, the North Star, which is almost directly above Earth's North Pole. Because of Earth's rotation, all of the stars in the sky appear to make one complete circle around Polaris every 24 h.

The Actual Motion of Stars

You now know that the apparent motion of the sun and stars in our sky is due to Earth's rotation. But each star is also moving in space. Because stars are so distant, however, their actual motion is hard to see. If you could put thousands of years into one hour, a star's movement would be obvious. **Figure 8** shows how familiar star patterns slowly change their shapes.

✓ **Reading Check** Why is the actual motion of stars hard to see?

Figure 8 *Over time, the shapes of star patterns, such as the Big Dipper and other groups, change.*

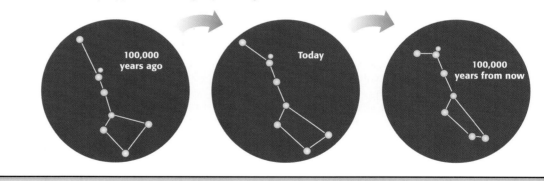

100,000 years ago

Today

100,000 years from now

Answer to Reading Check

The actual motion of stars is hard to see because the stars are so distant.

SECTION Review

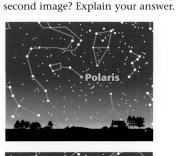

Summary

- The color of a star depends on its temperature. Hot stars are blue. Cool stars are red.
- The spectrum of a star shows the composition of a star.
- Scientists classify stars by temperature and brightness.
- Apparent magnitude is the brightness of a star as seen from Earth.

- Absolute magnitude is the measured brightness of a star at a distance of 32.6 light-years.
- Astronomers use parallax and trigonometry to measure distances from Earth to stars.
- Stars appear to move because of Earth's rotation. However, the actual motion of stars is very hard to see because stars are so distant.

Using Key Terms

1. Use the following terms in the same sentence: *apparent magnitude* and *absolute magnitude*.

2. Use each of the following terms in a separate sentence: *spectrum, light-year,* and *parallax*.

Understanding Key Ideas

3. When you look at white light through a glass prism, you see a rainbow of colors called a
   a. spectograph.
   b. spectrum.
   c. parallax.
   d. light-year.

4. Class F stars are
   a. blue.
   b. yellow.
   c. yellow-white.
   d. red.

5. Describe how scientists classify stars.

6. Explain how color indicates the temperature of a star.

Critical Thinking

7. **Applying Concepts** If a certain star displayed a large parallax, what could you say about the star's distance from Earth?

8. **Making Comparisons** Compare a continuous spectrum with an absorption spectrum. Then, explain how an absorption spectrum can identify a star's composition.

9. **Making Comparisons** Compare apparent motion with actual motion.

Interpreting Graphics

10. Look at the two figures below. How many hours passed between the first image and the second image? Explain your answer.

Polaris

Polaris

SC**LINKS**

NSTA
Developed and maintained by the
National Science Teachers Association

For a variety of links related to this chapter, go to www.scilinks.org

Topic: Stars
SciLinks code: HSM1448

CHAPTER RESOURCES

Chapter Resource File

- Section Quiz **GENERAL**
- Section Review **GENERAL**
- Vocabulary and Section Summary **GENERAL**
- Critical Thinking **ADVANCED**
- SciLinks Activity **GENERAL**
- Datasheet for Quick Lab

Answers to Section Review

1. Sample answer: Apparent magnitude is the brightness of a star as seen from Earth, whereas absolute magnitude is the measured brightness of a star at a distance of 32.6 light-years.

2. Sample answer: A spectrum is the band of color produced when white light passes through a prism. A light-year is the distance that light travels in one year. Parallax is an apparent shift in the position of an object when the object is viewed from different locations.

3. b

4. c

5. Scientists classify stars by temperature and brightness.

6. The color of a star indicates the star's temperature because the color of a star is determined by its temperature. For example, hot stars are blue, and cool stars are red.

7. The star would be relatively close to Earth.

8. A continuous spectrum is a spectrum that shows all colors. An absorption spectrum is produced when light from a hot solid or dense gas passes through a cooler gas. Stars give off an absorption spectrum. The pattern of lines in a star's absorption spectrum shows some of the elements that are in the star's atmosphere and therefore identifies the star's composition.

9. The apparent motion of stars is due to Earth's rotation. The actual motion, or true motion of stars, is very hard to see because stars are so distant from Earth.

10. About 6 hours have passed. The stars would make a complete circle (360°) in 24 h. In the figures, they have turned 90°, which is 1/4 of 360°. Therefore, 24 h × 1/4 = 6 h.

Focus

Overview

This section discusses the life cycle of stars. It also explores how the H-R diagram shows the relationship between a star's surface temperature and absolute magnitude. Finally, this section discusses how stars can become supernovas, neutron stars, pulsars, or black holes.

🔊 Bellringer

Display photographs of Supernova 1987A and a photograph of the Large Magellanic Cloud taken before the explosion. Explain that supernovas represent the "death" of stars that exceed a certain mass. In a few seconds, a supernova can release more energy than it previously released in its entire existence.

READING WARM-UP

Objectives

● Describe different types of stars.
● Describe the quantities that are plotted in the H-R diagram.
● Explain how stars at different stages in their life cycle appear on the H-R diagram.

Terms to Learn

red giant	supernova
white dwarf	neutron star
H-R diagram	pulsar
main sequence	black hole

READING STRATEGY

Paired Summarizing Read this section silently. In pairs, take turns summarizing the material. Stop to discuss ideas that seem confusing.

The Life Cycle of Stars

Stars exist for billions of years. But how are they born? And what happens when a star dies?

Because stars exist for billions of years, scientists cannot observe a star throughout its entire life. Therefore, scientists have developed theories about the life cycle of stars by studying them in different stages of development.

The Beginning and End of Stars

A star enters the first stage of its life cycle as a ball of gas and dust. Gravity pulls the gas and dust together into a sphere. As the sphere becomes denser, it gets hotter and the hydrogen changes to helium in a process called *nuclear fusion*.

As stars get older, they lose some of their material. Stars usually lose material slowly, but sometimes they can lose material in a big explosion. Either way, when a star dies, much of its material returns to space. In space, some of the material combines with more gas and dust to form new stars.

Different Types of Stars

Stars can be classified by their size, mass, brightness, color, temperature, spectrum, and age. Some types of stars include *main-sequence stars*, *giants*, *supergiants*, and *white dwarf stars*. A star can be classified as one type of star early in its life cycle and then can be classified as another star when it gets older. For example, the star shown in **Figure 1** has reached the final stage in its life cycle. It has run out of fuel, which has caused the central parts of the star to collapse inward.

Figure 1 *This star has entered the last stage of its life cycle.*

CHAPTER RESOURCES

Chapter Resource File

- **Lesson Plan**
- **Directed Reading A** BASIC
- **Directed Reading B** SPECIAL NEEDS

Technology

Transparencies
- Bellringer

CONNECTION to Astronomy ——— ADVANCED

The *Hubble Space Telescope* From its orbit around the Earth, the *Hubble Space Telescope* has photographed newborn stars emerging from huge pillars of dense gas and dust. Some of these events occurred about 7,000 light-years away in the Eagle Nebula. The largest of these pillars photographed by the telescope is an estimated 10 trillion kilometers high. Have students find these images on the Internet and discuss them during class. **LS Verbal**

Main-Sequence Stars

After a star forms, it enters the second and longest stage of its life cycle known as the main sequence. During this stage, energy is generated in the core of the star as hydrogen atoms fuse into helium atoms. This process releases an enormous amount of energy. The size of a main-sequence star will change very little as long as the star has a continuous supply of hydrogen atoms to fuse into helium atoms.

Giants and Supergiants

After the main-sequence stage, a star can enter the third stage of its life cycle. In this third stage, a star can become a red giant. A **red giant** is a star that expands and cools once it uses all of its hydrogen. Eventually, the loss of hydrogen causes the center of the star to shrink. As the center of the star shrinks, the atmosphere of the star grows very large and cools to form a red giant or a red supergiant, as shown in **Figure 2.** Red giants can be 10 or more times bigger than the sun. Supergiants are at least 100 times bigger than the sun.

✓ **Reading Check** What is the difference between a red giant star and a red supergiant star? *(See the Appendix for answers to Reading Checks.)*

White Dwarfs

In the final stages of a star's life cycle, a star the size of the sun or smaller, can be classified as a white dwarf. A **white dwarf** is a small, hot star that is the leftover center of an older star. A white dwarf has no hydrogen left and can no longer generate energy by fusing hydrogen atoms into helium atoms. White dwarfs can shine for billions of years before they cool completely.

Figure 2 *The red supergiant star Antares is shown above. Antares is located in the constellation of Scorpius.*

red giant a large, reddish star late in its life cycle

white dwarf a small, hot, dim star that is the leftover center of an old star

CONNECTION TO Astronomy

WRITING SKILL **Long Live the Sun** Our sun probably took about 10 million years to become a main-sequence star. It has been shining for about 5 billion years. In another 5 billion years, our sun will burn up all of its hydrogen and change into a red giant. When this change happens, the sun's diameter will increase. How will this change affect Earth and our solar system? Use the Internet or library resources to find out what might happen as the sun gets older and how the changes in the sun might affect our solar system. Gather your findings, and write a report on what you find out about the life cycle the sun.

Answer to Reading Check

A red giant star is a star that expands and cools once it uses all of its hydrogen. As the center of a star continues to shrink, a red giant star can become a red supergiant star.

Using the Figure — BASIC

Star Magnitudes Remind students that the lower the magnitude of a star is, the brighter the star is. By looking at the H-R diagram, students should be able to identify the sun as a main-sequence, yellowish dwarf star with medium brightness and a surface temperature of almost 6,000°C. Have students describe other stars in the diagram in a similar manner. **LS** Visual/Logical

Group ACTIVITY — GENERAL

Star Cycle Divide the class into small groups, and provide each group with a piece of newsprint paper and markers. Direct each group to use these materials to create a flowchart describing the life of a star. Encourage students to refer to the H-R diagram as they work. Their chart should indicate that stars (1) form when gas and dust are drawn together by gravity and nuclear fusion begin, (2) enter the main sequence when they mature, and (3) may then become red giants, supergiants, or eventually white dwarfs. Have students label their charts and write a descriptive caption for each stage. **LS** Interpersonal/Visual

A Tool for Studying Stars

In 1911, a Danish astronomer named Ejnar Hertzsprung (IE nawr HUHRTS sproong) compared the brightness and temperature of stars on a graph. Two years later, American astronomer Henry Norris Russell made some similar graphs. Although these astronomers used different data, they had similar results. The combination of their ideas is now called the Hertzsprung-Russell diagram, or H-R diagram. The **H-R diagram** is a graph that shows the relationship between a star's surface temperature and its absolute magnitude. Over the years, the H-R diagram has become a tool for studying the lives of stars. It shows not only how stars are classified by brightness and temperature but also how stars change over time.

H-R diagram Hertzsprung-Russell diagram, a graph that shows the relationship between a star's surface temperature and absolute magnitude

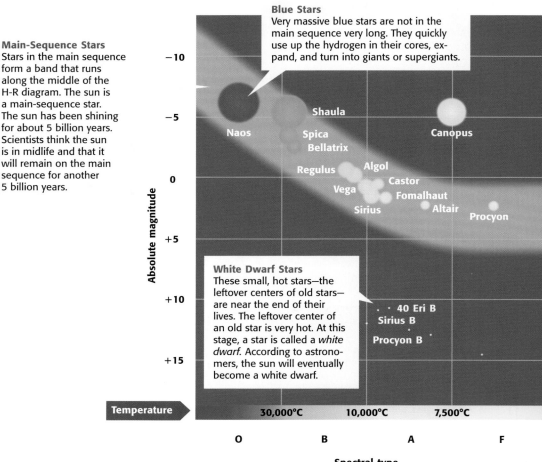

Main-Sequence Stars Stars in the main sequence form a band that runs along the middle of the H-R diagram. The sun is a main-sequence star. The sun has been shining for about 5 billion years. Scientists think the sun is in midlife and that it will remain on the main sequence for another 5 billion years.

Blue Stars Very massive blue stars are not in the main sequence very long. They quickly use up the hydrogen in their cores, expand, and turn into giants or supergiants.

White Dwarf Stars These small, hot stars—the leftover centers of old stars—are near the end of their lives. The leftover center of an old star is very hot. At this stage, a star is called a *white dwarf*. According to astronomers, the sun will eventually become a white dwarf.

Naos   Shaula   Spica   Bellatrix   Canopus   Regulus   Algol   Vega   Castor   Fomalhaut   Sirius   Altair   Procyon   40 Eri B   Sirius B   Procyon B

Absolute magnitude: −10, −5, 0, +5, +10, +15

Temperature: 30,000°C   10,000°C   7,500°C

Spectral type: O   B   A   F

WEIRD SCIENCE

When the core from a star that is the size of the sun becomes a white dwarf that is the size of Earth, the white dwarf is much denser than our planet. In fact, a teaspoon of the matter that makes up a white dwarf would weigh several metric tons on Earth!

Is That a Fact!

Most of the stars near our solar system are not as bright as the sun. How do we know this fact? When the 100 stars nearest to Earth are arranged on the H-R diagram, we can see that almost all of them fall in the region of the red dwarfs. The sun is a brighter, type G main-sequence star.

Reading the H-R Diagram

The modern H-R diagram is shown below. Temperature is given along the bottom of the diagram and absolute magnitude, or brightness, is given along the left side. Hot (blue) stars are located on the left, and cool (red) stars are on the right. Bright stars are at the top, and dim stars are at the bottom. The brightest stars are 1 million times brighter than the sun. The dimmest stars are 1/10,000 as bright as the sun. The diagonal pattern on the H-R diagram where most stars lie, is called the **main sequence.** A star spends most of its lifetime in the main sequence. As main-sequence stars age, they move up and to the right on the H-R diagram to become giants or supergiants and then down and to the left to become white dwarfs.

main sequence the location on the H-R diagram where most stars lie

Giants and Supergiants

When a star runs out of hydrogen in its core, the center of the star shrinks inward and the outer parts expand outward. For a star the size of our sun, the star's atmosphere will grow very large and become cool. When this change happens, the star becomes a *red giant.* If the star is very massive, it becomes a supergiant.

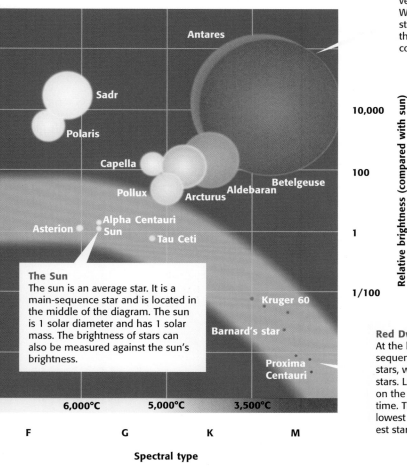

The Sun
The sun is an average star. It is a main-sequence star and is located in the middle of the diagram. The sun is 1 solar diameter and has 1 solar mass. The brightness of stars can also be measured against the sun's brightness.

Red Dwarf Stars

At the lower end of the main sequence are the red dwarf stars, which are low-mass stars. Low-mass stars remain on the main sequence a long time. The stars that have the lowest mass may be the oldest stars in the universe.

ACTIVITY — BASIC

Using the H-R Diagram Have students locate where on the H-R diagram each of the stars in the table below would be found.

	Magnitude	Temperature
Star A	+10	10,000°C
Star B	−2	5,000°C
Star C	+3	7,000°C
Star D	−9	3,500°C

Ask students the following questions:

• Which star is a giant? (B)

• Which star is a white dwarf? (A)

• Which star is a supergiant? (D)

• Which star is most like the sun? (C)

LS Verbal/Visual

CONNECTION ACTIVITY
Language Arts — BASIC

Star Crossword Puzzles Divide the class into small groups, and challenge each group to create a crossword puzzle using the vocabulary and concepts from this section. Have students in each group work together to write clues and to construct the puzzle. Then allow groups to exchange and solve the puzzles.

LS Interpersonal

English Language Learners

INTERNET ACTIVITY
Sequence Board — GENERAL

For an Internet activity related to this chapter, have students go to **go.hrw.com** and type in the keyword **HZ5UNVW.**

CHAPTER RESOURCES

Technology

Transparencies
• The H-R Diagram: A
• The H-R Diagram: B

Reteaching — BASIC

Autobiography of the Stars

Have students write a brief autobiography of each type of star and then read their autobiography to the class. **LS** Verbal

Quiz — GENERAL

1. What information does the H-R diagram give us? (It indicates the relationship between a star's temperature and the star's brightness, which indicates the star's age.)

2. What is a supernova? (the explosion of a massive star at the end of its life)

3. What is a neutron star? How is it different from a pulsar? (A neutron star is the compressed core of a star that became a supernova. A pulsar is a spinning neutron star.)

Alternative Assessment — GENERAL

Writing **1987A** Have students research the astral events of 1987, known as the "year of the supernova." Have them prepare a brief report and share their findings with the class. (Reports should include the following information: In 1987, for the first time in almost 400 years, people on Earth witnessed the death of a star without using a telescope. The supernova was located in a satellite galaxy of the Milky Way called the *Large Magellanic Cloud* and was visible only from the Southern Hemisphere.) **LS** Interpersonal

When Stars Get Old

Although stars may stay on the main sequence for a long time, they don't stay there forever. Average stars, such as the sun, become red giants and then white dwarfs. However, stars that are more massive than the sun may explode with such intensity that they become a variety of strange objects such as supernovas, neutron stars, pulsars, and black holes.

Supernovas

Massive blue stars use their hydrogen much faster than stars like the sun do. Therefore, blue stars generate more energy than stars like the sun do, which makes blue stars very hot and blue! And compared with other stars, blue stars don't have long lives. At the end of its life, a blue star may explode in a large, bright flash called a *supernova*. A **supernova** is a gigantic explosion in which a massive star collapses. The explosion is so powerful that it can be brighter than an entire galaxy for several days. The ringed structure shown in **Figure 3** is the result of a supernova explosion.

Neutron Stars and Pulsars

After a supernova occurs, the materials in the center of a supernova are squeezed together to form a new star. This new star is about two times the mass of the sun. The particles inside the star's core lose their charge and become neutrons. A star that has collapsed under gravity to the point at which all of its particles are neutrons is called a **neutron star.**

If a neutron star is spinning, it is called a **pulsar.** A pulsar sends out beams of radiation that spin very rapidly. The beams are detected by radio telescopes as rapid clicks, or pulses.

supernova a gigantic explosion in which a massive star collapses and throws its outer layers into space

neutron star a star that has collapsed under gravity to the point that the electrons and protons have smashed together to form neutrons

pulsar a rapidly spinning neutron star that emits rapid pulses of radio and optical energy

Figure 3 Explosion of a Supernova

Supernova 1987A was the first supernova visible to the unaided eye in 400 years. The first image shows what the original star must have looked like only a few hours before the explosion. Today, the star's remains form a double ring of gas and dust, as shown at right.

Before (1984)

During (1987)

After
(Hubble Space Telescope close-up, 1994)

Black Holes

Sometimes the leftovers of a supernova are so massive that they collapse to form a black hole. A **black hole** is an object that is so massive that even light cannot escape its gravity. So, it is called a *black hole*. A black hole doesn't gobble up other stars like some movies show. Because black holes do not give off light, locating them is difficult. If a star is nearby, some gas or dust from the star will spiral into the black hole and give off X rays. These X rays allow astronomers to detect the existence of black holes.

black hole an object so massive and dense that even light cannot escape its gravity

✔ **Reading Check** What is a black hole? How do astronomers detect the presence of black holes?

SECTION Review

Summary

- New stars form from the material of old stars that have gone through their lives.
- Types of stars include main-sequence stars, giants and supergiants, and white dwarf stars.
- The H-R diagram shows the brightness of a star in relation to the temperature of a star. It also shows the life cycle of stars.
- Most stars are main-sequence stars.
- Massive stars become supernovas. Their cores can change into neutron stars or black holes.

Using Key Terms

For each pair of terms, explain how the meanings of the terms differ.

1. *white dwarf* and *red giant*
2. *supernova* and *neutron star*
3. *pulsar* and *black hole*

Understanding Key Ideas

4. The sun is a
   a. white dwarf.
   b. main-sequence star.
   c. red giant.
   d. red dwarf.

5. A star begins as a ball of gas and dust pulled together by
   a. black holes.
   b. electrons and protons.
   c. heavy metals.
   d. gravity.

6. Are blue stars young or old? How can you tell?

7. In main-sequence stars, what is the relationship between brightness and temperature?

8. Arrange the following stages in order of their appearance in the life cycle of a star: white dwarf, red giant, and main-sequence star. Explain your answer.

Math Skills

9. The sun's present diameter is 700,000 km. If the sun's diameter increased by 150 times, what would its diameter be?

Critical Thinking

10. **Applying Concepts** Given that there are more low-mass stars than high-mass stars in the universe, do you think there are more white dwarfs or more black holes in the universe? Explain.

11. **Analyzing Processes** Describe what might happen to a star after it becomes a supernova.

12. **Evaluating Data** How does the H-R diagram explain the life cycle of a star?

Developed and maintained by the National Science Teachers Association

For a variety of links related to this chapter, go to www.scilinks.org

Topic: Supernova
SciLinks code: HSM1482

Answer to Reading Check

A black hole is an object that is so massive that even light cannot escape its gravity. A black hole can be detected when it gives off X rays.

Answers to Section Review

1. Sample answer: White dwarfs are small, hot, dim stars that are leftover centers of old stars. Red giants are large, reddish stars that are in the third stage of their life cycles.

CHAPTER RESOURCES

Chapter Resource File

- Section Quiz GENERAL
- Section Review GENERAL
- Vocabulary and Section Summary GENERAL
- Reinforcement Worksheet BASIC

2. Sample answer: A supernova is a gigantic explosion in which a massive star collapses and throws its outer layers into space. A neutron star is a massive star that has collapsed under gravity to the point that the electrons and protons have become neutrons.

3. Sample answer: A pulsar is a rapidly spinning neutron star that emits rapid pulses of radio and optical energy. A black hole is an object that is so massive and dense that even light cannot escape its gravity.

4. b

5. d

6. Blue stars are young. They use up their hydrogen quickly and become supernovas before they get old.

7. In the main sequence, hotter stars are usually brighter.

8. The order would be main-sequence star, red giant, and white dwarf. As a main-sequence star runs out of fuel, its core shrinks and its atmosphere expands. It then becomes a red giant. After the red giant loses its outer layers, its core remains as a white dwarf.

9. 700,000 km × 150 = 105,000,000 km

10. There are more white dwarfs. White dwarfs are the remains of average-sized stars that grow old. Only very massive stars become black holes.

11. After a star becomes a supernova, it could become a neutron star, a pulsar, or a black hole.

12. The H-R diagram explains the life cycle of a star, because a star can be plotted on the diagram at any point during its life cycle. The H-R diagram plots stars according to their magnitude and temperature. Therefore, as a star goes through its life cycle, it can be plotted in different locations as its magnitude and temperature change.

Focus

Overview

This section discusses the differences between the three types of galaxies: spiral, elliptical, and irregular. Students will learn that galaxies have features known as *nebulas, open clusters,* and *globular clusters*. Finally, students will learn how scientists study the origin of galaxies.

🔔 Bellringer

Show students a photograph of a spiral galaxy. Discuss the evidence that indicates that the galaxy is rotating. Ask students the following questions: "What other objects that look similar have you seen? Do they rotate?" **English Language Learners**
LS Visual/Verbal

Motivate

Discussion ——— GENERAL

Star Factory Galaxies can be thought of as star factories. Ask students to identify the raw materials used by the "factory" to produce stars. (clouds of gas and dust) Have students describe how stars are assembled. (The gases and dust are drawn together by gravity.) Discuss whether star formation is an ongoing process. (Establish that the process is ongoing because of the abundance of raw materials.) **LS Verbal**

SECTION
3

READING WARM-UP

Objectives
- Identify three types of galaxies.
- Describe the contents and characteristics of galaxies.
- Explain why looking at distant galaxies reveals what young galaxies looked like.

Terms to Learn
galaxy
nebula
globular cluster
open cluster
quasar

READING STRATEGY

Reading Organizer As you read this section, make a table comparing the different types of galaxies.

galaxy a collection of stars, dust, and gas bound together by gravity

---

Galaxies

Your complete address is part of a much larger system than your street, city, state, country, and even the planet Earth. You also live in the Milky Way galaxy.

Large groups of stars, dust, and gas are called **galaxies.** Galaxies come in a variety of sizes and shapes. The largest galaxies contain more than a trillion stars. Astronomers don't count the stars, of course. They estimate how many sun-sized stars the galaxy might have by studying the size and brightness of the galaxy.

Types of Galaxies

There are many different types of galaxies. Edwin Hubble, the astronomer for whom the *Hubble Space Telescope* is named, began to classify galaxies, mostly by their shapes, in the 1920s. Astronomers still use the galaxy classification that Hubble developed.

Spiral Galaxies

When someone says the word *galaxy*, most people probably think of a spiral galaxy. *Spiral galaxies,* such as the one shown in **Figure 1,** have a bulge at the center and spiral arms. The spiral arms are made up of gas, dust, and new stars that have formed in these denser regions of gas and dust.

✔ **Reading Check** What are two characteristics of spiral galaxies? What makes up the arms of a spiral galaxy? (*See the Appendix for answers to Reading Checks.*)

Figure 1 Types of Galaxies

▼ **Spiral Galaxy**
The Andromeda galaxy is a spiral galaxy that looks similar to what our galaxy, the Milky Way, is thought to look like.

Answer to Reading Check

Spiral galaxies have a bulge at the center and spiral arms. The arms of spiral galaxies are made up of gas, dust, and new stars.

CHAPTER RESOURCES

Chapter Resource File
- Lesson Plan
- Directed Reading A BASIC
- Directed Reading B SPECIAL NEEDS

Technology
- Transparencies
  - Bellringer

The Milky Way

It is hard to tell what type of galaxy we live in because the gas, dust, and stars keep astronomers from having a good view of our galaxy. Observing other galaxies and making measurements inside our galaxy, the Milky Way, has led astronomers to think that our solar system is in a spiral galaxy.

Elliptical Galaxies

About one-third of all galaxies are simply massive blobs of stars. Many look like spheres, and others are more stretched out. Because we don't know how they are oriented, some of these galaxies could be cucumber shaped, with the round end facing our galaxy. These galaxies are called *elliptical galaxies*. Elliptical galaxies usually have very bright centers and very little dust and gas. Elliptical galaxies contain mostly old stars. Because there is so little free-flowing gas in an elliptical galaxy, few new stars form. Some elliptical galaxies, such as M87, shown in **Figure 1,** are huge and are called *giant elliptical galaxies*. Other elliptical galaxies are much smaller and are called *dwarf elliptical galaxies*.

Irregular Galaxies

When Hubble first classified galaxies, he had a group of leftovers. He named the leftovers "irregulars." *Irregular galaxies* are galaxies that don't fit into any other class. As their name suggests, their shape is irregular. Many of these galaxies, such as the Large Magellanic Cloud, shown in **Figure 1,** are close companions of large spiral galaxies. The large spiral galaxies may be distorting the shape of these irregular galaxies.

CONNECTION TO Language Arts

WRITING SKILL **Alien Observer** As you read earlier, it's hard to tell what type of galaxy we live in because the gas, dust, and stars keep us from having a good view. But our galaxy might look different to an alien observer. Write a short story describing how our galaxy would look to an alien observer in another galaxy.

▼ **Elliptical Galaxy**
Unlike the Milky Way, the galaxy known as M87 has no spiral arms.

▼ **Irregular Galaxy**
The Large Magellanic Cloud, an irregular galaxy, is located within our galactic neighborhood.

SCIENTISTS AT ODDS

The Shapley-Curtis Debate When the American astronomer Harlow Shapley mapped globular clusters, he found that they form an enormous spherical system surrounding the Milky Way and that Earth was not at the center of the galaxy. This discovery sparked a heated astronomical debate over whether other spiral nebulas in the distant universe are part of our galaxy or are separate "island universes," or distant galaxies. This discussion led to the 1920 Shapley-Curtis debate at the National Academy of Sciences. The debate was resolved in 1924, when observations by Edwin Hubble showed that these nebulas were so far away that they must be distinct galaxies.

Close

Reteaching — BASIC

Types of Galaxies Help students differentiate between the types of galaxies by having them divide a large sheet of paper into thirds and draw a spiral, an ellipse, and an irregular shape. Have students describe the galaxy type under each drawing. **LS** Visual   English Language Learners

Quiz — GENERAL

1. What is a nebula? (A nebula is an enormous cloud of gas and dust in space.)

2. What are open clusters? (They are groups of a few hundred to a few thousand stars that form when great amounts of gas and dust come together.)

Alternative Assessment — ADVANCED

Interstellar Medium Tell students that the material that exists between a galaxy's stars is called i*nterstellar medium*. Ask them to hypothesize what makes up interstellar medium. (Students should recognize that nebulas—giant clouds of gas and dust—make up interstellar medium.)

Challenge students to compare irregular, spiral, and elliptical galaxies and to determine how much interstellar medium each galaxy has. (Irregular and spiral galaxies have more interstellar medium than elliptical galaxies have.) **LS** Verbal

nebula a large cloud of dust and gas in interstellar space; a region in space where stars are born or where stars explode at the end of their lives

globular cluster a tight group of stars that looks like a ball and contains up to 1 million stars

open cluster a group of stars that are close together relative to surrounding stars

Contents of Galaxies

Galaxies are composed of billions of stars and some planetary systems, too. Some of these stars form large features, such as gas clouds and star clusters, as shown in **Figure 2.**

Gas Clouds

The Latin word for "cloud" is *nebula*. In space, **nebulas** (or nebulae) are large clouds of gas and dust. Some types of nebulas glow, while others absorb light and hide stars. Still, other nebulas reflect starlight and produce some amazing images. Some nebulas are regions in which new stars form. **Figure 2** shows part of the Eagle nebula. Spiral galaxies usually contain nebulas, but elliptical galaxies contain very few.

Star Clusters

Globular clusters are groups of older stars. A **globular cluster** is a group of stars that looks like a ball, as shown in **Figure 2.** There may be up to one million stars in a globular cluster. Globular clusters are located in a spherical *halo* that surrounds spiral galaxies such as the Milky Way. Globular clusters are also common near giant elliptical galaxies.

Open clusters are groups of closely grouped stars that are usually located along the spiral disk of a galaxy. Newly formed open clusters have many bright blue stars, as shown in **Figure 2.** There may be a few hundred to a few thousand stars in an open cluster.

Reading Check What is the difference between a globular cluster and an open cluster?

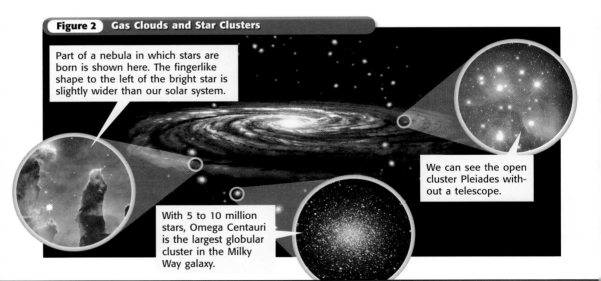

Figure 2 Gas Clouds and Star Clusters

Part of a nebula in which stars are born is shown here. The fingerlike shape to the left of the bright star is slightly wider than our solar system.

We can see the open cluster Pleiades without a telescope.

With 5 to 10 million stars, Omega Centauri is the largest globular cluster in the Milky Way galaxy.

Answer to Reading Check

A globular cluster is a tight group of up to 1 million stars that looks like a ball. An open cluster is a group of closely grouped stars that are usually located along the spiral disk of a galaxy.

Is That a Fact!

Earth is about two-thirds of the distance from the center of the Milky Way to the edge of the Milky Way. Our solar system revolves around the galaxy every 200 million years. The last time the solar system was in its current position was during the Triassic period, when dinosaurs first appeared on Earth!

Origin of Galaxies

Scientists investigate the early universe by observing objects that are extremely far away in space. Because it takes time for light to travel through space, looking through a telescope is like looking back in time. Looking at distant galaxies reveals what early galaxies looked like. This information gives scientists an idea of how galaxies change over time and may give them insight about what caused the galaxies to form.

Quasars

Among the most distant objects are quasars. **Quasars** are starlike sources of light that are extremely far away. They are among the most powerful energy sources in the universe. Some scientists think that quasars may be the core of young galaxies that are in the process of forming. **Figure 3** shows a quasar that is 6 billion light-years away.

✓ Reading Check What are quasars? What do some scientists think quasars might be?

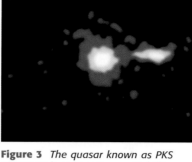

Figure 3 *The quasar known as PKS 0637-752 is as massive as 10 billion suns.*

quasar a very luminous, starlike object that generates energy at a high rate; quasars are thought to be the most distant objects in the universe

---

SECTION Review

Summary

- Edwin Hubble classified galaxies according to their shape including spiral, elliptical, and irregular galaxies.

- Some galaxies consist of nebulas and star clusters.

- Nebulas are large clouds of gas and dust. Globular clusters are tightly grouped stars. Open clusters are closely grouped stars.

- Scientists look at distant galaxies to learn what early galaxies looked like.

Using Key Terms

1. Use the following terms in the same sentence: *nebula, globular cluster,* and *open cluster.*

Understanding Key Ideas

2. Arrange the following galaxies in order of decreasing size: spiral, giant elliptical, dwarf elliptical, and irregular.

3. All of the following are shapes used to classify galaxies EXCEPT
   a. elliptical.
   b. irregular.
   c. spiral.
   d. triangular.

Critical Thinking

4. **Making Comparisons** Describe the difference between an elliptical galaxy and a globular cluster.

5. **Identifying Relationships** Explain how looking through a telescope is like looking back in time.

Math Skills

6. The quasar known as PKS 0637-752 is 6 billion light-years away from Earth. The North Star is 431 light-years away from Earth. What is the ratio of the distances these two celestial objects are from Earth? (Hint: One light-year is equal to 9.46 trillion kilometers.)

SciLINKS® NSTA
Developed and maintained by the National Science Teachers Association

For a variety of links related to this chapter, go to www.scilinks.org

Topic: Galaxies
SciLinks code: HSM0632

Overview

This section discusses the study of cosmology. Students will learn about the big bang theory and about evidence that supports the theory. Finally, they will learn about the structure of the universe.

🔊 Bellringer

Have students examine **Figure 1** and describe, in writing, the differences between the images. The first image represents the initial explosion, and the following images represent the expansion of the universe and the formation of the galaxies.

ACTIVITY ——————— GENERAL

The Expanding Universe Have students draw several dots on an uninflated balloon with a permanent marker and label the dots with letters. Ask students to measure the distances between the dots. Then, have students blow up their balloon and tie the end. Have students measure the distances between the dots again. Ask students to explain how this model represents the expansion of the universe. (As the universe expands, the distance between every star and galaxy increases.) **LS Visual/Verbal**

SECTION

4

READING WARM-UP

Objectives
- Describe the big bang theory.
- Explain evidence used to support the big bang theory.
- Describe the structure of the universe.
- Describe two ways scientists calculate the age of the universe.
- Explain what will happen if the universe expands forever.

Terms to Learn
cosmology
big bang theory

READING STRATEGY

Prediction Guide Before reading this section, write the title of each heading in this section. Next, under each heading, write what you think you will learn.

cosmology the study of the origin, properties, processes, and evolution of the universe

Figure 1 *Some astronomers think the big bang caused the universe to expand in all directions.*

Formation of the Universe

Imagine explosions, bright lights, and intense energy. Does that scene sound like an action movie? This scene could also describe a theory about the formation of the universe.

The study of the origin, structure, and future of the universe is called **cosmology.** Like other scientific theories, theories about the beginning and end of the universe must be tested by observations or experiments.

Universal Expansion

To understand how the universe formed, scientists study the movement of galaxies. Careful measurements have shown that most galaxies are moving apart.

A Raisin-Bread Model

To understand how the galaxies are moving, imagine a loaf of raisin bread before it is baked. Inside the dough, each raisin is a certain distance from every other raisin. As the dough gets warm and rises, it expands and all of the raisins begin to move apart. No matter which raisin you observe, the other raisins are moving farther away from it. The universe, like the rising bread dough, is expanding. Think of the raisins as galaxies. As the universe expands, the galaxies move farther apart.

The Big Bang Theory

With the discovery that the universe is expanding, scientists began to wonder what it would be like to watch the formation of the universe in reverse. The universe would appear to be contracting, not expanding. All matter would eventually come together at a single point. Thinking about what would happen if all of the matter in the universe were squeezed into such a small space led scientists to the big bang theory.

CHAPTER RESOURCES

Chapter Resource File
- Lesson Plan
- Directed Reading A **BASIC**
- Directed Reading B **SPECIAL NEEDS**

Technology
- Transparencies
  - Bellringer
  - *LINK TO PHYSICAL SCIENCE* The Doppler Effect
  - The Big Bang Theory

MISCONCEPTION ///ALERT\\\

Distances in Space Students may have difficulty understanding that because of the great distances between Earth and stars, the light that they see was emitted from the stars in the distant past. If a star is 100 light-years away from Earth, the star's light takes 100 years to travel from the star to Earth.

A Tremendous Explosion

The theory that the universe began with a tremendous explosion is called the **big bang theory.** According to the theory, 13.7 billion years ago all the contents of the universe was compressed under extreme pressure, temperature, and density in a very tiny spot. Then, the universe rapidly expanded, and matter began to come together and form galaxies. **Figure 1** illustrates what the big bang might have looked like.

Cosmic Background Radiation

In 1964, two scientists using a huge antenna accidentally found radiation coming from all directions in space. One explanation for this radiation is that it is *cosmic background radiation* left over from the big bang. To understand the connection between the big bang theory and cosmic background radiation, think about a kitchen oven. When an oven door is left open after the oven has been used, thermal energy is transferred throughout the kitchen and the oven cools. Eventually, the room and the oven are the same temperature. According to the big bang theory, the thermal energy from the original explosion was distributed in every direction as the universe expanded. This cosmic background radiation now fills all of space.

Reading Check Explain the relationship between cosmic background radiation and the big bang theory. (*See the Appendix for answers to Reading Checks.*)

big bang theory the theory that states the universe began with a tremendous explosion 13.7 billion years ago

CONNECTION to Physical Science — GENERAL

Red Shift Ask students if they have ever noticed how the siren of an ambulance sounds higher in pitch when the ambulance is approaching than when it is receding. Students may be surprised to learn that a similar effect occurs with light. Scientists can observe that the light emitted by galaxies that are moving away from us is shifted to the red end of the spectrum. This shift occurs because as galaxies move away, the waves appear to have longer wavelengths. This phenomenon is caused by the Doppler effect. Because the light from all distant galaxies is red shifted, scientists can conclude that the universe is expanding. Use the teaching transparency called "The Doppler Effect" to show how the Doppler effect works with sound waves.
LS Visual

Answer to Reading Check

Cosmic background radiation is radiation that is left over from the big bang. After the big bang, cosmic background radiation was distributed everywhere and filled all of space.

MISCONCEPTION ALERT

The Big Bang Theory Make sure that students realize that according to the big bang theory, the big bang was not an explosion that happened "somewhere in space." Space and time did not exist before the big bang;—they came into being with the big bang. Just before expansion, the universe was compressed into an infinitely dense mass. There was no "space" outside this mass. Thus, we are not receding away from the bang; rather, the explosion continues to expand. We aren't moving away from the point of the big bang because the big bang is happening everywhere.

Structure of the Universe

From our home on Earth, the universe stretches out farther than astronomers can see with their most advanced instruments. The universe contains a variety of objects. But these objects in the universe are not simply scattered through the universe in a random pattern. The universe has a structure that is loosely repeated over and over again.

A Cosmic Repetition

Every object in the universe is part of a larger system. As illustrated in **Figure 2,** a cluster or group of galaxies can be made up of smaller star clusters and galaxies. Galaxies, such as the Milky Way, can include planetary systems, such as our solar system. Earth is part of our solar system. Although our solar system is the planetary system that we are most familiar with, other planets have been detected in orbit around other stars. Scientists think that planetary systems are common in the universe.

How Old Is the Universe?

One way scientists can calculate the age of the universe is to measure the distance from Earth to various galaxies. By using these distances, scientists can calculate the age of the universe and predict its rate of expansion.

Another way to estimate the age of the universe is to calculate the ages of old, nearby stars. Because the universe must be at least as old as the oldest stars it contains, the ages of the stars provide a clue to the age of the universe.

✓ Reading Check What is one way that scientists calculate the age of the universe?

Figure 2 *Every object in the universe is part of a larger system. Earth is part of our solar system, which is in turn part of the Milky Way galaxy.*

A Forever Expanding Universe

What will happen to the universe? As the galaxies move farther apart, they get older and stop forming stars. The farther galaxies move apart from each other, the less visible to us they will become. The expansion of the universe depends on how much matter the universe contains. Scientists predict that if there is enough matter, gravity could eventually stop the expansion of the universe. If the universe stops expanding, it could start collapsing to its original state. This process would be a reverse of what might have happened during the big bang.

However, scientists now think that there may not be enough matter in the universe, so the universe will continue to expand forever. Therefore, stars will age and die, and the universe will probably become cold and dark after many billions of years. Even after the universe becomes cold and dark, it will continue to expand forever.

Reading Check If the universe expanded to the point at which gravity stopped the expansion, what would happen? What will happen if the expansion of the universe continues forever?

CONNECTION TO Physics

WRITING SKILL **Origin of the Universe** The big bang theory is one scientific theory about the origin of the universe. Use library resources to research these other scientific theories. In your **science journal**, describe in your own words the different theories of the origin of the universe. Use charts or tables to examine and evaluate these differences.

SECTION Review

Summary

- Observations show that the universe is expanding.
- The big bang theory states that the universe began with an explosion about 13.7 billion years ago.
- Cosmic background radiation helps support the big bang theory.
- Scientists use different ways to calculate the age of the universe.
- Scientists think that the universe may expand forever.

Using Key Terms

1. In your own words, write a definition for the following terms: *cosmology* and *big bang theory*.

Understanding Key Ideas

2. Describe two ways scientists calculate the age of the universe.

3. The expansion of the universe can be compared to
   a. cosmology.
   b. raisin bread baking in an oven.
   c. thermal energy leaving an oven as the oven cools.
   d. bread pudding.

4. How does cosmic background radiation support the big bang theory?

5. What do scientists think will eventually happen to the universe?

Math Skills

6. The distance from Earth to the North Star is 4.08×10^{12} km. What is this number written in its long form?

Critical Thinking

7. **Applying Concepts** Explain how every object in the universe is part of a larger system.

8. **Analyzing Ideas** Why do scientists think that the universe will expand forever?

SCILINKS **NSTA** Developed and maintained by the National Science Teachers Association

For a variety of links related to this chapter, go to www.scilinks.org

Topic: Structure of the Universe
SciLinks code: HSM1469

Homework ——— ADVANCED

Age of the Universe Remind students that one way to determine the age of the universe is to divide the distance to other galaxies by the speed at which those galaxies appear to be moving away from us. Scientists now think that the universe is 13.7 billion years old. Have students consider how many light-years away a celestial object would be if it existed during the formation of the universe. (It would be 13.7 billion light-years away.) **LS Logical**

CHAPTER RESOURCES

Chapter Resource File

- Section Quiz **GENERAL**
- Section Review **GENERAL**
- Vocabulary and Section Summary **GENERAL**
- Critical Thinking **ADVANCED**

Answers to Section Review

1. Sample answer: The study of the origin, structure, and future of the universe is called *cosmology*. The theory that the universe began with a tremendous explosion is called the *big bang theory*.

2. One way that scientists calculate the age of the universe is by measuring the distance from Earth to various galaxies. Another way to estimate the age of the universe is to calculate the ages of old, nearby stars.

3. b

4. Cosmic background radiation supports the big bang theory because measurements show that cosmic background radiation is distributed everywhere in the universe, which is what the big bang theory predicts.

5. Scientists think that the universe may continue to expand forever. Therefore, stars will die and the universe will eventually become a cold and dark place after billions of years, because the fusion process will gradually cause stars to die out.

6. 4,080,000,000,000 km

7. Sample answer: Every object in the universe is part of a larger system because the universe has a structure that is loosely repeated. For example, a cluster or group of galaxies can be made up of smaller star clusters and galaxies.

8. Scientists think that the universe will expand forever because they think that there may not be enough matter to produce the gravitational attraction needed to counteract expansion.

Answer to Reading Check

If gravity stops the expansion of the universe, the universe might collapse. If the expansion of the universe continues forever, stars will age and die and the universe will eventually become cold and dark.

Red Hot, or Not?

Teacher's Notes

Time Required

One 45-minute class period

Lab Ratings

EASY ———————————→ HARD

Teacher Prep 🧪🧪
Student Set-Up 🧪🧪
Concept Level 🧪
Clean Up 🧪

MATERIALS

The materials listed on this page are enough for a group of 3 to 4 students.

Safety Caution

Remind students to review all safety cautions and icons before beginning this lab activity. Be sure that the students disconnect the wires at each step. If left connected, the wires can get very hot.

Lab Notes

Students may find that holding the wires to the light bulb is difficult. If so, you may use a light socket. Any miniature incandescent light bulb can be used as the flashlight bulb.

Red Hot, or Not?

When you look at the night sky, some stars are brighter than others. Some are even different colors. For example, Betelgeuse, a bright star in the constellation Orion, glows red. Sirius, one of the brightest stars in the sky, glows bluish white. Astronomers use color to estimate the temperature of stars. In this activity, you will experiment with a light bulb and some batteries to discover what the color of a glowing object reveals about the temperature of the object.

OBJECTIVES

Discover what the color of a glowing object reveals about the temperature of the object.

Describe how the color and temperature of a star are related.

MATERIALS

- battery, D cell (2)
- battery, D cell, weak
- flashlight bulb
- tape, electrical
- wire, insulated copper, with ends stripped, 20 cm long (2)

SAFETY

Ask a Question

1 How are the color and temperature of a star related?

Form a Hypothesis

2 On a sheet of paper, change the question above into a statement that gives your best guess about the relationship between a star's color and temperature.

Kathy McKee
Hoyt Middle School
Des Moines, Iowa

CHAPTER RESOURCES

Chapter Resource File

- **Datasheet for Chapter Lab**
- **Lab Notes and Answers**

Technology

Classroom Videos
- Lab Video

LabBook

- I See the Light!

Test the Hypothesis

3 Tape one end of an insulated copper wire to the positive pole of the weak D cell. Tape one end of the second wire to the negative pole.

4 Touch the free end of each wire to the light bulb. Hold one of the wires against the bottom tip of the light bulb. Hold the second wire against the side of the metal portion of the bulb. The bulb should light.

5 Record the color of the filament in the light bulb. Carefully touch your hand to the bulb. Observe the temperature of the bulb. Record your observations.

6 Repeat steps 3–5 with one of the two fresh D cells.

7 Use the electrical tape to connect two fresh D cells so that the positive pole of the first cell is connected to the negative pole of the second cell.

8 Repeat steps 3–5 using the fresh D cells that are taped together.

Analyze the Results

1 **Describing Events** What was the color of the filament in each of the three trials? For each trial, compare the bulb temperature to the temperature of the bulb in the other two trials.

2 **Analyzing Results** What information does the color of a star tell you about the star?

3 **Classifying** What color are stars that have relatively high surface temperatures? What color are stars that have relatively low surface temperatures?

Draw Conclusions

4 **Applying Conclusions** Arrange the following stars in order from highest to lowest surface temperature: Sirius, which is bluish white; Aldebaran, which is orange; Procyon, which is yellow-white; Capella, which is yellow; and Betelgeuse, which is red.

Analyze the Results

1. With the weaker cell, the filament should glow with a dull red color. The filament glows bright red or orange with the stronger cell. With two fresh D cells, the filament becomes almost white. The temperature increases as more cells—or if fresh cells—are used.

2. Cooler objects emit red light. As an object becomes hotter, its color gradually changes from red to orange to white. Therefore, the color of the light emitted from the stars helps scientists determine the surface temperatures of stars.

3. Stars that have relatively high surface temperatures are white or blue. Stars that have relatively low surface temperatures are red or orange.

Draw Conclusions

4. The order of the stars from highest to lowest surface temperature is as follows: Sirius, Procyon, Capella, Aldebaran, and Betelgeuse.

Chapter Review

Assignment Guide

Section	Questions
1	1, 4, 5, 7–10, 13–14, 17, 19, 21
2	12, 16, 18
3	2, 3
4	6, 11, 15, 20
1, 3	22–24

ANSWERS

Using Key Terms

1. a light-year
2. Open clusters
3. elliptical galaxies
4. spectrum

Understanding Key Ideas

5. c
6. d
7. c
8. a
9. d
10. Scientists classify stars by temperature and magnitude.
11. Sample answer: The universe has a structure that is loosely repeated over and over again. Every object in the universe is part of a larger system. For example, a group of galaxies can include planetary systems, such as our solar system.

USING KEY TERMS

The statements below are false. For each statement, replace the underlined term to make a true statement.

1. The distance that light travels in space in 1 year is called <u>apparent magnitude</u>.

2. <u>Globular clusters</u> are groups of stars that are usually located along the spiral disk of a galaxy.

3. Galaxies that have very bright centers and very little dust and gas are called <u>spiral galaxies</u>.

4. When you look at white light through a glass prism, you see a rainbow of colors called a <u>supernova</u>.

UNDERSTANDING KEY IDEAS

Multiple Choice

5. A scientist can identify a star's composition by looking at
   a. the star's prism.
   b. the star's continuous spectrum.
   c. the star's absorption spectrum.
   d. the star's color.

6. If the universe expands forever,
   a. the universe will collapse.
   b. the universe will repeat itself.
   c. the universe will remain just as it is today.
   d. stars will age and die and the universe will become cold and dark.

7. The majority of stars in our galaxy are
   a. blue stars.
   b. white dwarfs.
   c. main-sequence stars.
   d. red giants.

8. Which of the following is used to measure the distance between objects in space?
   a. parallax   c. zenith
   b. magnitude   d. altitude

9. Which of the following stars would be seen as the brightest star?
   a. Alcyone, which has an apparent magnitude of 3
   b. Alpheratz, which has an apparent magnitude of 2
   c. Deneb, which has an apparent magnitude of 1
   d. Rigel, which has an apparent magnitude of 0

Short Answer

10. Describe how scientists classify stars.

11. Describe the structure of the universe.

12. Explain how stars at different stages in their life cycle appear on the H-R diagram.

13. Explain the difference between the apparent motion and actual motion of stars.

14. Describe how color indicates the temperature of a star.

15. Describe two ways that scientists calculate the age of the universe.

12. Sample answer: After a star forms, it enters the second and longest stage of its life cycle, known as the *main sequence*. The main sequence is the diagonal pattern on the H-R diagram where most stars lie. After the main sequence, a star can enter the third stage of its life cycle and become a red giant star. Red giants are located on the upper-right part of the H-R diagram because these stars are very massive. In the final stages of a star's life cycle, a star can become a white dwarf. White dwarfs are located in the lower-left part of the H-R diagram because they are extremely hot but dim.

13. The apparent motion of stars, or how the stars appear to move, is due to Earth's rotation. The actual motion of stars is very hard to see because the stars are very distant.

14. Stars that have different temperatures have different colors. For example, blue stars are much hotter than red stars are.

15. Two ways that scientists calculate the age of the universe are by measuring the distance from Earth to various galaxies and by calculating the ages of old, nearby stars.

CRITICAL THINKING

16 Concept Mapping Use the following terms to create a concept map: *main-sequence star, nebula, red giant, white dwarf, neutron star,* and *black hole.*

17 Evaluating Conclusions While looking through a telescope, you see a galaxy that doesn't appear to contain any blue stars. What kind of galaxy is it most likely to be? Explain your answer.

18 Making Comparisons Explain the differences between main-sequence stars, giant stars, supergiant stars, and white dwarfs.

19 Evaluating Data Why do astronomers use absolute magnitudes to plot stars? Why don't astronomers use apparent magnitudes to plot stars?

20 Evaluating Sources According to the big bang theory, how did the universe begin? What evidence supports this theory?

21 Evaluating Data If a certain star displayed a large parallax, what could you say about the star's distance from Earth?

INTERPRETING GRAPHICS

The graph below shows Hubble's law, which relates how far galaxies are from Earth and how fast they are moving away from Earth. Use the graph below to answer the questions that follow.

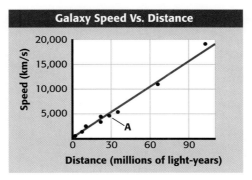

Galaxy Speed Vs. Distance

y-axis: Speed (km/s) — 5,000, 10,000, 15,000, 20,000
x-axis: Distance (millions of light-years) — 0, 30, 60, 90

A

22 Look at the point that represents galaxy A in the graph. How far is galaxy A from Earth, and how fast is it moving away from Earth?

23 If a galaxy is moving away from Earth at 15,000 km/s, how far is the galaxy from Earth?

24 If a galaxy is 90,000,000 light-years from Earth, how fast is it moving away from Earth?

Critical Thinking

16. An answer to this exercise can be found at the end of this book.

17. The galaxy is most likely an elliptical galaxy because it lacks the gas and dust needed for star formation. Blue stars are young stars.

18. Sample answer: Main-sequence stars are stars that are in the second stage of their life cycle. Red giant stars are large, reddish stars that are late in their life cycle. Red super-giant stars are similar to red giant stars but are much larger. White dwarf stars are small, hot, dim stars that are the left-over centers of old stars.

19. Absolute magnitude is a physical property of the star. Apparent magnitude varies according to a star's distance and absolute magnitude.

20. According to the big bang theory, the universe began with an explosion of matter and energy. Scientists can support this theory because they have shown that all distant galaxies are moving apart from all other galaxies.

21. The star would be relatively close to Earth.

Interpreting Graphics

22. The galaxy is about 30 million light-years away from Earth, and its speed is about 5,000 km/s.

23. The galaxy is almost 90 million light-years away from Earth.

24. The galaxy is moving at about 15,000 km/s.

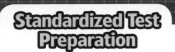

Standardized Test Preparation

Teacher's Note

To provide practice under more realistic testing conditions, give students 20 minutes to answer all of the questions in this Standardized Test Preparation.

MISCONCEPTION ALERT

Answers to the standardized test preparation can help you identify student misconceptions and misunderstandings.

READING

Passage 1

1. C
2. G
3. C

➕ TEST DOCTOR

Question 2: Answer G is correct because according to the passage, quasars appear as tiny pinpoints of light, but they emit a large amount of energy. Answer F is incorrect because quasars are not the same as galaxies. Answer H is incorrect because the passage does not state that quasars can be viewed only by using an optical telescope. Although scientists do not yet understand exactly how quasars can emit so much energy, answer I is incorrect because the passage does not state that quasars will never be understood.

READING

Read each of the passages below. Then, answer the questions that follow each passage.

Passage 1 Quasars are some of the most puzzling objects in the sky. If viewed through an optical telescope, a quasar appears as a small, dim star. Quasars are the most distant objects that have been observed from Earth. But many quasars are hundreds of times brighter than the brightest galaxy. Because quasars are so far away from Earth and yet are very bright, they most likely emit a large amount of energy. Scientists do not yet understand exactly how quasars can emit so much energy.

1. Based on the passage, which of the following statements is a fact?
   - **A** Quasars, unlike galaxies, include billions of bright objects.
   - **B** Galaxies are brighter than quasars.
   - **C** Quasars are hundreds of times brighter than the brightest galaxy.
   - **D** Galaxies are the most distant objects observed from Earth.

2. Based on the information in the passage, what can the reader conclude?
   - **F** Quasars are the same as galaxies.
   - **G** Quasars appear as small, dim stars, but they emit a large amount of energy.
   - **H** Quasars can be viewed only by using an optical telescope.
   - **I** Quasars will never be understood.

3. Why do scientists think that quasars emit a large amount of energy?
   - **A** because quasars are the brightest stars in the universe
   - **B** because quasars can be viewed only through an optical telescope
   - **C** because quasars are very far away and are still bright
   - **D** because quasars are larger than galaxies

Passage 2 If you live away from bright outdoor lights, you may be able to see a faint, narrow band of light and dark patches across the sky. This band is called the Milky Way. Our galaxy, the Milky Way, consists of stars, gases, and dust. Between the stars of the Milky Way are clouds of gas and dust called <u>interstellar matter</u>. These clouds provide materials that form new stars.

Every star that you can see in the night sky is a part of the Milky Way, because our solar system is inside the Milky Way. Because we are inside the galaxy, we cannot see the entire galaxy. But scientists can use astronomical data to create a picture of the Milky Way.

1. In the passage, what does the term *interstellar matter* mean?
   - **A** stars in the Milky Way
   - **B** the Milky Way
   - **C** a narrow band of light and dark patches across the sky
   - **D** the clouds of gas and dust between the stars in the Milky Way

2. Based on the information in the passage, what can the reader conclude?
   - **F** The Milky Way can be seen in the night sky near a large city.
   - **G** The entire Milky Way can be seen all at once.
   - **H** Every star that is seen in the night sky is a part of the Milky Way.
   - **I** Scientists have no idea what the entire Milky Way looks like.

Passage 2

1. D
2. H

➕ TEST DOCTOR

Question 2: Answer F is incorrect because according to the passage, the Milky Way galaxy can usually be seen in the night sky if you are away from bright outdoor lights. Answer G is incorrect because the second paragraph indicates that we cannot see the entire Milky Way galaxy. Answer H is correct because this fact is mentioned in the second paragraph. Answer I is incorrect because the second paragraph mentions that scientists use astronomical data to create pictures of the Milky Way.

The graph below shows the relationship between a star's age and mass. Use the graph below to answer the questions that follow.

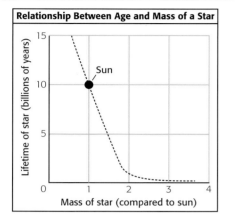

Relationship Between Age and Mass of a Star

1. How long does a star that has 1.2 times the mass of the sun live?
   - **A** 10 billion years
   - **B** 8 billion years
   - **C** 6 billion years
   - **D** 5 billion years

2. How long does a star that has 2 times the mass of the sun live?
   - **F** 4 billion years
   - **G** 1 billion years
   - **H** 10 billion years
   - **I** 5 billion years

3. If the sun's mass was reduced by half, how long would the sun live?
   - **A** 2 billion years
   - **B** 8 billion years
   - **C** 10 billion years
   - **D** more than 15 billion years

4. According to the graph, how long is the sun predicted to live?
   - **F** 15 billion years
   - **G** 10 billion years
   - **H** 5 billion years
   - **I** 2 billion years

Read each question below, and choose the best answer.

1. How many kilometers away from Earth is an object that is 8 light-years away from Earth? (Hint: One light-year is equal to 9.46 trillion kilometers.)
   - **A** 77 trillion kilometers
   - **B** 76 trillion kilometers
   - **C** 7.66 trillion kilometers
   - **D** 7.6 trillion kilometers

2. An astronomer observes two stars of about the same temperature and size. Alpha Centauri B is about 4 light-years away from Earth, and Sigma 2 Eridani A is about 16 light-years away from Earth. How many times as bright as Sigma 2 Eridani A does Alpha Centauri B appear? (Hint: One light-year is equal to 9.46 trillion kilometers.)
   - **F** 2 times as bright
   - **G** 4 times as bright
   - **H** 16 times as bright
   - **I** 32 times as bright

3. Star A is 5 million kilometers from Star B. What is this distance expressed in meters?
   - **A** 0.5 m
   - **B** 5,000 m
   - **C** 5×10^6 m
   - **D** 5×10^9 m

4. In the vacuum of space, light travels 3×10^8 m/s. How far does light travel in 1 h in space?
   - **F** 3,600 m
   - **G** 1.80×10^{10} m
   - **H** 1.08×10^{12} m
   - **I** 1.08×10^{16} m

5. The mass of the known universe is about 10^{23} solar masses, which is 10^{50} metric tons. How many metric tons is one solar mass?
   - **A** 10^{27} solar masses
   - **B** 10^{27} metric tons
   - **C** 10^{73} solar masses
   - **D** 10^{73} metric tons

Standardized Test Preparation

1. B
2. G
3. D
4. G

TEST DOCTOR

Question 1: Answer B is correct because when 1.2 is located on the x-axis, the student will find that the star will live for approximately 8 billion years. Answer A is incorrect because the star would have to be almost the same mass as the sun to live for approximately 10 billion years. Answers C and D are incorrect because they apply to a star that is closer to 1.5 times the mass of the sun.

1. B
2. H
3. D
4. H
5. B

TEST DOCTOR

Question 1: When 9.46 trillion kilometers is multiplied by 8 light-years, the product is 75.68 trillion kilometers. When this answer is rounded to the nearest whole number, the answer equals 76 trillion kilometers. Therefore, answer B is correct. Answer A is too high, and answers C and D are 10 times too low.

CHAPTER RESOURCES

Chapter Resource File

- • Standardized Test Preparation GENERAL

State Resources

 For specific resources for your state, visit **go.hrw.com** and type in the keyword **HSMSTR**.

Weird Science

Background

Scientists theorize that two types of stars can turn into black holes. When an extremely large star (about 8 to 25 times as massive as the sun) runs out of fuel and dies, it usually explodes as a supernova. A star that is more than 25 times as massive as the sun may collapse without exploding. If the core of either type of star is at least 3 times as massive as the sun, it is predicted that the core will collapse under its own gravity and become a black hole.

Scientific Discoveries

Discussion ——— GENERAL

Scientists think that Eta Carinae could become a supernova at anytime. Review the definition of *supernova* with students, and then ask them, "If Eta Carinae becomes a supernova, would life on this planet be threatened?" Then, ask them why or why not. (Answers may vary. At a distance of more than 8,000 light-years from our solar system, Eta Carinae is not close enough to threaten us. However, it's explosion would be seen from Earth.)

Science in Action

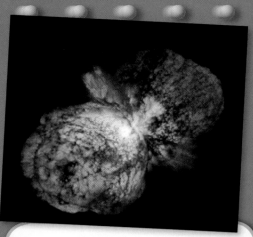

Weird Science

Holes Where Stars Once Were

An invisible phantom lurks in space, ready to swallow everything that comes near it. Once trapped in its grasp, matter is stretched, torn, and crushed into oblivion. Does this tale sound like a horror story? Guess again! Scientists call this phantom a *black hole*. As a star runs out of fuel, it cools and eventually collapses under the force of its own gravity. If the collapsing star is massive enough, it may shrink to become a black hole. The resulting gravitational attraction is so strong that even light cannot escape! Many astronomers think that black holes lie at the heart of many galaxies. Some scientists suggest that there is a giant black hole at the center of our own Milky Way.

Scientific Discoveries

Eta Carinae: The Biggest Star Ever Discovered

In 1841, Eta Carinae was the second-brightest star in the night sky. Why is this observation a part of history? Eta Carinae's brightness is historic because before 1837, Eta Carinae wasn't even visible to the naked eye! Strangely, a few years later Eta Carinae faded again and disappeared from the night sky. Something unusual was happening to Eta Carinae, and scientists wanted to know what it was. As soon as scientists had telescopes with which they could see far into space, they took a closer look at Eta Carinae. Scientists discovered that this star is highly unstable and prone to violent outbursts. These outbursts, the last of which was seen in 1841, can be seen on Earth. Scientists also discovered that Eta Carinae is 150 times as big as our sun and about 4 million times as bright. Eta Carinae is the biggest and brightest star ever found!

Language Arts ACTIVITY

WRITING SKILL Can you imagine traveling through a black hole? Write a short story that describes what you would see if you led a space mission to a black hole.

Math ACTIVITY

If Eta Carinae is 8,000 light-years from our solar system, how many kilometers is Eta Carinae from our solar system? (Hint: One light-year is equal to 9.46 trillion kilometers.)

Answer to Language Arts Activity

To help students with their short story, suggest that they use the Internet or library resources to read more about black holes.

Answer to Math Activity

9,460,000,000,000 km $\times$ 8,000 light-years = approximately 7.57×10^{16} km away from our solar system

Jocelyn Bell-Burnell

Astrophysicist Imagine getting a signal from far out in space and not knowing what or whom it's coming from. That's what happened to astrophysicist Jocelyn Bell-Burnell. Bell-Burnell is known for discovering pulsars, objects in space that emit radio waves at short, regular intervals. But before she and her advisor discovered that the signals came from pulsars, they thought that the signals may have come from aliens!

Born in 1943 in Belfast, Northern Ireland, Jocelyn Bell-Burnell became interested in astronomy at an early age. At Cambridge University in 1967, Bell-Burnell, who was a graduate student, and her advisor, Anthony Hewish, completed work on a huge radio telescope designed to pick up signals from quasars. Bell-Burnell's job was to operate the telescope and analyze its chart paper recordings on a graph. Each day, the telescope recordings used 29.2 m of chart paper! After a month, Bell-Burnell noticed that the recordings showed a few "bits of scruff"—very short, pulsating radio signals—that she could not explain. Bell-Burnell and Hewish struggled to find the source of the mysterious signal. They checked the equipment and began eliminating possible sources of the signal, such as satellites, television, and radar. Shortly after finding the first signal, Bell-Burnell discovered a second. The second signal was similar to the first but came from a different position in the sky. By January 1968, Bell-Burnell had discovered two more pulsating signals. In March of 1968, her findings that the signals were from a new kind of star were published and amazed the scientific community. The scientific press named the newly discovered stars *pulsars*.

Today, Bell-Burnell is a leading expert in the field of astrophysics and the study of stars. She is currently head of the physics department at the Open University, in Milton Keynes, England.

Social Studies ACTiViTy

Use the Internet or library resources to research historical events that occurred during 1967 and 1968. Find out if the prediction that the signals from pulsars were coming from aliens affected historical events during this time.

go.hrw.com

To learn more about these Science in Action topics, visit **go.hrw.com** and type in the keyword **HZ5UNVF**.

Current Science

Check out Current Science® articles related to this chapter by visiting go.hrw.com. Just type in the keyword **HZ5CS19**.

Careers

Background

One of the most fascinating properties of neutron stars is their incredible mass. Most neutron stars are only about 10 to 16 km in diameter, but their mass can be equal to the mass of our sun. The mass of a neutron star is so great that if you could stand on its surface and drop a coin, the coin would hit the ground at half the speed of light.

Discussion ———— GENERAL

Controversial Nobel Prize The discovery of the pulsar is a very important part of the history of astrophysics. In 1967, Sir Martin Ryle and Tony Hewish, from the Cavendish Laboratory in Cambridge, England, were jointly awarded the Nobel Prize in physics. Hewish was recognized for his discovery of pulsars. This Nobel Prize announcement sparked a public controversy because some people thought that Hewish should have shared the Nobel Prize with Bell-Burnell. Have students do some research to find out what Bell-Burnell's reaction was to the controversy. Then, have a class discussion on the topic. Encourage students to share their opinions about the controversy.

Answer to Social Studies Activity

To help students conduct research, suggest that they look at the front-page sections or science sections of major newspapers during 1967 and 1968. Students may also find information or news stories that were featured in mainstream news and science magazines during 1967 and 1968.

Formation of the Solar System
Chapter Planning Guide

Compression guide:
To shorten instruction because of time limitations, omit the Chapter Lab.

OBJECTIVES	LABS, DEMONSTRATIONS, AND ACTIVITIES	TECHNOLOGY RESOURCES
PACING • 90 min pp. 62–67 **Chapter Opener**	SE **Start-up Activity**, p. 63 (GENERAL)	OSP **Parent Letter** ■ (GENERAL) CD **Student Edition on CD-ROM** CD **Guided Reading Audio CD** ■ TR **Chapter Starter Transparency*** VID **Brain Food Video Quiz**
Section 1 A Solar System Is Born • Explain the relationship between gravity and pressure in a nebula. • Describe how the solar system formed.	TE **Group Activity** The Solar System, p. 64 (GENERAL) TE **Activity** Modeling Planetesimal Formation, p. 66 ◆ (BASIC) SE **Connection to Language Arts** Eyewitness Account, p. 67 (GENERAL) LB **Whiz-Bang Demonstrations** Can You Vote on Venus?* (GENERAL)	CRF **Lesson Plans*** TR **Bellringer Transparency***
PACING • 45 min pp. 68–73 **Section 2 The Sun: Our Very Own Star** • Describe the basic structure and composition of the sun. • Explain how the sun generates energy. • Describe the surface activity of the sun, and identify how this activity affects Earth.	TE **Demonstration** Observing Sunspots, p. 68 ◆ (GENERAL) TE **Connection Activity** Math, p. 69 (GENERAL) SE **Connection to Chemistry** Atoms, p. 70 (GENERAL) TE **Group Activity** Escape from the Sun, p. 70 (ADVANCED) TE **Connection Activity** Math, p. 71 (BASIC) TE **Connection Activity** History, p. 71 (GENERAL) SE **Skills Practice Lab** How Far Is the Sun?, p. 84 ◆ (GENERAL) CRF **Datasheet for Chapter Lab***	CRF **Lesson Plans*** TR **Bellringer Transparency*** TR The Structure and Atmosphere of the Sun* TR Fusion of Hydrogen in the Sun* TR **LINK TO PHYSICAL SCIENCE** The Periodic Table of the Elements* CRF **SciLinks Activity*** (GENERAL) VID **Lab Videos for Earth Science**
PACING • 45 min pp. 74–79 **Section 3 The Earth Takes Shape** • Describe the formation of the solid Earth. • Describe the structure of the Earth. • Explain the development of Earth's atmosphere and the influence of early life on the atmosphere. • Describe how the Earth's oceans and continents formed.	TE **Discussion** Earth's Atmosphere, p. 74 (GENERAL) SE **Connection to Environmental Science** The Greenhouse Effect, p. 76 (GENERAL) TE **Activity** Rings Around Our Planet, p. 76 (GENERAL) TE **Internet Activity** Researching New Planets, p. 76 (ADVANCED) SE **School-to-Home Activity** Comets and Meteors, p. 77 (GENERAL)	CRF **Lesson Plans*** TR **Bellringer Transparency*** TR Formation of Earth's Layers* SE **Internet Activity**, p. 79 (GENERAL)
PACING • 45 min pp. 80–83 **Section 4 Planetary Motion** • Explain the difference between rotation and revolution. • Describe three laws of planetary motion. • Describe how distance and mass affect gravitational attraction.	TE **Activity** Measuring Ellipses, p. 80 (GENERAL) TE **Connection Activity** Language Arts, p. 81 (GENERAL) SE **Quick Lab** Staying in Focus, p. 82 (GENERAL) CRF **Datasheet for Quick Lab*** LB **Long-Term Projects & Research Ideas** A Two-Sun Solar System* (ADVANCED)	CRF **Lesson Plans*** TR **Bellringer Transparency*** TR Earth's Rotation and Revolution* TR Ellipse* TR Gravity and the Motion of the Moon*

PACING • 90 min

CHAPTER REVIEW, ASSESSMENT, AND STANDARDIZED TEST PREPARATION

CRF **Vocabulary Activity*** (GENERAL)
SE **Chapter Review**, pp. 86–87 (GENERAL)
CRF **Chapter Review*** ■ (GENERAL)
CRF **Chapter Tests A*** ■ (GENERAL), **B*** (ADVANCED), **C*** (SPECIAL NEEDS)
SE **Standardized Test Preparation**, pp. 88–89 (GENERAL)
CRF **Standardized Test Preparation*** (GENERAL)
CRF **Performance-Based Assessment*** (GENERAL)
OSP **Test Generator** (GENERAL)
CRF **Test Item Listing*** (GENERAL)

Online and Technology Resources

Visit **go.hrw.com** for a variety of free resources related to this textbook. Enter the keyword **HZ5SOL**.

Holt Online Learning

Students can access interactive problem-solving help and active visual concept development with the *Holt Science and Technology* Online Edition available at **www.hrw.com**.

Guided Reading Audio CD

A direct reading of each chapter using instructional visuals as guideposts. For auditory learners, reluctant readers, and Spanish-speaking students. Available in English and Spanish.

SKILLS DEVELOPMENT RESOURCES	SECTION REVIEW AND ASSESSMENT	STANDARDS CORRELATIONS
SE Pre-Reading Activity, p. 62 `GENERAL` **OSP** Science Puzzlers, Twisters & Teasers* `GENERAL`		National Science Education Standards UCP 1, 2; SAI 2; ES 3b
CRF Directed Reading A* ■ `BASIC`, B* `SPECIAL NEEDS` **CRF** Vocabulary and Section Summary* ■ `GENERAL` **SE** Reading Strategy Reading Organizer, p. 64 `GENERAL` **TE** Inclusion Strategies, p. 65 ◆ **MS** Math Skills for Science Density* `GENERAL` **CRF** Critical Thinking A Balooney Universe* `ADVANCED`	**SE** Reading Checks, pp. 65, 67 `GENERAL` **TE** Reteaching, p. 66 `BASIC` **TE** Quiz, p. 66 `GENERAL` **TE** Alternative Assessment, p. 66 `GENERAL` **TE** Homework, p. 66 `ADVANCED` **SE** Section Review,* p. 67 ■ `GENERAL` **CRF** Section Quiz* ■ `GENERAL`	UCP 1, 2, 4; ST 2; HNS 2, 3; SPSP 5; ES 3a, 3b, 3c
CRF Directed Reading A* ■ `BASIC`, B* `SPECIAL NEEDS` **CRF** Vocabulary and Section Summary* ■ `GENERAL` **SE** Reading Strategy Reading Organizer, p. 68 `GENERAL` **CRF** Reinforcement Worksheet Stay on the Sunny Side* `GENERAL`	**SE** Reading Checks, pp. 69, 71, 72 `GENERAL` **TE** Reteaching, p. 72 `BASIC` **TE** Quiz, p. 72 `GENERAL` **TE** Alternative Assessment, p. 72 `ADVANCED` **SE** Section Review,* p. 73 ■ `GENERAL` **CRF** Section Quiz* ■ `GENERAL`	ST 2; SPSP 5; HNS 1, 2, 3; ES 3a; *Chapter Lab:* UCP 2, 3, 5; SAI 1, 2; ST 1; SPSP 5; HNS 1; ES 3a
CRF Directed Reading A* ■ `BASIC`, B* `SPECIAL NEEDS` **CRF** Vocabulary and Section Summary* ■ `GENERAL` **SE** Reading Strategy Discussion, p. 74 `GENERAL` **TE** Inclusion Strategies, p. 75 **MS** Math Skills for Science Reducing Fraction to Lowest Terms* `GENERAL` **CRF** Reinforcement Worksheet Third Rock from the Sun* `GENERAL`	**SE** Reading Checks, pp. 74, 76 78 `GENERAL` **TE** Homework, p. 77 `GENERAL` **TE** Reteaching, p. 78 `BASIC` **TE** Quiz, p. 78 `GENERAL` **TE** Alternative Assessment, p. 78 `GENERAL` **SE** Section Review,* p. 79 ■ `GENERAL` **CRF** Section Quiz* ■ `GENERAL`	UCP 2, 4; SAI 1; ES 2b
CRF Directed Reading A* ■ `BASIC`, B* `SPECIAL NEEDS` **CRF** Vocabulary and Section Summary* ■ `GENERAL` **SE** Reading Strategy Paired Summarizing, p. 80 `GENERAL` **SE** Math Practice Kepler's Formula, p. 81 `GENERAL`	**SE** Reading Checks, pp. 81, 82 `GENERAL` **TE** Reteaching, p. 82 `BASIC` **TE** Quiz, p. 82 `GENERAL` **TE** Alternative Assessment, p. 82 `GENERAL` **SE** Section Review,* p. 83 ■ `GENERAL` **CRF** Section Quiz* ■ `GENERAL`	UCP 1, 2, 3; SAI 1, 2; ST 2; SPSP 5; HNS 1, 2, 3; ES 3b

One-Stop Planner® CD-ROM

This convenient CD-ROM includes:
- Lab Materials QuickList Software
- Holt Calendar Planner
- Customizable Lesson Plans
- Printable Worksheets
- ExamView® Test Generator

cnnstudentnews.com

Find the latest news, lesson plans, and activities related to important scientific events.

www.scilinks.org

Maintained by the **National Science Teachers Association.** See Chapter Enrichment pages for a complete list of topics.

Current Science®

Check out *Current Science* articles and activities by visiting the HRW Web site at **go.hrw.com.** Just type in the keyword **HZ5CS20T.**

Classroom Videos

- **Lab Videos** demonstrate the chapter lab.
- **Brain Food Video Quizzes** help students review the chapter material.

Visual Resources

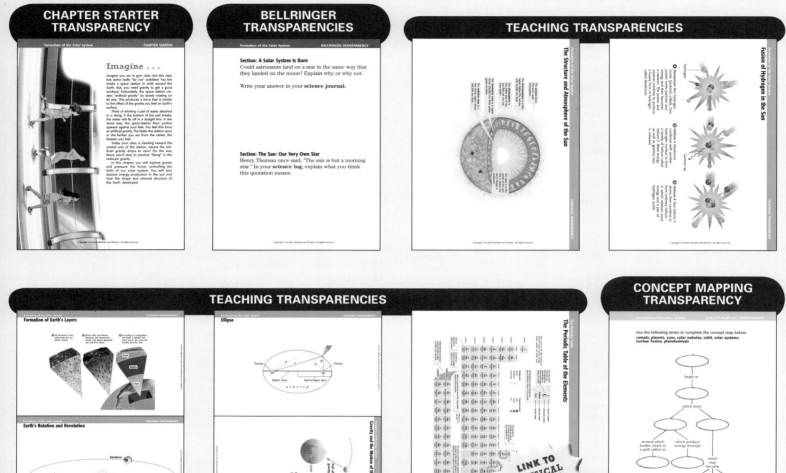

CHAPTER STARTER TRANSPARENCY

Imagine . . .

BELLRINGER TRANSPARENCIES

Section: A Solar System Is Born
Could astronauts land on a star in the same way that they landed on the moon? Explain why or why not.

Write your answer in your **science journal.**

Section: The Sun: Our Very Own Star
Henry Thoreau once said, "The sun is but a morning star." In your **science log,** explain what you think this quotation means.

TEACHING TRANSPARENCIES

The Structure and Atmosphere of the Sun

Fusion of Hydrogen in the Sun

TEACHING TRANSPARENCIES

Formation of Earth's Layers

Earth's Rotation and Revolution

Ellipse

Gravity and the Motion of the Moon

The Periodic Table of the Elements

LINK TO PHYSICAL SCIENCE

Chapter: The Periodic Table

CONCEPT MAPPING TRANSPARENCY

Use the following terms to complete the concept map below: comets, planets, suns, solar nebulas, orbit, solar systems, nuclear fusion, planetesimals

Planning Resources

LESSON PLANS

Lesson Plan   SAMPLE

Section: Waves

Pacing
Regular Schedule: with lab(s)2 days   without lab(s)2 days
Block Schedule: with lab(s) 1 1/2 days   without lab(s)1 day

Objectives
1. Relate the seven properties of life to a living organism.
2. Describe seven themes that can help you to organize what you learn about biology.
3. Identify the tiny structures that make up all living organisms.
4. Differentiate between reproduction and heredity and between metabolism and homeostasis.

National Science Education Standards Covered
LSInter6: Cells have particular structures that underlie their functions.
LSMat1: Most cell functions involve chemical reactions.
LSBeh1: Cells store and use information to guide their functions.
UCP1: Cell functions are regulated.
SI1: Cells can differentiate and form complete multicellular organisms.
SI1: Species evolve over time.
ESS1: The great diversity of organisms is the result of more than 3.5 billion years of evolution.
ESS2: Natural selection and its evolutionary consequences provide a scientific explanation for the fossil record of ancient life forms as well as for the striking molecular similarities observed among the diverse species of living organisms.
ST1: The millions of different species of plants, animals, and microorganisms that live on Earth today are related by descent from common ancestors.
ST2: The energy for life primarily comes from the sun.
SPSP1: The complexity and organization of organisms accommodate the need for obtaining, transforming, transporting, releasing, and eliminating the matter and energy used to sustain the organism.
SPSP6: As matter and energy flows through different levels of organization of living systems—cells, organs, communities—and between living systems and the physical environment, chemical elements are recombined in different ways.
HNS1: Organisms have behavioral responses to internal changes and to external stimuli.

PARENT LETTER

    SAMPLE
Dear Parent,

Your son's or daughter's science class will soon begin exploring the chapter entitled "The World of Physical Science." In this chapter, students will learn about how the scientific method applies to the world of physical science and the role of physical science in the world. By the end of the chapter, students should demonstrate a clear understanding of the chapter's main ideas and be able to discuss the following topics:

1. physical science as the study of energy and matter (Section 1)
2. the role of physical science in the world around them (Section 1)
3. careers that rely on physical science (Section 1)
4. the steps used in the scientific method (Section 2)
5. examples of technology (Section 2)
6. how the scientific method is used to answer questions and solve problems (Section 2)
7. how our knowledge of science changes over time (Section 2)
8. how models represent real objects or systems (Section 3)
9. examples of different ways models are used in science (Section 3)
10. the importance of the International System of Units (Section 4)
11. the appropriate units to use for particular measurements (Section 4)
12. how area and density are derived quantities (Section 4)

Questions to Ask Along the Way

You can help your son or daughter learn about these topics by asking interesting questions such as the following:

• What are some surprising careers that use physical science?
• What is a characteristic of a good hypothesis?
• When is it a good idea to use a model?
• Why do Americans measure things in terms of inches and yards and feet and meters ?

ALSO IN SPANISH

TEST ITEM LISTING

TEST ITEM LISTING
The World of Earth Science   SAMPLE

MULTIPLE CHOICE

1. A limitation of models is that
   a. they are large enough to use.
   b. they do not act exactly like the things that they model.
   c. they are smaller than the things that they model.
   d. they model unfamiliar things.
   Answer: B   Difficulty: 1   Section: 3   Objective: 2

2. The length 10 m is equal to
   a. 100 cm.   c. 10,000 mm.
   b. 1,000 cm.   d. Both (b) and (c)
   Answer: D   Difficulty: 1   Section: 3   Objective: 2

3. To be valid, a hypothesis must be
   a. testable.   c. made into a law.
   b. supported by evidence.   d. Both (a) and (b)
   Answer: B   Difficulty: 1   Section: 3   Objective: 2 1

4. The statement "Sheila has a stain on her shirt" is an example of a(n)
   a. law.   c. observation.
   b. hypothesis.   d. prediction.
   Answer: B   Difficulty: 3   Section: 3   Objective: 2

5. A hypothesis is often developed out of
   a. observations.   c. laws.
   b. experiments.   d. Both (a) and (b)
   Answer: B   Difficulty: 1   Section: 3   Objective: 2

6. How many milliliters are in 3.5 kL?
   a. 3,500 mL   c. 3,500, 000 mL
   b. 0.0035 mL   d. 35,000 mL
   Answer: B   Difficulty: 1   Section: 3   Objective: 2

7. A map of Seattle is an example of a
   a. law.   c. model.
   b. theory.   d. unit.
   Answer: C   Difficulty: 1   Section: 3   Objective: 2

8. A lab has the safety icons shown below. These icons mean that you should wear
   a. only safety goggles.   c. safety goggles and a lab apron.
   b. only a lab apron.   d. safety goggles, a lab apron, and gloves.
   Answer: C   Difficulty: 3   Section: 3   Objective: 2

9. The law of conservation of mass says the tot al mass before a chemical change is
   a. more than the total mass after the change.
   b. less than the total mass after the change.
   c. the same as the total mass after the change.
   d. not the same as the total mass after the change.
   Answer: C   Difficulty: 3   Section: 3   Objective: 2

10. In which of the following areas might you find a geochemist at work?
    a. studying the chemistry of rocks   c. studying fishes
    b. studying forestry   d. studying the atmosphere
    Answer: D   Difficulty: 1   Section: 3   Objective: 2

 **One-Stop Planner® CD-ROM**

This CD-ROM includes all of the resources shown here and the following time-saving tools:

• *Lab Materials QuickList Software*
• *Customizable lesson plans*
• *Holt Calendar Planner*
• *The powerful ExamView® Test Generator*

For a preview of available worksheets covering math and science skills, see pages T12–T19. All of these resources are also on the One-Stop Planner®.

Meeting Individual Needs

DIRECTED READING A
Skills Worksheet
Directed Reading A SAMPLE

Section:
THAT'S SCIENCE!

1. How did James Czarnowski get his idea for the penguin boat?
Explain.

ALSO IN SPANISH

BASIC

DIRECTED READING B
Skills Worksheet
Directed Reading B SAMPLE

Section:
THAT'S SCIENCE!

1. How did James Czarnowski get his idea for the penguin boat, Proteus?
Explain.

2. What is unusual about the way that Proteus moves through the water?

SPECIAL NEEDS

VOCABULARY ACTIVITY
Activity
Vocabulary Activity SAMPLE

Getting the Dirt on the Soil
After you finish reading Chapter [Unique Title], try this puzzle! Use the clues below to unscramble the vocabulary words. Write your answer in the space provided.

9. the chemical breakdown of rocks and minerals into new substances: CAMILCHE THEARIGWEN

GENERAL

VOCABULARY AND SECTION SUMMARY
Skills Worksheet
Vocabulary & Notes SAMPLE

Section:
VOCABULARY
In your own words, write a definition of the following term in the space provided.

1. scientific method

2. technology

ALSO IN SPANISH

GENERAL

REINFORCEMENT
Skills Worksheet
Reinforcement SAMPLE

The Plane Truth
Complete this worksheet after you finish reading the Section: [Unique Section Title]

You plan to enter a paper airplane contest sponsored by Talkin' Physical Science magazine. The person whose airplane flies the farthest wins a lifetime subscription to the magazine! The week before the contest, you watch an airplane landing at a nearby

You notice that the wings of the airplane have flaps, as shown in the illustration at right. The paper airplanes you've been testing do not have wing flaps. What question would you ask yourself based on these observations? Write your

BASIC

CRITICAL THINKING
Skills Worksheet
Critical Thinking SAMPLE

A Solar Solution

ADVANCED

SCILINKS ACTIVITY
Activity
SciLinks Activity SAMPLE

MARINE ECOSYSTEMS
Go to www.scilinks.org. To find links related to marine ecosystems, type in the keyword HL5495. Then, use the links to answer the questions about marine ecosystems.

percentage of the Earth's surface is covered by water?

GENERAL

SCIENCE PUZZLERS, TWISTERS & TEASERS
CHAPTER
19 SCIENCE PUZZLERS, TWISTERS & TEASERS
Formation of the Solar System

Find the Oddballs and Decode the Message
1. Each group of terms below contains an unrelated oddball. Circle the term that doesn't belong and explain why it doesn't. Then arrange the first letters of the oddballs to discover something central to the study of stars and planets.

a.	mantle	orbit	crust	core
b.	radiative zone	photosphere	crust	sunspot
		planetesimal	radiative zone	planet
	chromosphere	corona	radiative zone

GENERAL

Labs and Activities

LONG-TERM PROJECTS & RESEARCH IDEAS
Name _____ Date _____ Class _____
PROJECT
47 STUDENT WORKSHEET
A Two-Sun Solar System? DESIGN YOUR OWN

It's late on a Saturday night and you and a friend are watching a science fiction movie on television. It's your favorite part, when the hero finds himself stranded in a desert after his ship has crashed on a distant planet. Without water he may never make it to the nearest scientific outpost. The blazing sun hangs high overhead, hurting his eyes and burning his parched lips. Suddenly he sees a second sun rising. Our hero is doomed! The heat from the second sun will surely seal his fate.

A Star Is Born
1. A solar system with two suns? Is that possible? Some people believe that in the dawning of our solar system Jupiter almost became a star. Find out why Jupiter is sometimes called a "failed star" or a "near sun." What are the characteristics of a star? What are brown dwarfs? What would have had to happen during the formation of our solar system in order for Jupiter to become a star? Could this still happen? Pretend that you and a partner are film critics who understand the history of planets, as described above. Videotape a show in which you debate whether this scene could actually take place in our solar system.

Research Ideas
2. Solar wind particles can disturb the Earth's ionosphere. A strong solar wind can disturb the Earth's magnetic field, interfering with radio and television transmissions as well as microwave communications. A strong solar wind can also endanger astronauts in space. What are solar winds? What causes them? What problems do they cause for some technologies, and why? Make a poster display of your findings.

3. Are there clues about Mars on Earth? Research how information on the changing composition of Earth's atmosphere might help scientists understand the history of other planets. What is the atmosphere of Mars like now? What was the composition of its original atmosphere? Write an article about your findings.

4. Meteorites give us clues about the formation of the solar system. By using radiometric dating methods, scientists have determined the age of all meteorites to be about 4.5 billion years, about the same age as Earth. What are some scientific theories explaining this phenomenon? Write a report on the competing theories. Decide which one you support and explain why.

ADVANCED

WHIZ-BANG DEMONSTRATIONS
DEMO
32 TEACHER-LED DEMONSTRATION DISCOVERY LAB
Can You Vote on Venus?

Purpose
Students calculate their age in years or other planets and develop a better understanding of planetary motion around the sun.

Time Required
10–15 minutes

Lab Ratings
TEACHER PREP
CONCEPT LEVEL
CLEAN UP

What to Do
1. Ask students what it means to say that someone is 10 years old. (It means that a person has lived 10 Earth years.)
2. Point out to students that a year on Earth is the time it takes for the Earth to complete one revolution around the sun. Other planets revolve around the sun at different speeds, so a year is not the same on every planet.
3. Tell students they will calculate their age in years on the other planets. First have them multiply their age on Earth by 365 to find their age in Earth days. Then have them divide their age in Earth days by the number of Earth days that make up a year on a particular planet. The following is an example for a 25-year-old person on Mercury:

25 Earth years × 365 Earth days / 1 Earth year
= 9,125 Earth days

Mercury completes one revolution around the sun in 88 Earth days.

9,125 Earth days / 88 Earth days = 1 Mercury year
104 Mercury years

A person who is 25 years old on Earth is 104 on Mercury!

The number of Earth days or Earth years for each planet is given in the chart below. Have students record their answers in their ScienceLogs.

Table of Planetary Years

Planet	Length of year
Mercury	88 Earth days (0.24 Earth year)
Venus	225 Earth days (0.62 Earth year)
Earth	365 Earth days (1.00 Earth year)
Mars	687 Earth days (1.88 Earth years)
Jupiter	4,383 Earth days (12 Earth years)
Saturn	10,775 Earth days (29.5 Earth years)
Uranus	30,681 Earth days (84 Earth years)
Neptune	60,266 Earth days (165 Earth years)
Pluto	90,582 Earth days (248 Earth years)

Discussion
After completing the activity, ask students to look for trends in their responses. Point out that they would be older on Mercury and Venus and younger than on all the other planets. This is because Mercury and Venus revolve around the sun faster than the Earth does, while Earth revolves around the sun faster than all of the remaining planets do. Students should be led to discover that the farther a planet is from the sun, the longer it takes the planet to revolve around the sun.

continued...

GENERAL

DATASHEETS FOR QUICKLABS
TEACHER RESOURCE PAGE
Quick Lab
Reaction to Stress DATASHEET FOR QUICK LAB SAMPLE

Background
The graph below illustrates changes that occur in the membrane potential of a neuron during an action potential. Use the graph to answer the following questions. Refer to Figure 3 as needed.

DATASHEETS FOR CHAPTER LABS
TEACHER RESOURCE PAGE
Skills Practice Lab
Using Scientific Methods DATASHEET FOR CHAPTER LAB SAMPLE

Teacher's Notes
TIME REQUIRED
One 45-minute class period.

DATASHEETS FOR LABBOOK
TEACHER RESOURCE PAGE
Skills Practice Lab
Does It All Add Up? DATASHEET FOR LABBOOK LAB SAMPLE

Teacher's Notes
TIME REQUIRED
One 45-minute class period.

Review and Assessments

SECTION QUIZ
Assessment
Section Quiz SAMPLE

Section:
In the space provided, write the letter of the description that best matches the term or phrase.

_____ 1. building molecules that can be used as an energy source, or breaking down molecules in which energy is stored
_____ the process by which light energy is converted to chemical energy
_____ an organism that uses sunlight or inorganic substances to make organic compounds

f. cellular respiration

ALSO IN SPANISH

GENERAL

SECTION REVIEW
Skills Worksheet
Section Review SAMPLE

Section:
KEY TERMS
1. What is paleontologist study?

2. How does a trace fossil differ from petrified wood?

3. Define fossil.

ALSO IN SPANISH

GENERAL

CHAPTER REVIEW
Skills Worksheet
Chapter Review SAMPLE

USING VOCABULARY
1. Define biome in your own words.

2. Describe the characteristics of a savanna and a desert.

3. Identify the relationship between tundra and permafrost.

ALSO IN SPANISH

GENERAL

CHAPTER TEST A
Assessment
Chapter Test A SAMPLE

MULTIPLE CHOICE
In the space provided, write the letter of the term or phrase that best completes each statement or best answers each question.

_____ 1. Surface currents are formed by
a. the moon's gravity.   c. wind.
b. the sun's gravity.   d. increased water density.

_____ 2. When waves come near the shore,
a. they speed up.   c. their wavelength increases.
b. they maintain their speed. d. their wave height increases.

_____ 3. Longshore currents transport sediment
a. out to the open ocean.   c. only during low tide.
b. along the shore.   d. only during high tide.

_____ 4. Which of the following does NOT control surface currents?

ALSO IN SPANISH

GENERAL

CHAPTER TEST B
Assessment
Chapter Test B SAMPLE

MULTIPLE CHOICE
In the space provided, write the letter of the term or phrase that best completes each statement or best answers each question.

_____ 1. Surface currents are formed by
a. the moon's gravity.   c. wind.
b. the sun's gravity.   d. increased water density.

_____ 2. When waves come near the shore,
a. they speed up.   c. their wavelength increases.
b. they maintain their speed. d. their wave height increases.

ADVANCED

CHAPTER TEST C
Assessment
Chapter Test C SAMPLE

MULTIPLE CHOICE
In the space provided, write the letter of the term or phrase that best completes each statement or best answers each question.

_____ 1. Surface currents are formed by
a. the moon's gravity.   c. wind.
b. the sun's gravity.   d. increased water density.

_____ 2. When waves come near the shore,
a. they speed up.   c. their wavelength increases.
b. they maintain their speed. d. their wave height increases.

_____ 3. Longshore currents transport sediment
a. out to the open ocean.   c. only during low tide.
b. along the shore.   d. only during high tide.

_____ 4. Which of the following does NOT control surface currents?

SPECIAL NEEDS

STANDARDIZED TEST PREPARATION
Assessment
Standardized Test Preparation SAMPLE

READING
Read the passages below. Then, read each question that follows the passage. Decide which is the best answer to each question.

summer camp in the world, Billy can't to head for the outdoors. Billy checked the recommended apply light, summer clothes, sunscreen, rain gear, heavy, non-filled jacket, ski mask, and thick gloves. Wait a minute! Billy thought he was traveling to only one destination, so why does he need to bring such a wide variety of clothes? On further investigation

GENERAL

PERFORMANCE-BASED ASSESSMENT
Assessment
Performanced-Based Assessment SKILL BUILDER SAMPLE

OBJECTIVE
Determine which factors cause some sugar shapes to break down faster than others.

KNOW THE SCORE!
As you work through the activity, keep in mind that you will be earning a grade for the following:
• how you form and test the hypotheses (30%)
• the quality of your analysis (40%)
• the clarity of your conclusions (30%)

Using Scientific Methods
QUESTIONS
some sugar shapes erode more rapidly than others?

MATERIALS AND EQUIPMENT
• 1 regular sugar cube   • 90 mL of water

GENERAL

This Chapter Enrichment provides relevant and interesting information to expand and enhance your presentation of the chapter material.

Section 1

A Solar System Is Born

A Computer Model of Planet Building

- The Planetary Science Institute in Tucson, Arizona, produced a computer program in the late 1970s to test hypotheses about the formation of planetesimals and planets. The program has given credence to the theory that particle collisions within a swirling, collapsing nebula could have led to the creation of our solar system. The program simulates the motion of particles at various distances from the sun and tracks the results of collisions based on actual physical and mechanical properties, such as gas drag and particle speed. Run with various starting conditions, the program shows small particles aggregating into numbers of larger bodies and the eventual production of a system of planets.

Is That a Fact!

- ◆ A star begins life when fusion reactions start (at about 10 million degrees Celsius). It can burn steadily for billions of years by converting hydrogen to helium in its core. As the hydrogen runs out, the core collapses and heats up. The star's atmosphere expands and cools, and the star becomes a red giant.

High-Mass Stars: Live Fast and Die Young

- Stars that have a mass similar to that of our sun have a life cycle of 10 billion to 11 billion years. High-mass stars, which have a mass at least 10 times that of our sun, actually burn up much quicker than the sun— in 50 million to 100 million years—and burn much brighter than the sun. When a high-mass star runs out of fuel, its core collapses, and the star becomes a supernova. A supernova explosion is one of the most spectacular events in the universe.

Section 2

The Sun: Our Very Own Star

The Sunspot Cycle

- The increasing and decreasing number of sunspots in an 11-year cycle appears to be driven by the magnetic field in the sun's surface layers. The field seems to "wind up" much as a rubber band does (perhaps because of the difference in the rate of rotation between the sun's poles and its equator). This process intensifies the magnetic field; therefore, more sunspots appear, and the sun becomes much more active. At places on the sun that are most active, giant explosions called *solar flares* occur.

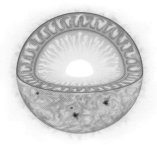

- Solar flares spew out electrically charged particles and enormous bursts of X rays that affect human technology on Earth. Despite the great distance between the sun and Earth, solar flares can disrupt TV programs, damage satellites, and endanger astronauts.

Section 3

The Earth Takes Shape

Evidence of Earth's Origins

- Many different sciences have contributed to our understanding of Earth's origin. Much remains to be discovered through computer models and the study of other planets. The following evidence has shaped the current scientific theories about the formation of Earth and its atmosphere:

- The oldest rocks on Earth are about 4 billion years old. Some of the oldest rocks are sedimentary in origin, so we know that oceans must have existed early in the history of Earth.

- The sun was about 30% less luminous when Earth was forming. We know this from the study of how hydrogen fusion reactions work.

- Earth must have had a dense atmosphere with greenhouse gases early in its life, or Earth would have been too cold to have liquid oceans.

- The oldest fossils of primitive life are stromatolites, blue-green algae colonies that originated between 3.7 billion and 3.4 billion years ago. Simple life-forms may have appeared on Earth before this time.

- Blue-green algae release oxygen as a byproduct of photosynthesis. Evidence from the oxidation of minerals in the rock record indicates that oxygen started to appear in significant concentrations in Earth's atmosphere between 2.5 billion and 2.0 billion years ago.

- Some of the water that formed the oceans came from the early Earth's interior and was released by outgassing during the differentiation process. As Earth heated and differentiated, water that was chemically bound in minerals was carried to the surface of the planet along with magma. Water vapor was then released to form the early atmosphere. Even at current rates of volcanism, the water vapor released from lava flows would be more than enough to fill Earth's oceans in 500 million years.

Section 4

Planetary Motion

The Orbit of Comets

- Orbits represent the entire path of an orbiting body. A planet or asteroid orbit has an elliptical shape. Scientists observe that near the sun comets seem to have a parabolic orbit (shaped like an open-ended ellipse). This shape may indicate that comets come from the outer reaches of the solar system. Scientists theorize that some comets originate in an enormous spherical cloud surrounding the solar system. This region, named the *Oort cloud*, may contain 100 billion comets. The gravitational pull of another object can knock a comet out of the Oort cloud, after which the comet's orbit brings it into the inner solar system.

Is That a Fact!

◆ Although Kepler's laws of motion were formulated almost 400 years ago, some space scientists and engineers still use the laws to plan and calculate the flight paths of artificial satellites orbiting the Earth.

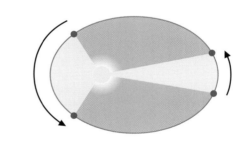

SciLINKS.

NSTA
Developed and maintained by the
National Science Teachers Association

SciLinks is maintained by the National Science Teachers Association to provide you and your students with interesting, up-to-date links that will enrich your classroom presentation of the chapter.

Visit www.scilinks.org and enter the SciLinks code for more information about the topic listed.

Topic: Planets
SciLinks code: HSM1152

Topic: The Oceans
SciLinks code: HSM1069

Topic: The Sun
SciLinks code: HSM1477

Topic: Kepler's Laws
SciLinks code: HSM0827

Topic: Layers of the Earth
SciLinks code: HSM0862

Overview

Tell students that this chapter will help them learn about how the solar system formed. The chapter also describes the processes that formed the sun and Earth. Finally, the chapter discusses the laws related to planetary motion.

Assessing Prior Knowledge

Students should be familiar with the following topics:

• the scale of the universe

• the life cycle of stars

• the contents of galaxies

Identifying Misconceptions

As students learn about the action of gravity, some of them may characterize it as an action of "holding." Students may also confuse gravity with atmospheric pressure, and they may describe gravity as something that keeps things from floating away. Remind students that gravity does not require air and that gravity does not stop acting on an object when the object has finished falling. Also, when teaching about the formation of the solar system, remind students that gravity is present in space.

3

Formation of the Solar System

About the PHOTO

The Orion Nebula, a vast cloud of dust and gas that is 35 trillion miles wide, is part of the familiar Orion constellation. Here, swirling clouds of dust and gas give birth to systems like our own solar system.

PRE-READING ACTIVITY

Graphic Organizer

Chain-of-Events Chart Before you read the chapter, create the graphic organizer entitled "Chain-of-Events Chart" described in the **Study Skills** section of the Appendix. As you read the chapter, fill in the chart with details about each step of the formation of the solar system.

Standards Correlations

National Science Education Standards

The following codes indicate the National Science Education Standards that correlate to this chapter. The full text of the standards is at the front of the book.

Chapter Opener
UCP 1, 2; SAI 2; ES 3b

Section 1 A Solar System Is Born
UCP 1, 2, 4; ST 2; HNS 2, 3; SPSP 5; ES 3a, 3b, 3c

Section 2 The Sun: Our Very Own Star
ST 2; SPSP 5; HNS 1, 2, 3; ES 3a

Section 3 The Earth Takes Shape
UCP 2, 4; SAI 1; ES 2b

Section 4 Planetary Motion
UCP 1, 2, 3; SAI 1, 2; ST 2; SPSP 5; HNS 1, 2, 3; ES 3b

Chapter Lab
UCP 2, 3, 5; SAI 1, 2; ST 1; SPSP 5; HNS 1; ES 3a

Chapter Review
UCP 1, 2, 4; SAI 1, 2; SPSP 5; HNS 2, 3; ES 2b; 3a, 3b, 3c

Science in Action
UCP 1, 2; SAI 2; ST 2; SPSP 5; HNS 1, 2, 3

Teacher's Notes: You might point out to students that when David Scott performed his experiment on the moon's surface, he paid homage to Galileo. While Galileo did predict that the mass of an object does not affect the rate at which the object falls, it is uncertain whether Galileo demonstrated this theory by dropping cannonballs of different masses from the Leaning Tower of Pisa.

Answers

1. Sample answer: Both pieces of paper should reach the bottom at the same time. They should fall at the same rate as the book. Gravity causes all objects to fall at the same rate regardless of their mass.

2. Sample answer: The crumpled piece of paper should reach the floor first. Although gravity pulled both pieces toward the floor, the crumpled piece of paper hit the ground first because it fell with less air resistance than the flat piece of paper did.

START-UP ACTIVITY

Strange Gravity

If you drop a heavy object, will it fall faster than a lighter one? According to the law of gravity, the answer is no. In 1971, *Apollo 15* astronaut David Scott stood on the moon and dropped a feather and a hammer. Television audiences were amazed to see both objects strike the moon's surface at the same time. Now, you can perform a similar experiment.

Procedure

1. Select **two pieces of identical notebook paper.** Crumple one piece of paper into a ball.

2. Place the flat piece of paper on top of a **book** and the paper ball on top of the flat piece of paper.

3. Hold the book waist high, and then drop it to the floor.

Analysis

1. Which piece of paper reached the bottom first? Did either piece of paper fall slower than the book? Explain your observations.

2. Now, hold the crumpled paper in one hand and the flat piece of paper in the other. Drop both pieces of paper at the same time. Besides gravity, what affected the speed of the falling paper? Record your observations.

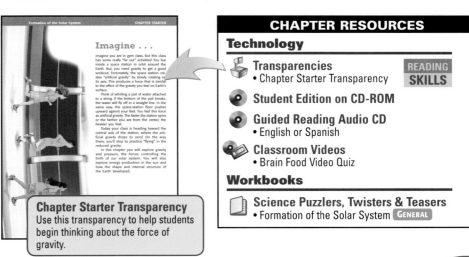

Chapter Starter Transparency
Use this transparency to help students begin thinking about the force of gravity.

CHAPTER RESOURCES

Technology

Transparencies
- Chapter Starter Transparency

READING SKILLS

Student Edition on CD-ROM

Guided Reading Audio CD
- English or Spanish

Classroom Videos
- Brain Food Video Quiz

Workbooks

Science Puzzlers, Twisters & Teasers
- Formation of the Solar System GENERAL

Focus

Overview

This section describes the formation of the solar system. It also describes the role that gravity and pressure played in the formation of the solar system.

🎧 Bellringer

Write the following question on the board: "Could astronauts land on a star in the same way that they landed on the moon?" (Sample answer: No, stars are composed of gas, not solid rock like the moon is. Stars are also a lot hotter than the moon is!)

Motivate

Group ACTiViTy — GENERAL

The Solar System Display a poster of the solar system. Have student groups brainstorm a list of facts they know about the solar system and some questions that they want to answer. Ask each group to study the display and note the following: what a planet's orbit looks like and how planets close to the sun differ from those far away from the sun. Discuss students' observations and hypotheses.

LS Visual

English Language Learners

READING WARM-UP

Objectives
- Explain the relationship between gravity and pressure in a nebula.
- Describe how the solar system formed.

Terms to Learn
nebula
solar nebula

READING STRATEGY

Reading Organizer As you read this section, make a flowchart of the steps of the formation of a solar system.

nebula a large cloud of gas and dust in interstellar space; a region in space where stars are born or where stars explode at the end of their lives

Figure 1 *The Horsehead Nebula is a cold, dark cloud of gas and dust. But observations suggest that it is also a site where stars form.*

A Solar System Is Born

As you read this sentence, you are traveling at a speed of about 30 km/s around an incredibly hot star shining in the vastness of space!

Earth is not the only planet orbiting the sun. In fact, Earth has eight fellow travelers in its cosmic neighborhood. The solar system includes a star we call the sun, nine planets, and many moons and small bodies that travel around the sun. For almost 5 billion years, planets have been orbiting the sun. But how did the solar system come to be?

The Solar Nebula

All of the ingredients for building planets, moons, and stars are found in the vast, seemingly empty regions of space between the stars. Just as there are clouds in the sky, there are clouds in space. These clouds are called nebulas. **Nebulas** (or nebulae) are mixtures of gases—mainly hydrogen and helium—and dust made of elements such as carbon and iron. Although nebulas are normally dark and invisible to telescopes, they can be seen when nearby stars illuminate them. So, how can a cloud of gas and dust such as the Horsehead Nebula, shown in **Figure 1,** form planets and stars? To answer this question, you must explore two forces that interact in nebulas—gravity and pressure.

Gravity Pulls Matter Together

The gas and dust that make up nebulas are made of matter. The matter of a nebula is held together by the force of gravity. In most nebulas, there is a lot of space between the particles. In fact, nebulas are less dense than air! Thus, the gravitational attraction between the particles in a nebula is very weak. The force is just enough to keep the nebula from drifting apart.

CHAPTER RESOURCES

Chapter Resource File

- Lesson Plan
- Directed Reading A **BASIC**
- Directed Reading B **SPECIAL NEEDS**

Technology

Transparencies
- Bellringer

Answer to Reading Check

The solar nebula is the cloud of gas and dust that formed our solar system.

Figure 2 Gravity and Pressure in a Nebula

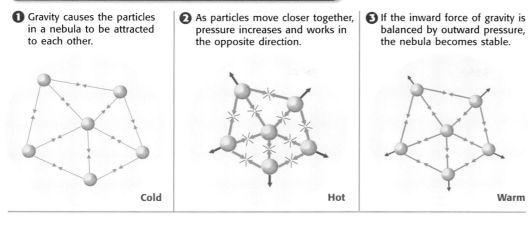

❶ Gravity causes the particles in a nebula to be attracted to each other.

❷ As particles move closer together, pressure increases and works in the opposite direction.

❸ If the inward force of gravity is balanced by outward pressure, the nebula becomes stable.

Cold

Hot

Warm

Pressure Pushes Matter Apart

If gravity pulls on all of the particles in a nebula, why don't nebulas slowly collapse? The answer has to do with the relationship between temperature and pressure in a nebula. *Temperature* is a measure of the average kinetic energy, or the energy of motion, of the particles in an object. If the particles in a nebula have little kinetic energy, they move slowly and the temperature of the cloud is very low. If the particles move fast, the temperature of the cloud is high. As particles move around, they sometimes crash into each other. As shown in **Figure 2,** these collisions cause particles to push away from each other, which creates *pressure*. If you have ever blown up a balloon, you understand how pressure works—pressure keeps a balloon from collapsing. In a nebula, outward pressure balances the inward gravitational pull and keeps the cloud from collapsing.

Upsetting the Balance

The balance between gravity and pressure in a nebula can be upset if two nebulas collide or a nearby star explodes. These events compress, or push together, small regions of a nebula called *globules,* or gas clouds. Globules can become so dense that they contract under their own gravity. As the matter in a globule collapses inward, the temperature increases and the stage is set for stars to form. The **solar nebula**—the cloud of gas and dust that formed our solar system—may have formed in this way.

solar nebula the cloud of gas and dust that formed our solar system

☑ **Reading Check** What is the solar nebula? (*See the Appendix for answers to Reading Checks.*)

Teach

Using the Figure — BASIC

Forces in a Nebula Have students refer to the three steps in **Figure 2** as they explain the force that pulls particles together (gravity) and the force that pushes particles apart (pressure due to the collision of particles). Be sure that students understand the role of temperature in a nebula. (As temperature increases, particles speed up, collisions increase, and pressure increases.) **LS Visual**

Discussion — BASIC

Reaching Equilibrium Have volunteers describe in their own words an example of a system that is in equilibrium because opposing forces of gravity (pulling) and pressure (pushing) balance one another. (Sample answer: One example is a person sitting in a chair. Gravity pulls down with a force equal to the force with which the chair pushes up.) **LS Verbal**

CONNECTION to Physical Science — GENERAL

Writing **Gas Laws** Discuss with students how Boyle's and Charles's laws help us understand the formation of the solar system.

• Boyle's law: At a constant temperature, the volume of a gas is inversely proportional to the pressure.

• Charles's law: At constant pressure, the volume of a gas is directly proportional to the temperature.

Review with students the difference between inverse and direct relationships, and have them give examples. Ask students to explain, in writing, why the temperature of a nebula increases as the nebula becomes denser. **LS Verbal**

Close

Reteaching ——— BASIC

Stages of Formation Draw four large squares on the board, and label them as follows: "The solar nebula," "The nebula collapsing," "The planetesimals form," and "The sun and planets form." Sketch each stage of solar system formation in the appropriate square. Have students use arrows and phrases to indicate changes in temperature and the balance between gravity and pressure.
LS Visual

Quiz ——— GENERAL

1. Why does the center of a collapsing nebula form a star? (Pressure is so intense between the crowded particles that atoms fuse and give off large amounts of energy.)

2. How do planets form? (Particles swirling in a cloud of dust and gas stick together and form planetesimals, which accumulate more matter and eventually form planets.)

Alternative Assessment ——— GENERAL

Explaining How the Solar System Formed Have students write a story in their **science journal** that would explain to a 17th-century astronomer how the sun and the planets formed. A 17th-century astronomer would not know nebulas or planetesimals by their current names. Students should share their stories with the class.
LS Verbal/Interpersonal

Figure 3 **The Formation of the Solar System**

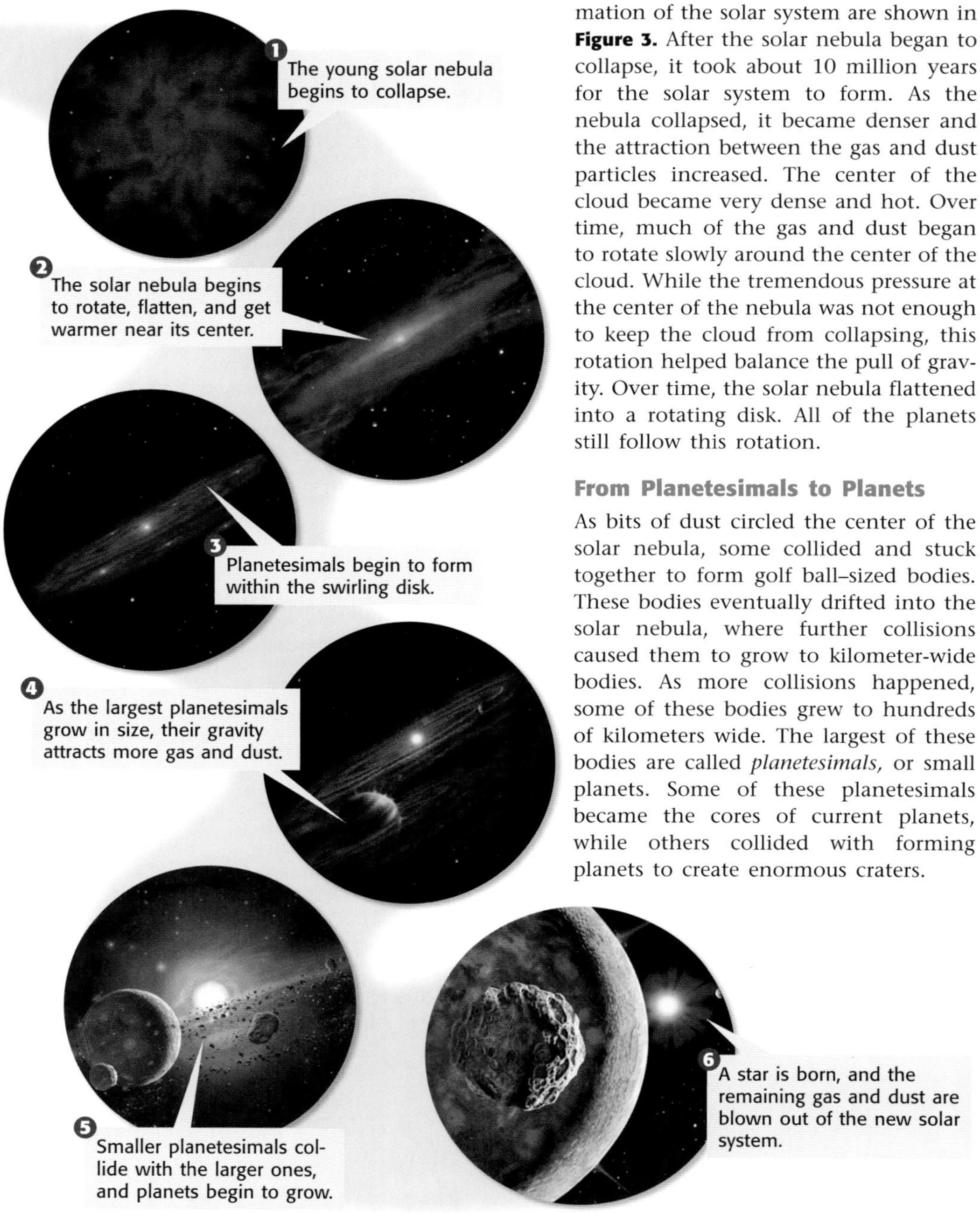

1. The young solar nebula begins to collapse.

2. The solar nebula begins to rotate, flatten, and get warmer near its center.

3. Planetesimals begin to form within the swirling disk.

4. As the largest planetesimals grow in size, their gravity attracts more gas and dust.

5. Smaller planetesimals collide with the larger ones, and planets begin to grow.

6. A star is born, and the remaining gas and dust are blown out of the new solar system.

How the Solar System Formed

The events that may have led to the formation of the solar system are shown in **Figure 3.** After the solar nebula began to collapse, it took about 10 million years for the solar system to form. As the nebula collapsed, it became denser and the attraction between the gas and dust particles increased. The center of the cloud became very dense and hot. Over time, much of the gas and dust began to rotate slowly around the center of the cloud. While the tremendous pressure at the center of the nebula was not enough to keep the cloud from collapsing, this rotation helped balance the pull of gravity. Over time, the solar nebula flattened into a rotating disk. All of the planets still follow this rotation.

From Planetesimals to Planets

As bits of dust circled the center of the solar nebula, some collided and stuck together to form golf ball–sized bodies. These bodies eventually drifted into the solar nebula, where further collisions caused them to grow to kilometer-wide bodies. As more collisions happened, some of these bodies grew to hundreds of kilometers wide. The largest of these bodies are called *planetesimals,* or small planets. Some of these planetesimals became the cores of current planets, while others collided with forming planets to create enormous craters.

ACTIVITY ——— BASIC

Modeling Planetesimal Formation Have students work in pairs. Each pair will need a couple of sheets of wax paper and a small spray bottle containing some water tinted by food coloring. Have students spray a little water on a sheet of wax paper. As students observe the wax paper after each spray, they will see large drops form. Discuss whether this model of planetesimal formation is accurate, and note how gravity, rather than surface tension, causes planetesimals to form. **LS** Visual

Homework ——— ADVANCED

Preparing a Presentation Have students find information about the asteroid belt, the Kuiper belt, the Oort cloud, and comets. Ask students to explain one or more of these phenomena using the steps shown in **Figure 3.** Encourage students to find a creative way to present their findings.
LS Visual/Interpersonal

Gas Giant or Rocky Planet?

The largest planetesimals formed near the outside of the rotating solar disk, where hydrogen and helium were abundant. These planetesimals were far enough from the solar disk that their gravity could attract the nebula gases. These outer planets grew to huge sizes and became the gas giants—Jupiter, Saturn, Uranus, and Neptune. Closer to the center of the nebula, where Mercury, Venus, Earth, and Mars formed, temperatures were too hot for gases to remain. Therefore, the inner planets in our solar system are made mostly of rocky material.

Reading Check Which planets are gas giants?

The Birth of a Star

As the planets were forming, nearly all of the extra matter in the solar nebula was traveling toward the center. The center became so dense and hot that hydrogen atoms began to fuse, or join, to form helium. Fusion released huge amounts of energy and created enough outward pressure to balance the inward pull of gravity. At this point, when the gas stopped collapsing, our sun was born and the new solar system was complete!

CONNECTION TO Language Arts

WRITING SKILL **Eyewitness Account** Research information on the formation of the outer planets, inner planets, and the sun. Then, imagine that you witnessed the formation of the planets and sun. Write a short story describing your experience.

SECTION Review

Summary

- The solar system formed out of a vast cloud of gas and dust called the *nebula*.
- Gravity and pressure were balanced until something upset the balance. Then, the nebula began to collapse.
- Collapse of the solar nebula caused heating in the center, which caused planetesimals to form.
- The central mass of the nebula became the sun. Planets formed from the surrounding materials.

Using Key Terms

1. In your own words, write a definition for each of the following terms: *nebula* and *solar nebula*.

Understanding Key Ideas

2. What is the relationship between gravity and pressure in a nebula?
   a. Gravity reduces pressure.
   b. Pressure balances gravity.
   c. Pressure increases gravity.
   d. None of the above

3. Describe how our solar system formed.

4. Compare the inner planets with the outer planets.

Math Skills

5. If the planets, moons, and other bodies make up 0.15% of the solar system's mass, what percentage does the sun make up?

Critical Thinking

6. **Evaluating Hypotheses** Pluto, the outermost planet, is small and rocky. Some scientists argue that Pluto is a captured asteroid, not a planet. Use what you know about how solar systems form to evaluate this hypothesis.

7. **Making Inferences** Why do all of the planets go around the sun in the same direction, and why do the planets lie on a relatively flat plane?

For a variety of links related to this chapter, go to www.scilinks.org

Topic: The Planets
SciLinks code: HSM1152

Answer to Reading Check

Jupiter, Saturn, Uranus, and Neptune

CHAPTER RESOURCES

Chapter Resource File

- Section Quiz GENERAL
- Section Review GENERAL
- Vocabulary and Section Summary GENERAL
- Critical Thinking ADVANCED

Answers to Section Review

1. Sample answer: A nebula is a large cloud of gas and dust in interstellar space. The solar nebula is the cloud of gas and dust that formed our solar system.

2. b

3. Sample answer: After the balance between gravity and pressure became unbalanced in the solar nebula, it began to collapse. The solar nebula became denser, and the attraction between the gas and dust particles increased. This attraction caused the center of the nebula to become dense and hot. As bits of dust circled the center, some collided to form planetesimals. The central mass of the nebula became the sun, and the planetesimals that continued to circle the sun eventually formed the planets.

4. The inner planets, Mercury, Venus, Earth, and Mars, formed closer to the sun where temperatures were too hot for gases to remain. So, the inner planets are made of mostly rocky material. The outer planets, Jupiter, Saturn, Uranus, and Neptune, formed farther away from the sun and are made of mostly gases. Pluto, the outermost planet, is not a gas giant.

5. 100% − 0.15% = 99.85%

6. Answers may vary.

7. Sample answer: The planets formed from the flattened disk of the nebula, which rotated. The dust and gas that formed the planets rotated in the same direction that the nebula was spinning.

Focus

Overview

This section describes the structure of the sun. It discusses the early theories about the source of the sun's energy and why nuclear fusion is the accepted theory today.

Bellringer

Have students write about the following quotation by Henry Thoreau: "The sun is but a morning star."

Motivate

Demonstration ——— GENERAL

Observing Sunspots Clamp a pair of binoculars in a ring stand, and cut a hole in a piece of cardboard that fits around the eyepiece of one binocular lens. Darken the classroom, and orient the binoculars toward the sun. Hold a mirror in the shadow of the cardboard, and project an image of the sun onto a wall. Focus the image, and have students identify sunspots and other features of the sun.

Safety Caution: Make sure that students do not look at the sun through the binoculars. Also, caution students never to look directly at the sun. **LS** Visual

READING WARM-UP

Objectives
● Describe the basic structure and composition of the sun.
● Explain how the sun generates energy.
● Describe the surface activity of the sun, and identify how this activity affects Earth.

Terms to Learn
nuclear fusion
sunspot

READING STRATEGY

Reading Organizer As you read this section, create an outline of the section. Use the headings from the section in your outline.

The Sun: Our Very Own Star

Can you imagine what life on Earth would be like if there were no sun? Without the sun, life on Earth would be impossible!

Energy from the sun lights and heats Earth's surface. Energy from the sun even drives the weather. Making up more than 99% of the solar system's mass, the sun is the dominant member of our solar system. The sun is basically a large ball of gas made mostly of hydrogen and helium held together by gravity. But what does the inside of the sun look like?

The Structure of the Sun

Although the sun may appear to have a solid surface, it does not. When you see a picture of the sun, you are really seeing through the sun's outer atmosphere. The visible surface of the sun starts at the point where the gas becomes so thick that you cannot see through it. As **Figure 1** shows, the sun is made of several layers.

Figure 1 **The Structure and Atmosphere of the Sun**

The **corona** forms the sun's outer atmosphere.

The **chromosphere** is a thin region below the corona, only 30,000 km thick.

The **photosphere** is the visible part of the sun that we can see from Earth.

The **convective zone** is a region about 200,000 km thick where gases circulate.

The **radiative zone** is a very dense region about 300,000 km thick.

The **core** is at the center of the sun. This is where the sun's energy is produced.

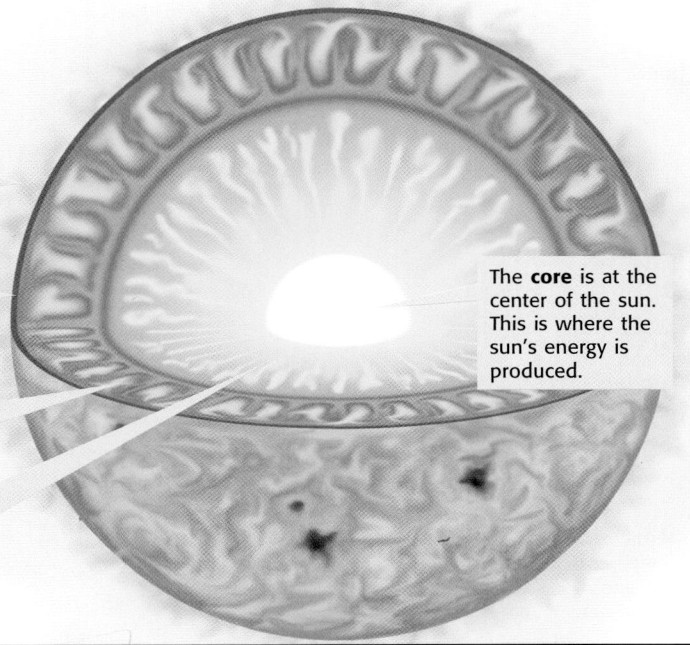

CHAPTER RESOURCES

Chapter Resource File

- Lesson Plan
- Directed Reading A **BASIC**
- Directed Reading B **SPECIAL NEEDS**

Technology

Transparencies
- Bellringer
- The Structure and Atmosphere of the Sun

Is That a Fact!

During an eclipse in 1868, a French astronomer named Pierre Janssen detected in the chromosphere of the sun a new element that was unknown on Earth. The new element, called helium (named after *helios,* the Greek word for "sun"), was not discovered on Earth until 1895.

At first, some type of burning fuel was thought to be the source of the sun's energy.

A shrinking sun was another explanation for solar energy.

Figure 2 *Ideas about the source of the sun's energy have changed over time.*

Energy Production in the Sun

The sun has been shining on Earth for about 4.6 billion years. How can the sun stay hot for so long? And what makes it shine? **Figure 2** shows two theories that were proposed to answer these questions. Many scientists thought that the sun burned fuel to generate its energy. But the amount of energy that is released by burning would not be enough to power the sun. If the sun were simply burning, it would last for only 10,000 years.

Burning or Shrinking?

It eventually became clear to scientists that burning wouldn't last long enough to keep the sun shining. Then, scientists began to think that gravity was causing the sun to slowly shrink. They thought that perhaps gravity would release enough energy to heat the sun. While the release of gravitational energy is more powerful than burning, it is not enough to power the sun. If all of the sun's gravitational energy were released, the sun would last for only 45 million years. However, fossils that have been discovered prove that dinosaurs roamed the Earth more than 65 million years ago, so this couldn't be the case. Therefore, something even more powerful than gravity was needed.

✔ Reading Check Why isn't energy from gravity enough to power the sun? (*See the Appendix for answers to Reading Checks.*)

Answer to Connection to Chemistry

nitrogen: 7 protons, 7 neutrons, and 7 electrons; oxygen: 8 protons, 8 neutrons, and 8 electrons; and carbon: 6 protons, 5 neutrons, and 6 electrons

Group ACTiViTy —ADVANCED

Escape from the Sun Have groups of students create a board game to model the movement of energy in the sun. Students should begin by making a game board out of poster board. The board should accurately illustrate the structure of the sun from the core to the corona. Then, have students create a series of index cards containing questions and answers about the sun. Explain to students the following objectives of the game: "A player begins the game as a proton and attempts to "collide" with another proton in the sun's core. A player "collides" when he or she correctly answers a question. Players will continue through each step of the fusion process by correctly answering more questions. Players then leave the core as energy and continue traveling to the convective and radiative zones by correctly answering more questions. The object of the game is to reach Earth's surface as infrared, UV, or visible light energy."

LS Verbal/Interpersonal

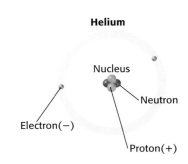

Helium

Nucleus

Neutron

Electron(−)

Proton(+)

nuclear fusion the combination of the nuclei of small atoms to form a larger nucleus; releases energy

CONNECTION TO Chemistry

Atoms An atom consists of a nucleus surrounded by one or more electrons. Electrons have a negative charge. In most elements, the atom's nucleus is made up of two types of particles: *protons,* which have a positive charge, and *neutrons,* which have no charge. The protons in the nucleus are usually balanced by an equal number of electrons. The number of protons and electrons gives the atom its chemical identity. A helium atom, shown at left, has two protons, two neutrons, and two electrons. Use a Periodic Table to find the chemical identity of the following atoms: nitrogen, oxygen, and carbon.

Nuclear Fusion

At the beginning of the 20th century, Albert Einstein showed that matter and energy are interchangeable. Matter can change into energy according to his famous formula: $E = mc^2$. (E is energy, m is mass, and c is the speed of light.) Because c is such a large number, tiny amounts of matter can produce a huge amount of energy. With this idea, scientists began to understand a very powerful source of energy.

Nuclear fusion is the process by which two or more low-mass nuclei join together, or fuse, to form a more massive nucleus. In this way, two hydrogen nuclei can fuse to form a single nucleus of helium. During the process, a lot of energy is produced. Scientists now know that the sun gets its energy from nuclear fusion. Einstein's equation, shown in **Figure 3,** changed ideas about the sun's energy source by equating mass and energy.

$$E = mc^2$$

Figure 3 Einstein's equation changed ideas about the sun's energy source by equating mass and energy.

CONNECTION to Physical Science ——— GENERAL

Star Stuff Stars are the crucibles in which the heavy elements of the universe are forged. As generations of stars are born, grow old, and re-form, heavy elements have become more abundant in the universe. The calcium in our bones and the iron in our blood originated in stars—we are made of "star stuff." The big bang produced mainly hydrogen and helium, the fuel that powers stars. All other elements in the universe are produced during the life cycle of stars. Our sun is massive enough to create elements as heavy as oxygen, and more-massive stars can produce elements as heavy as sodium. Elements heavier than iron are synthesized only when extremely massive supergiants become supernovas. Use the teaching transparency entitled "The Periodic Table of the Elements" to discuss the types of elements produced by stars. **LS** Verbal/Visual

Fusion in the Sun

During fusion, under normal conditions, the nuclei of hydrogen atoms never get close enough to combine. The reason is that they are positively charged. Like charges repel each other, as shown in **Figure 4**. In the center of the sun, however, the temperature and pressure are very high. As a result, the hydrogen nuclei have enough energy to overcome the repulsive force, and hydrogen fuses into helium, as shown in **Figure 5**.

The energy produced in the center, or core of the sun takes millions of years to reach the sun's surface. The energy passes from the core through a very dense region called the *radiative zone*. The matter in the radiative zone is so crowded that the light and energy are blocked and sent in different directions. Eventually, the energy reaches the *convective zone*. Gases circulate in the convective zone, which is about 200,000 km thick. Hot gases in the convective zone carry the energy up to the *photosphere,* the visible surface of the sun. From there, the energy leaves the sun as light, which takes only 8.3 min to reach Earth.

Figure 4 *Like charges repel just as similar poles on a pair of magnets do.*

Reading Check What causes the nuclei of hydrogen atoms to repel each other?

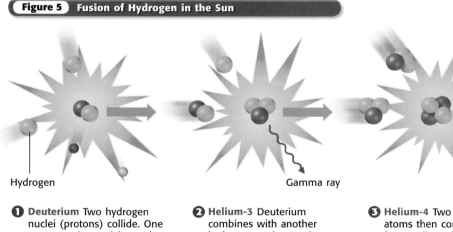

Figure 5 Fusion of Hydrogen in the Sun

Hydrogen

Gamma ray

❶ **Deuterium** Two hydrogen nuclei (protons) collide. One proton emits particles and energy and then becomes a neutron. The proton and neutron combine to produce a heavy form of hydrogen called *deuterium*.

❷ **Helium-3** Deuterium combines with another hydrogen nucleus to form a variety of helium called *helium*-3. More energy, as well as gamma rays, is released.

❸ **Helium-4** Two helium-3 atoms then combine to form ordinary helium-4, which releases more energy and a pair of hydrogen nuclei.

Close

Reteaching — Basic

The Layers of the Sun's Atmosphere Have students create a mnemonic device to help them remember the layers of the sun's atmosphere. LS **Verbal**

Quiz — General

1. How do you know that gravity does not produce the sun's energy? (If all of the sun's gravitational energy were released, the sun would last only 45 million years. The solar system is at least 4.6 billion years old.)
2. How does energy produced by nuclear fusion move from the sun's core to space? (It moves very slowly through the radiative zone, circulates through the convective zone, passes through the photosphere, and leaves the sun as light.)

Alternative Assessment — Advanced

Energy Transfer Have students explain how energy released from the collision of two protons in the sun's core warms a car seat on Earth. Students should account for the following: nuclear fusion, the movement of energy through the radiative and convective zones, Earth's atmosphere, and the amount of time this process takes. LS **Verbal**

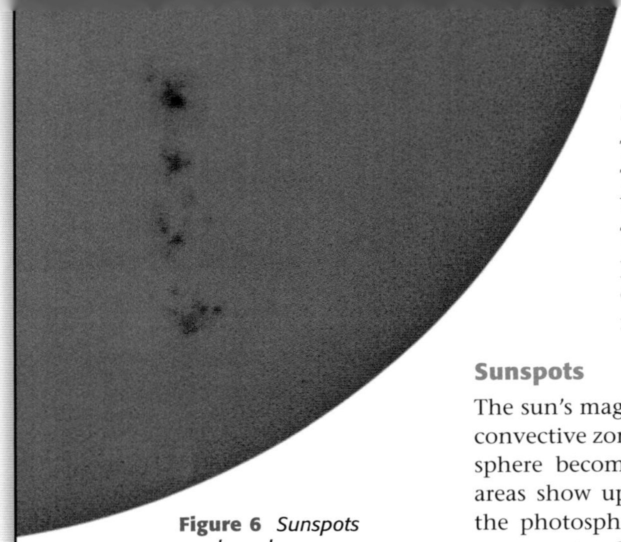

Figure 6 *Sunspots mark cooler areas on the sun's surface. They are related to changes in the magnetic properties of the sun.*

Solar Activity

The photosphere is an ever-changing place. Thermal energy moves from the sun's interior by the circulation of gases in the convective zone. This movement of energy causes the gas in the photosphere to boil and churn. This circulation, combined with the sun's rotation, creates magnetic fields that reach far out into space.

Sunspots

The sun's magnetic fields tend to slow down the activity in the convective zone. When activity slows down, areas of the photosphere become cooler than surrounding areas. These cooler areas show up as sunspots. **Sunspots** are cooler, dark spots of the photosphere of the sun, as shown in **Figure 6.** Sunspots can vary in shape and size. Some sunspots can be as large as 50,000 miles in diameter.

The numbers and locations of sunspots on the sun change in a regular cycle. Scientists have found that the sunspot cycle lasts about 11 years. Every 11 years, the amount of sunspot activity in the sun reaches a peak intensity and then decreases. **Figure 7** shows the sunspot cycle since 1610, excluding the years 1645–1715, which was a period of unusually low sunspot activity.

sunspot a dark area of the photosphere of the sun that is cooler than the surrounding areas and that has a strong magnetic field

Reading Check What are sunspots? What causes sunspots to occur?

Climate Confusion

Scientists have found that sunspot activity can affect the Earth. For example, some scientists have linked the period of low sunspot activity, 1645–1715, with the very low temperatures that Europe experienced during that time. This period is known as the "Little Ice Age." Most scientists, however, think that more research is needed to fully understand the possible connection between sunspots and Earth's climate.

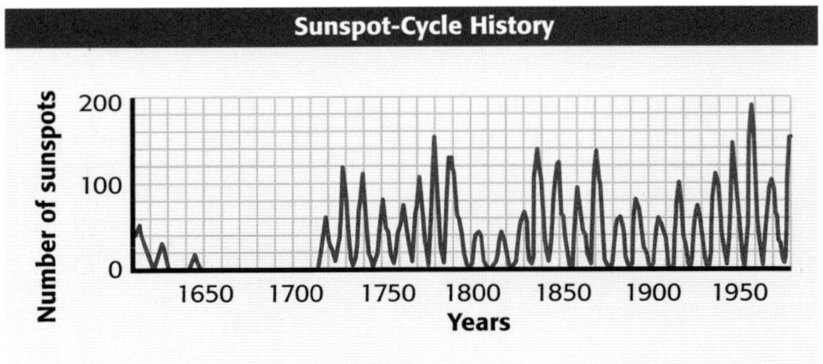

Figure 7 *This graph shows the number of sunspots that have occurred each year since Galileo's first observation in 1610.*

Answer to Reading Check

Sunspots are cooler, dark spots on the sun. Sunspots occur because when activity slows down in the convective zone, areas of the photosphere become cooler.

Solar Flares

The magnetic fields that cause sunspots also cause solar flares. *Solar flares,* as shown in **Figure 8,** are regions of extremely high temperature and brightness that develop on the sun's surface. When a solar flare erupts, it sends huge streams of electrically charged particles into the solar system. Solar flares can extend upward several thousand kilometers within minutes. Solar flares usually occur near sunspots and can interrupt radio communications on Earth and in orbit. Scientists are trying to find ways to give advance warning of solar flares.

Figure 8 *Solar flares are giant eruptions on the sun's surface.*

SECTION Review

Summary

- The sun is a large ball of gas made mostly of hydrogen and helium. The sun consists of many layers.
- The sun's energy comes from nuclear fusion that takes place in the center of the sun.
- The visible surface of the sun, or the photosphere, is very active.
- Sunspots and solar flares are the result of the sun's magnetic fields that reach space.
- Sunspot activity may affect Earth's climate, and solar flares can interact with Earth's atmosphere.

Using Key Terms

1. In your own words, write a definition for each of the following terms: *sunspot* and *nuclear fusion*.

Understanding Key Ideas

2. Which of the following statements describes how energy is produced in the sun?
   a. The sun burns fuels to generate energy.
   b. As hydrogen changes into helium deep inside the sun, a great deal of energy is made.
   c. Energy is released as the sun shrinks because of gravity.
   d. None of the above

3. Describe the composition of the sun.

4. Name and describe the layers of the sun.

5. In which area of the sun do sunspots appear?

6. Explain how sunspots form.

7. Describe how sunspots can affect the Earth.

8. What are solar flares, and how do they form?

Math Skills

9. If the equatorial diameter of the sun is 1.39 million kilometers, how many kilometers is the sun's radius?

Critical Thinking

10. **Applying Concepts** If nuclear fusion in the sun's core suddenly stopped today, would the sky be dark in the daytime tomorrow? Explain.

11. **Making Comparisons** Compare the theories that scientists proposed about the source of the sun's energy with the process of nuclear fusion in the sun.

SCI LINKS.

NSTA
Developed and maintained by the
National Science Teachers Association

For a variety of links related to this chapter, go to www.scilinks.org

Topic: The Sun
SciLinks code: HSM1477

4. The corona forms the sun's outer atmosphere. The chromosphere is below the corona and is only 30,000 km thick. The photosphere is the visible part of the sun that we can see from Earth. The convective zone is a region about 200,000 km thick where gases circulate. The radiative zone is a very dense region about 300,000 km thick. The core is at the center of the sun, and it is where the sun's energy is produced.

5. Sunspots appear as cool, dark spots of the photosphere of the sun.

6. The sun's magnetic fields slow down the activity in the convective zone, which causes areas of the photosphere to become cooler. Sunspots are the dark, cool spots on the sun.

7. Scientists think that a period of less sunspot activity may cause lower temperatures on Earth.

8. Solar flares are regions of extremely high temperature and brightness that develop on the sun's surface. Solar flares are caused by the sun's magnetic fields, which are caused by the movement of energy in the sun.

9. 1,390,000 km ÷ 2 = 695,000 km

10. No, it would take millions of years for the last energy made in the core to reach the surface of the sun.

11. In the 19th century, some scientists thought that the sun burned fuel to generate its energy. Other scientists thought that gravity was causing the sun to slowly shrink and release energy. Finally, with the help of Albert Einstein's equation, $E = mc^2$, the process of nuclear fusion was defined. Nuclear fusion is the combination of the nuclei of small atoms to form a large nucleus. The result of nuclear fusion is the release of a large amount of energy. The fusion of hydrogen into helium in the sun generates a large amount of energy and therefore is the source of the sun's energy.

Answers to Section Review

1. Sample answer: A sunspot is a dark area of the photosphere of the sun that is cooler than the surrounding areas. Nuclear fusion occurs when the nuclei of small atoms combine to form a larger nucleus.

2. b

3. The sun is a large ball of gas made mostly of hydrogen and helium held together by gravity.

CHAPTER RESOURCES

Chapter Resource File

- **Section Quiz** GENERAL
- **Section Review** GENERAL
- **Vocabulary and Section Summary** GENERAL
- **Reinforcement Worksheet** BASIC
- **SciLinks Activity** GENERAL

Focus

Overview

This section explores Earth's formation. Students will learn how Earth's atmosphere, oceans, and continents developed and how our atmosphere sustains life today.

Bellringer

Tell students that Earth is about 4.6 billion years old. The first fossil evidence of life on Earth has been dated to nearly 3.5 billion years ago. Have students write a paragraph describing what Earth might have been like during the first billion years of its existence.

Motivate

Discussion —— GENERAL

Earth's Atmosphere Tell students that gases released from Earth's molten surface helped create Earth's first atmosphere. Explain that the gases in our atmosphere are held by gravity, and ask students to speculate why there is very little hydrogen or helium in our atmosphere if these are the most abundant elements in the universe. (These gases are so light that Earth's gravity cannot trap them.) **LS** Visual

READING WARM-UP

Objectives

- Describe the formation of the solid Earth.
- Describe the structure of the Earth.
- Explain the development of Earth's atmosphere and the influence of early life on the atmosphere.
- Describe how the Earth's oceans and continents formed.

Terms to Learn

crust
mantle
core

READING STRATEGY

Discussion Read this section silently. Write down questions that you have about this section. Discuss your questions in a small group.

Figure 1 When Earth is seen from space, one of its unique features—the presence of water—is apparent.

The Earth Takes Shape

In many ways, Earth seems to be a perfect place for life.

We live on the third planet from the sun. The Earth, shown in **Figure 1,** is mostly made of rock, and nearly three-fourths of its surface is covered with water. It is surrounded by a protective atmosphere of mostly nitrogen and oxygen and smaller amounts of other gases. But Earth has not always been such an oasis in the solar system.

Formation of the Solid Earth

The Earth formed as planetesimals in the solar system collided and combined. From what scientists can tell, the Earth formed within the first 10 million years of the collapse of the solar nebula!

The Effects of Gravity

When a young planet is still small, it can have an irregular shape, somewhat like a potato. But as the planet gains more matter, the force of gravity increases. When a rocky planet, such as Earth, reaches a diameter of about 350 km, the force of gravity becomes greater than the strength of the rock. As the Earth grew to this size, the rock at its center was crushed by gravity and the planet started to become round.

The Effects of Heat

As the Earth was changing shape, it was also heating up. Planetesimals continued to collide with the Earth, and the energy of their motion heated the planet. Radioactive material, which was present in the Earth as it formed, also heated the young planet. After Earth reached a certain size, the temperature rose faster than the interior could cool, and the rocky material inside began to melt. Today, the Earth is still cooling from the energy that was generated when it formed. Volcanoes, earthquakes, and hot springs are effects of this energy trapped inside the Earth. As you will learn later, the effects of heat and gravity also helped form the Earth's layers when the Earth was very young.

✓ **Reading Check** What factors heated the Earth during its early formation? (*See the Appendix for answers to Reading Checks.*)

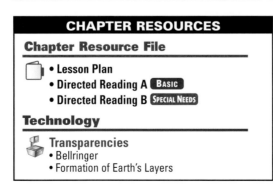

CHAPTER RESOURCES

Chapter Resource File

- Lesson Plan
- Directed Reading A **BASIC**
- Directed Reading B **SPECIAL NEEDS**

Technology

Transparencies
- Bellringer
- Formation of Earth's Layers

Answer to Reading Check

During Earth's early formation, planetesimals collided with the Earth. The energy of their motion heated the planet.

How the Earth's Layers Formed

Have you ever watched the oil separate from vinegar in a bottle of salad dressing? The vinegar sinks because it is denser than oil. The Earth's layers formed in much the same way. As rocks melted, denser elements, such as nickel and iron, sank to the center of the Earth and formed the core. Less dense elements floated to the surface and became the crust. This process is shown in **Figure 2.**

The **crust** is the thin, outermost layer of the Earth. It is 5 to 100 km thick. Crustal rock is made of elements that have low densities, such as oxygen, silicon, and aluminum. The **mantle** is the layer of Earth beneath the crust. It extends 2,900 km below the surface. Mantle rock is made of elements such as magnesium and iron and is denser than crustal rock. The **core** is the central part of the Earth below the mantle. It contains the densest elements (nickel and iron) and extends to the center of the Earth—almost 6,400 km below the surface.

crust the thin and solid outermost layer of the Earth above the mantle

mantle the layer of rock between the Earth's crust and core

core the central part of the Earth below the mantle

Figure 2 The Formation of Earth's Layers

❶ All elements in the early Earth are randomly mixed.

❷ Rocks melt, and denser elements sink toward the center. Less dense elements rise and form layers.

❸ According to composition, the Earth is divided into three layers: the crust, the mantle, and the core.

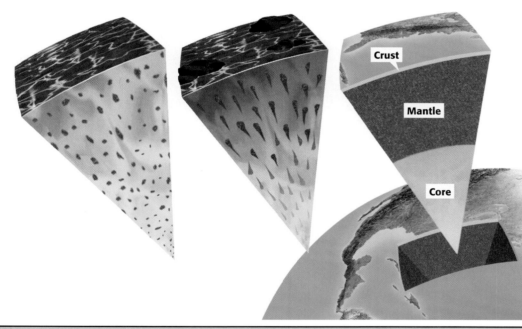

Crust

Mantle

Core

Science Bloopers

Impact Craters Impact craters have left scars on planets and moons throughout the solar system. In 1826, an eccentric Bavarian astronomer named Franz von Paula Gruithuisen was one of the first scientists to suggest that lunar craters were caused by meteorite impacts. However, he also asserted that other lunar features were built by a race of moon creatures called *Selenites*, and his theory of crater formation was not taken seriously. Students will enjoy reading and reporting on other imaginative descriptions of the moon by authors such as Jules Verne. Even astronomer Johannes Kepler wrote about creatures living on the moon in his book *The Dream*.

Teach

CONNECTION to Geology — ADVANCED

Comparing Mars and Earth's Core Earth's core is 33% of Earth's mass. Although Mars has approximately the same volume as Earth's core, the mass of Mars is only 11% of Earth's mass. Ask students to compare the density of Mars with the density of Earth's core and to explain the differences they note. (Mars is less dense than Earth's core. Mars is composed of a smaller core and lighter elements than Earth's core is.) **LS** Verbal

CONNECTION to Language Arts — GENERAL

Science Fiction Have students read "Desertion" by Clifford Simak in the *Holt Anthology of Science Fiction*. **LS** Intrapersonal

INCLUSION Strategies

- *Learning Disabled*
- *Attention Deficit Disorder*

Have students model the interior of the early Earth. Organize students into groups of three or four and give each group 50 mL of water, a 150 mL container, 50 mL of cooking oil, and a spoon. Ask students to pour 50 mL of water into the container. Then, have them add 50 mL of cooking oil to the water. Ask students to stir the mixture vigorously. Have students stop stirring, and ask them to let the mixture stand for a few minutes. Then, ask students how the mixture models the interior of the early Earth. **LS** Visual/Kinesthetic

Rings Around Our Planet The rings of ice and rock that circle Jupiter, Saturn, and Neptune may have formed when asteroids collided with the moons of these planets. But Earth is also developing a system of rings. These rings are not moon debris, however, but debris from satellites and space missions. Have interested students research the problem of orbiting space trash and the proposed solutions to clean it up. **LS** Intrapersonal

INTERNET ACTIVITY
Math ———— ADVANCED

Researching New Planets
In 1991, Alexander Wolszczan discovered the first extra-solar planets orbiting a pulsar called *PSR B1257+12* in the constellation Virgo. A pulsar is the remains of a star that has exploded as a supernova. The masses of the three planets are 0.02, 4.3, and 3.9 times the mass of Earth. Wolszczan carefully measured the rotation of the pulsar and detected slight changes caused by the gravity from the three planets. Astronomers do not know how these three planets formed or how the planets survived the explosion of their parent star. Have students do research on the Internet for explanations about these extra-solar planets and to report to the class any additional planets that have been discovered. **LS** Intrapersonal

CONNECTION TO
Environmental Science

WRITING SKILL **The Greenhouse Effect** Carbon dioxide is a greenhouse gas. Greenhouse gases are gases that trap thermal energy. When greenhouse gases trap thermal energy, the greenhouse effect occurs. The greenhouse effect is the natural heating process of a planet by which gases in the atmosphere trap thermal energy. Do research to find the percentage of carbon dioxide that is thought to make up Earth's early atmosphere. Write a report, and share your findings with your class.

Formation of the Earth's Atmosphere

Today, Earth's atmosphere is 78% nitrogen, 21% oxygen, and about 1% argon. (There are tiny amounts of many other gases.) Did you know that the Earth's atmosphere did not always contain the oxygen that you need to live? Scientists think that the Earth's atmosphere has changed several times since Earth formed.

Earth's First Atmosphere

Scientists think that Earth's first atmosphere was a mixture of gases that were released as Earth cooled. During the final stages of the Earth's formation, its surface was very hot—even molten in places—as shown in **Figure 3**. The molten rock released large amounts of carbon dioxide and water vapor. Therefore, scientists think that Earth's first atmosphere was a steamy mixture of carbon dioxide and water vapor.

✓ **Reading Check** Describe Earth's first atmosphere.

Figure 3 *This artwork is an artist's view of what Earth's surface may have looked like shortly after the Earth formed.*

SCIENCE HUMOR

Q: Have you heard about the new restaurant on the moon?

A: great food, lousy atmosphere

Answer to Reading Check
Scientists think that the Earth's first atmosphere was a steamy mixture of carbon dioxide and water vapor.

Figure 4 *As this volcano in Hawaii shows, a large amount of gas is released during an eruption.*

Earth's Second Atmosphere

As the Earth cooled and its layers formed, the Earth's second atmosphere was able to form. This atmosphere probably formed from volcanic gases. Volcanoes, such as the one in **Figure 4,** released chlorine, nitrogen, and sulfur in addition to large amounts of carbon dioxide and water vapor. Some of this water vapor may have condensed to form the Earth's first oceans.

Comets, which are planetesimals made of ice, also may have contributed to the Earth's second atmosphere. As comets crashed into the Earth, they brought in a range of elements, such as carbon, hydrogen, oxygen, and nitrogen. Comets also may have brought some of the water that helped form the oceans.

The Role of Life

How did the Earth's second atmosphere change to become the air you are breathing right now? The answer is related to the appearance of life on Earth.

Ultraviolet Radiation

Scientists think that ultraviolet (UV) radiation, the same radiation that causes sunburns, helped produce the conditions necessary for life. Because UV light has a lot of energy, it can break apart molecules in your skin and in the air. Today, we are shielded from most of the sun's UV rays by Earth's protective ozone layer. But Earth's early atmosphere probably did not have ozone, so many molecules in the air and at Earth's surface were broken apart. Over time, this material collected in the Earth's waters. Water offered protection from the effects of UV radiation. In these sheltered pools of water, chemicals may have combined to form the complex molecules that made life possible. The first life-forms were very simple and did not need oxygen to live.

Comets and Meteors

What is the difference between a comet and a meteor? With a parent, research the difference between comets and meteors. Then, find out if you can view meteor showers in your area!

ACTIVITY

Is That a Fact!

Earth is growing heavier by thousands of metric tons every year. Microscopic dust constantly filters through the atmosphere from space and lands on our planet. You can look at these so-called micrometeroids by collecting rainwater that has dropped directly into a glass. You can use a microscope to search and view round, dark particles from the rainwater sediment—more than likely, the particles are from space!

MISCONCEPTION ALERT

Meteors and Meteorites Students may think that large meteors threaten Earth. Movies and films have popularized the notion that large meteors threaten Earth and could cause widespread destruction. Although several tons of meteoroid material enters Earth's atmosphere each day, most pieces have a mass of only a few milligrams. Only the largest meteors reach Earth's surface and become meteorites.

Debate — GENERAL

Life on Earth: Could It Happen Again? Ask students to imagine that life on Earth is completely destroyed. Encourage students to debate whether life could evolve again in our current atmosphere. Students should consider conditions on primitive Earth as well as the requirements for life as we know it.  **Verbal**

CONNECTION to Life Science — GENERAL

The Presence of Oxygen If the early Earth had contained a lot of oxygen in its atmosphere, the chemical processes that gave rise to life probably would not have occurred! Oxygen is very reactive; it combines with other elements readily. High levels of oxygen would have prevented organic molecules from combining, which would have stopped the development of early forms of life.

Homework — GENERAL

Poster Project Have students create a series of drawings and captions to show how Earth's atmosphere first formed and how it changed over time. Encourage students to use creative approaches, such as comic-strips, to communicate the concepts. **Intrapersonal**

Answer to School-to-Home Activity

A comet is a small body of ice, rock, and cosmic dust that follows an elliptical orbit around the sun and that gives off gas and dust in the form of a tail as it passes close to the sun. A meteor is a bright streak of light that results when a meteoroid burns up in the Earth's atmosphere.

Reteaching — **BASIC**

Changing Earth's Composition
Ask students to choose one step in the development of Earth's composition and then eliminate it. Then, ask them to describe what the composition of Earth's atmosphere would be like if this step had not occurred. **LS** Verbal

Quiz — **GENERAL**

1. How did oxygen become abundant in Earth's atmosphere? (Millions of years ago, life-forms evolved that produced oxygen as a byproduct of photosynthesis; eventually oxygen levels increased.)

2. How has the relationship between ozone and life on Earth changed since Earth's early atmosphere? (The absence of ozone in Earth's early atmosphere allowed molecules to be broken apart by UV radiation. These molecules combined to form the complex molecules that gave rise to life. The ozone layer protects life on Earth from the harmful effects of UV radiation.)

Alternative Assessment — **GENERAL**

Earth Quiz Game Have students write facts about each stage of Earth's development on 3 in. × 5 in. cards. Mix the cards as a deck. Students should form teams and play a quiz game. Teams earn points by assigning the correct fact to the correct stage of Earth's development. **LS** Intrapersonal

The Source of Oxygen

Sometime before 3.4 billion years ago, organisms that produced food by photosynthesis appeared. *Photosynthesis* is the process of absorbing energy from the sun and carbon dioxide from the atmosphere to make food. During the process of making food, these organisms released oxygen—a gas that was not abundant in the atmosphere at that time. Scientists think that the descendants of these early life-forms are still around today, as shown in **Figure 5.**

Photosynthetic organisms played a major role in changing Earth's atmosphere to become the mixture of gases you breathe today. Over the next several million years, more and more oxygen was added to the atmosphere. At the same time, carbon dioxide was removed. As oxygen levels increased, some of the oxygen formed a layer of ozone in the upper atmosphere. This ozone blocked most of the UV radiation and made it possible for life, in the form of simple plants, to move onto land about 2.2 billion years ago.

Reading Check How did photosynthesis contribute to Earth's current atmosphere?

Formation of Oceans and Continents

Scientists think that the oceans probably formed during Earth's second atmosphere, when the Earth was cool enough for rain to fall and remain on the surface. After millions of years of rainfall, water began to cover the Earth. By 4 billion years ago, a global ocean covered the planet.

For the first few hundred million years of Earth's history, there weren't any continents. Given the composition of the rocks that make up the continents, scientists know that these rocks have melted and cooled many times in the past. Each time the rocks melted, the heavier elements sank and the lighter ones rose to the surface.

Figure 5 *Stromatolites, mats of fossilized algae (left), are among the earliest evidence of life. Blue-green algae (right) living today are thought to be similar to the first life-forms on Earth.*

Answer to Reading Check

When photosynthetic organisms appeared on Earth, they released oxygen into the Earth's atmosphere. Over several million years, more and more oxygen was added to the atmosphere, which helped form Earth's current atmosphere.

The Growth of Continents

After a while, some of the rocks were light enough to pile up on the surface. These rocks were the beginning of the earliest continents. The continents gradually thickened and slowly rose above the surface of the ocean. These scattered young continents did not stay in the same place, however. The slow transfer of thermal energy in the mantle pushed them around. Approximately 2.5 billion years ago, continents really started to grow. And by 1.5 billion years ago, the upper mantle had cooled and had become denser and heavier. At this time, it was easier for the cooler parts of the mantle to sink. These conditions made it easier for the continents to move in the same way that they do today.

INTERNET ACTIVITY

For another activity related to this chapter, go to **go.hrw.com** and type in the keyword **HZ5SOLW.**

SECTION Review

Summary

- The effects of gravity and heat created the shape and structure of Earth.
- The Earth is divided into three main layers based on composition: the crust, mantle, and core.
- The presence of life dramatically changed Earth's atmosphere by adding free oxygen.
- Earth's oceans formed shortly after the Earth did, when it had cooled off enough for rain to fall. Continents formed when lighter materials gathered on the surface and rose above sea level.

Using Key Terms

1. Use each of the following terms in a separate sentence: *crust*, *mantle*, and *core*.

Understanding Key Ideas

2. Earth's first atmosphere was mostly made of
   a. nitrogen and oxygen.
   b. chlorine, nitrogen, and sulfur.
   c. carbon dioxide and water vapor.
   d. water vapor and oxygen.

3. Describe the structure of the Earth.

4. Why did the Earth separate into distinct layers?

5. Describe the development of Earth's atmosphere. How did life affect Earth's atmosphere?

6. Explain how Earth's oceans and continents formed.

Critical Thinking

7. **Applying Concepts** How did the effects of gravity help shape the Earth?

8. **Making Inferences** How would the removal of forests affect the Earth's atmosphere?

Interpreting Graphics

Use the illustration below to answer the questions that follow.

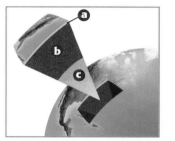

9. Which of the layers is composed mostly of the elements magnesium and iron?

10. Which of the layers is composed mostly of the elements iron and nickel?

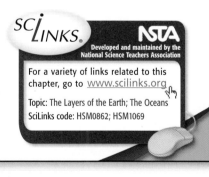

For a variety of links related to this chapter, go to www.scilinks.org
Topic: The Layers of the Earth; The Oceans
SciLinks code: HSM0862; HSM1069

Answers to Section Review

1. Sample answer: The crust is the thin and solid outermost layer of Earth above the mantle. The mantle is the layer of rock between Earth's crust and core. The core is below the mantle and is the central part of Earth.

2. c

3. Earth is divided into three layers: the crust, the mantle, and the core. The crust is the thin, outermost layer; the mantle is the layer of Earth beneath the crust; and the core is the central and densest part of Earth.

4. As rocks melted inside Earth, denser materials sank to the center of Earth, and less dense materials floated to the surface.

5. Sample answer: Scientists think that Earth's first atmosphere contained carbon dioxide and water vapor. Later, volcanoes added carbon dioxide, water vapor, chlorine, nitrogen, and sulfur. Comets brought water, carbon, hydrogen, nitrogen, and oxygen. Solar energy created new chemicals that led to the formation of living organisms. These organisms greatly changed the composition of the atmosphere by adding oxygen.

6. Scientists think that the oceans formed when Earth was cool enough for rain to fall and remain on the surface. After millions of years of rainfall, water began to cover Earth and eventually formed the global ocean. Earth's continents formed as heavy elements sank close to the core of Earth and light elements rose to Earth's surface. The light elements were light enough to pile up on the surface and began to form the earliest continents. These continents gradually thickened and slowly rose above the surface of the ocean.

7. When Earth was still a young planet, it had an irregular shape. But as Earth gained more matter, gravity became greater than the strength of the rock. Therefore, the rock at the center of Earth was crushed by gravity and Earth started to become round.

8. Answers may vary. Deforestation on a large scale would allow more carbon dioxide to accumulate in the atmosphere. As carbon dioxide increased, oxygen levels would likely decrease.

9. b

10. c

CHAPTER RESOURCES

Chapter Resource File

- **Section Quiz** GENERAL
- **Section Review** GENERAL
- **Vocabulary and Section Summary** GENERAL
- **Reinforcement Worksheet** BASIC

Focus

Overview

In this section, students will learn about planetary orbits and about Kepler's laws of planetary motion. The section discusses how Newton's law of universal gravitation helped explain the discoveries of Kepler.

Bellringer

Ask students to create a mnemonic device to help them differentiate between planetary rotation and revolution. For example, they might link the long *a* sound in *rotation* with the fact that Earth makes one rotation in a *day*. The following rhyme may also help them remember: As we rotate, we spin about our axis and live a day. A revolution is a voyage around the sun, and a year will pass away.

Motivate

ACTiViTY ——————— GENERAL

Measuring Ellipses Have students use a ruler to measure segments *a, b, c,* and *d* in **Figure 2** and then determine if $a + b = c + d$. The illustration shows the string at two distinct points in its description of an ellipse. **LS** Visual/Interpersonal

READING WARM-UP

Objectives

- Explain the difference between rotation and revolution.
- Describe three laws of planetary motion.
- Describe how distance and mass affect gravitational attraction.

Terms to Learn

rotation
orbit
revolution

READING STRATEGY

Paired Summarizing Read this section silently. In pairs, take turns summarizing the material. Stop to discuss ideas that seem confusing.

rotation the spin of a body on its axis

orbit the path that a body follows as it travels around another body in space

revolution the motion of a body that travels around another body in space; one complete trip along an orbit

Planetary Motion

Why do the planets revolve around the sun? Why don't they fly off into space? Does something hold them in their paths?

To answer these questions, you need to go back in time to look at the discoveries made by the scientists of the 1500s and 1600s. Danish astronomer Tycho Brahe (TIE koh BRAH uh) carefully observed the positions of planets for more than 25 years. When Brahe died in 1601, a German astronomer named Johannes Kepler (yoh HAHN uhs KEP luhr) continued Brahe's work. Kepler set out to understand the motions of planets and to describe the solar system.

A Revolution in Astronomy

Each planet spins on its axis. The spinning of a body, such as a planet, on its axis is called **rotation.** As the Earth rotates, only one-half of the Earth faces the sun. The half facing the sun is light (day). The half that faces away from the sun is dark (night).

The path that a body follows as it travels around another body in space is called the **orbit.** One complete trip along an orbit is called a **revolution.** The amount of time a planet takes to complete a single trip around the sun is called a *period of revolution.* Each planet takes a different amount of time to circle the sun. Earth's period of revolution is about 365.25 days (a year), but Mercury orbits the sun in only 88 days. **Figure 1** illustrates the orbit and revolution of the Earth around the sun as well as the rotation of the Earth on its axis.

Figure 1 *A planet rotates on its own axis and revolves around the sun in a path called an* orbit.

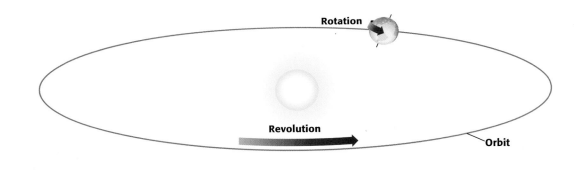

Rotation

Revolution

Orbit

CHAPTER RESOURCES

Chapter Resource File

- Lesson Plan
- Directed Reading A **BASIC**
- Directed Reading B **SPECIAL NEEDS**

Technology

Transparencies
- Bellringer
- Earth's Rotation and Revolution
- Ellipse

MISCONCEPTION ALERT

The Seasons Students may think that Earth's elliptical orbit brings it closer to the sun in the summer. The shape of Earth's orbit does not cause the seasons; the seasons are caused by Earth's tilt on its axis. In the summer, the Northern Hemisphere is tilted toward the sun. In the winter, the Northern Hemisphere is tilted away from the sun. Earth is actually closest to the sun during the Northern Hemisphere's winter.

Figure 2 Parts of an Ellipse

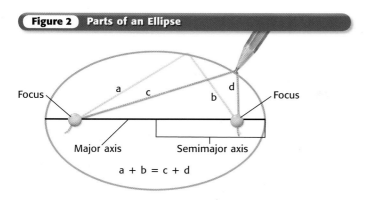

Focus — Major axis — Semimajor axis — Focus

a + b = c + d

Kepler's First Law of Motion

Kepler's first discovery came from his careful study of Mars. Kepler discovered that Mars did not move in a circle around the sun but moved in an elongated circle called an *ellipse*. This finding became Kepler's first law of motion. An ellipse is a closed curve in which the sum of the distances from the edge of the curve to two points inside the ellipse is always the same, as shown in **Figure 2**. An ellipse's maximum length is called its *major axis*. Half of this distance is the *semimajor axis,* which is usually used to describe the size of an ellipse. The semimajor axis of Earth's orbit—the maximum distance between Earth and the sun—is about 150 million kilometers.

Kepler's Second Law of Motion

Kepler's second discovery, or second law of motion, was that the planets seemed to move faster when they are close to the sun and slower when they are farther away. To understand this idea, imagine that a planet is attached to the sun by a string, as modeled in **Figure 3**. When the string is shorter, the planet must move faster to cover the same area.

Kepler's Third Law of Motion

Kepler noticed that planets that are more distant from the sun, such as Saturn, take longer to orbit the sun. This finding was Kepler's third law of motion, which explains the relationship between the period of a planet's revolution and its semimajor axis. Knowing how long a planet takes to orbit the sun, Kepler was able to calculate the planet's distance from the sun.

✓ Reading Check Describe Kepler's third law of motion. (*See the Appendix for answers to Reading Checks.*)

MATH PRACTICE

Kepler's Formula

Kepler's third law can be expressed with the formula

$$P^2 = a^3$$

where P is the period of revolution and a is the semimajor axis of an orbiting body. For example, Mars's period is 1.88 years, and its semimajor axis is 1.523 AU. Thus, $1.88^2 = 1.523^3 = 3.53$. Calculate a planet's period of revolution if the semimajor axis is 5.74 AU.

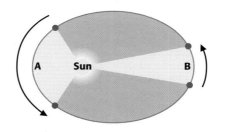

Figure 3 *According to Kepler's second law, to keep the area of* A *equal to the area of* B, *the planet must move faster in its orbit when it is closer to the sun.*

Science Bloopers

Discovery of Elliptical Orbits

Johannes Kepler was obsessed with trying to describe the geometric harmony of the universe. He believed that there were five perfect geometric solids that fit precisely between the six known planets and that this pattern contained the divine meaning of the solar system. Although Kepler was wrong, his efforts to prove this idea enabled him to discover the elliptical orbit of the planets.

Answer to Reading Check

Kepler's third law of motion states that planets that are farther away from the sun take longer to orbit the sun.

Teach

CONNECTION to Astronomy — GENERAL

The Moon's Orbit Every second, the moon travels 1 km in its orbit. During that second, it also falls about 1.4 mm toward Earth. Because of the moon's velocity and the pull of gravity, it travels along a path that follows the curved surface of Earth. This condition, known as *free fall*, keeps the moon in orbit around Earth. Explain to students that the condition of free fall does not mean that there is no gravity. Earth's gravity acts on the moon in the same way it acts on an apple that falls out of a tree. The difference is that the moon, unlike the apple, is moving forward much, faster than it is falling. **LS Verbal**

Answer to Math Practice

$a^3 = 5.74^3 = 189.12$; Because $a^3 = P^2$, the planet's period of revolution, P, can be found by taking the square root of a^3; $P^2 = 189.12 = 13.75$.

CONNECTION ACTIVITY Language Arts — GENERAL

Why Do Planets Orbit? Kepler described the orbits of planets, but Newton showed *why* planets orbit. Have students use the following laws to compose a letter to Kepler explaining planetary orbits.

- Every object in the universe attracts every other object in the universe with a force that is dependent on the object's mass and the inverse square of the distance between the objects.

- An object remains in a state of rest or motion unless acted on by an outside force.
LS Intrapersonal

Close

Reteaching — BASIC

Comparing Kepler and Newton
Have students compare the work of Kepler and Newton. Have students explain how Kepler's discoveries are related to Newton's discoveries and vice versa.
LS Verbal

Quiz — GENERAL

1. Does a car traveling on the outside lane of a racetrack take longer to complete one lap than a car traveling on the inside lane? How does this phenomenon relate to Kepler's third law? (Yes, the car on the outside takes longer to circle the track. This phenomenon relates to Kepler's third law because, according to the law, the farther a planet is from the sun, the longer the planet takes to orbit the sun.)

2. If the semimajor axis of Earth's orbit is 150 million kilometers, what is its major axis? (300 million kilometers)

Alternative Assessment — GENERAL

The Laws of Motion and Gravity Have students reproduce **Figures 2, 3, 4,** and **5** in their **science journal.** Students should write an explanation of each diagram in their own words. Have students exchange illustrations and quiz each other. **LS** Interpersonal

Quick Lab

Staying in Focus

1. Take a **short piece of string,** and pin both ends to a **piece of paper** by using **two thumbtacks.**

2. Keeping the string stretched tight at all times, use a **pencil** to trace the path of an ellipse.

3. Change the distance between the thumbtacks to change the shape of the ellipse.

4. How does the position of the thumbtacks (foci) affect the ellipse?

Newton to the Rescue!

Kepler wondered what caused the planets closest to the sun to move faster than the planets farther away. However, he never found an answer. Sir Isaac Newton finally put the puzzle together when he described the force of gravity. Newton didn't understand why gravity worked or what caused it. Even today, scientists do not fully understand gravity. But Newton combined the work of earlier scientists and used mathematics to explain the effects of gravity.

The Law of Universal Gravitation

Newton reasoned that an object falls toward Earth because Earth and the object are attracted to each other by gravity. He discovered that this attraction depends on the masses of the objects and the distance between the objects.

Newton's *law of universal gravitation* states that the force of gravity depends on the product of the masses of the objects divided by the square of the distance between the objects. The larger the masses of two objects and the closer together the objects are, the greater the force of gravity between the objects. For example, if two objects are moved twice as far apart, the gravitational attraction between them will decrease by 2×2 (a factor of 4), as shown in **Figure 4.** If two objects are moved 10 times as far apart, the gravitational attraction between them will decrease by 10×10 (a factor of 100).

Both Earth and the moon are attracted to each other. Although it may seem as if Earth does not orbit the moon, Earth and the moon actually orbit each other.

✓ **Reading Check** Explain Newton's law of universal gravitation.

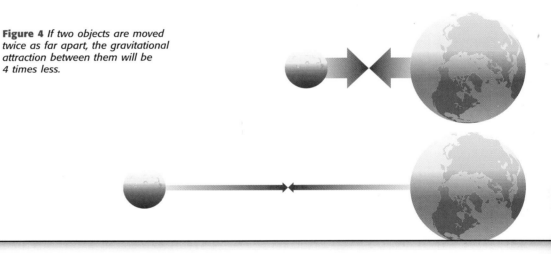

Figure 4 *If two objects are moved twice as far apart, the gravitational attraction between them will be 4 times less.*

Answer to Reading Check

Newton's law of universal gravitation states that the force of gravity depends on the product of the masses of the objects divided by the square of the distance between the objects.

Quick Lab

Safety Caution: Students should use care with thumbtacks to avoid injuring themselves or damaging the surface on which they work. Have them put a piece of cardboard under their sheet of paper.

Answers

4. The closer together the foci are, the more circular the ellipse is.

Orbits Falling Down and Around

If you drop a rock, it falls to the ground. So, why doesn't the moon come crashing into the Earth? The answer has to do with the moon's inertia. *Inertia* is an object's resistance in speed or direction until an outside force acts on the object. In space, there isn't any air to cause resistance and slow down the moving moon. Therefore, the moon continues to move, but gravity keeps the moon in orbit, as **Figure 5** shows.

Imagine twirling a ball on the end of a string. As long as you hold the string, the ball will orbit your hand. As soon as you let go of the string, the ball will fly off in a straight path. This same principle applies to the moon. Gravity keeps the moon from flying off in a straight path. This principle holds true for all bodies in orbit, including the Earth and other planets in our solar system.

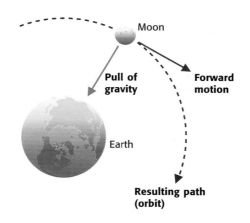

Figure 5 *Gravity causes the moon to fall toward the Earth and changes a straight-line path into a curved orbit.*

Answers to Section Review

1. Sample answer: Revolution is the motion of a body that travels around another body in space. Rotation is the spin of a body on its own axis.

2. b

3. mass and distance

4. The motion of a planet is balanced between falling toward the sun and moving in a straight line past the sun. The resultant path is a curved orbit around the sun.

5. 365.25 days $\times$ 24 h = 8,766 h

6. The closer moon would finish first. Kepler's third law states that period of revolution is related to the distance of an orbiting body from the object it orbits (its semimajor axis). A closer object has a shorter period of revolution.

7. Kepler's first law of motion states that orbits are elliptical. Kepler's second law of motion states that planets move faster when they are closer to the sun and slower when they are farther from the sun. Kepler's third law explains the relationship between the period of a planet's revolution and its semimajor axis. Kepler's three laws allowed him to understand how a planet orbits the sun and how to calculate its distance from the sun.

SECTION Review

Summary

- Rotation is the spinning of a planet on its axis, and revolution is one complete trip along an orbit.
- Planets move in an ellipse around the sun. The closer they are to the sun, the faster they move. The period of a planet's revolution depends on the planet's semimajor axis.
- Gravitational attraction decreases as distance increases and as mass decreases.

Using Key Terms

1. In your own words, write a definition for each of the following terms: *revolution* and *rotation*.

Understanding Key Ideas

2. Kepler discovered that planets move faster when they
 a. are farther from the sun.
 b. are closer to the sun.
 c. have more mass.
 d. rotate faster.

3. On what properties does the force of gravity between two objects depend?

4. How does gravity keep a planet moving in an orbit around the sun?

Math Skills

5. The Earth's period of revolution is 365.25 days. Convert this period of revolution into hours.

Critical Thinking

6. **Applying Concepts** If a planet had two moons and one moon was twice as far from the planet as the other, which moon would complete a revolution of the planet first? Explain your answer.

7. **Making Comparisons** Describe the three laws of planetary motion. How is each law related to the other laws?

Developed and maintained by the National Science Teachers Association

For a variety of links related to this chapter, go to www.scilinks.org

Topic: Kepler's Laws
SciLinks code: HSM0216

CHAPTER RESOURCES

Chapter Resource File

- Section Quiz GENERAL
- Section Review GENERAL
- Vocabulary and Section Summary GENERAL
- Datasheet for Quick Lab

Technology

Transparencies
- Gravity and Motion of the Moon

How Far Is the Sun?

Teacher's Notes

Time Required
One 45-minute class period

Lab Ratings

EASY ──────────→ HARD

Teacher Prep 🧪🧪
Student Set-Up 🧪
Concept Level 🧪🧪🧪🧪
Clean Up 🧪

MATERIALS
The materials listed on the student page are enough for a group of 2 to 3 students.

Safety Caution
Remind students to review all safety cautions and icons before beginning this lab activity. Also, caution students never to look directly at the sun.

Preparation Notes
Conduct this activity on a sunny day. This lab works best in the late afternoon because the sun is lower in the sky. The sunlight should shine through the window as close to perpendicular to the window as possible. You may want to lower the blinds so that the sunlight will pass through a narrow opening. Sample data are provided in the table at the top of the next page.

OBJECTIVES
Create a solar-distance measuring device.

Calculate the Earth's distance from the sun.

MATERIALS
- aluminum foil, 50 cm × 50 cm
- card, index
- meterstick
- poster board
- ruler, metric
- scissors
- tape, masking
- thumbtack

SAFETY

How Far Is the Sun?

It doesn't slice, it doesn't dice, but it can give you an idea of how big our universe is! You can build your very own solar-distance measuring device from household items. Amaze your friends by figuring out how many metersticks can be placed between the Earth and the sun.

Ask a Question

1 How many metersticks could I place between the Earth and the sun?

Form a Hypothesis

2 Write a hypothesis that answers the question above.

Test the Hypothesis

3 Measure and cut a 44 cm × 44 cm square from the middle of the poster board. Tape the foil square over the hole in the center of the poster board.

4 Using a thumbtack, carefully prick the foil to form a tiny hole in the center. Congratulations! You have just constructed your very own solar-distance measuring device!

5 Tape the device to a window facing the sun so that sunlight shines directly through the pinhole. **Caution:** Do not look directly into the sun.

6 Place one end of the meterstick against the window and beneath the foil square. Steady the meterstick with one hand.

7 With the other hand, hold the index card close to the pinhole. You should be able to see a circular image on the card. This image is an image of the sun.

8 Move the card back until the image is large enough to measure. Be sure to keep the image on the card sharply focused. Reposition the meterstick so that it touches the bottom of the card.

Daniel Bugenhagen
Yutan Jr.–Sr. High
Yutan, Nebraska

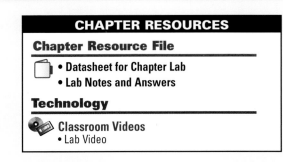

CHAPTER RESOURCES

Chapter Resource File
- • Datasheet for Chapter Lab
- • Lab Notes and Answers

Technology
- 💾 Classroom Videos
- • Lab Video

Analyze the Results

① **Analyzing Results** According to your calculations, how far from the Earth is the sun? Don't forget to convert your measurements to meters.

Draw Conclusions

② **Evaluating Data** You could put 150 billion metersticks between the Earth and the sun. Compare this information with your result in step 11. Do you think that this activity was a good way to measure the Earth's distance from the sun? Support your answer.

⑨ Ask your partner to measure the diameter of the image on the card by using the metric ruler. Record the diameter of the image in millimeters.

⑩ Record the distance between the window and the index card by reading the point at which the card rests on the meterstick.

⑪ Calculate the distance between Earth and the sun by using the following formula:

$$\text{distance between the sun and Earth} = \text{sun's diameter} \times \frac{\text{distance to the image}}{\text{image's diameter}}$$

(Hint: The sun's diameter is 1,392,000,000 m.)

1 cm = 10 mm
1 m = 100 cm
1 km = 1,000 m

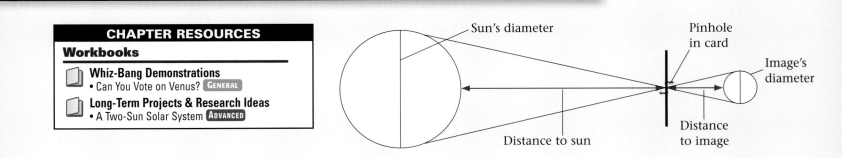

Background

On the board, draw the diagram at the bottom of this page and explain that this activity uses triangles and proportions to find the distance to the sun. Because the sun forms the image on the paper, there must be a proportionate relationship between the triangles in the diagram. The hole divides the proportions, so the distance between the image and the hole is related to the distance between the sun and the hole.

Analyze the Results

1. Answers may vary. Based on the sample data, the sun is 148,944,000,000 m from Earth. The sun is actually 149,600,000,000 m from Earth.

Draw Conclusions

2. Answers may vary. Accept all well-supported answers. According to the sample data, the calculated value was within 0.5% of the actual value. So, this activity is generally a good way to measure the distance from Earth to the sun.

CHAPTER RESOURCES

Workbooks

📕 **Whiz-Bang Demonstrations**
• Can You Vote on Venus? **GENERAL**

📕 **Long-Term Projects & Research Ideas**
• A Two-Sun Solar System **ADVANCED**

Sun's diameter

Pinhole in card

Image's diameter

Distance to sun

Distance to image

Chapter Review

Assignment Guide

SECTION	QUESTIONS
1	1, 3, 10, 15, 18
2	6, 16, 19
3	2, 4, 8, 11, 13–14, 17, 22
4	5, 7, 9, 12, 21, 23–26
1 and 2	20

ANSWERS

Using Key Terms

1. nebula
2. mantle
3. Sample answer: A nebula is a large cloud of gas and dust in interstellar space. The solar nebula is the cloud of gas and dust that formed our solar system.
4. Sample answer: The crust is the outermost layer of Earth. The mantle is the layer of rock between Earth's crust and core.
5. Sample answer: Rotation is the spin of a body on its axis. Revolution is the motion of a body that travels around another body in space.
6. Sample answer: Nuclear fusion is the combination of the nuclei of small atoms to form a large nucleus. A sunspot is a dark area of the photosphere of the sun that is cooler than the surrounding areas and that has a strong magnetic field.

USING KEY TERMS

Complete each of the following sentences by choosing the correct term from the word bank.

nebula	crust
mantle	solar nebula

1 A ___ is a large cloud of gas and dust in interstellar space.

2 The moon is made of rock similar to the rock in the Earth's ___.

For each pair of terms, explain how the meanings of the terms differ.

3 *nebula* and *solar nebula*

4 *crust* and *mantle*

5 *rotation* and *revolution*

6 *nuclear fusion* and *sunspot*

UNDERSTANDING KEY IDEAS

Multiple Choice

7 To determine a planet's period of revolution, you must know its
  a. size.
  b. distance from the sun.
  c. orbit.
  d. All of the above

8 During Earth's formation, materials such as nickel and iron sank to the
  a. mantle.
  b. core.
  c. crust.
  d. All of the above

9 Planetary orbits are shaped like
  a. orbits.
  b. spirals.
  c. ellipses.
  d. periods of revolution.

10 Impacts in the early solar system
  a. brought new materials to the planets.
  b. released energy.
  c. dug craters.
  d. All of the above

11 Organisms that photosynthesize get their energy from
  a. nitrogen.   c. the sun.
  b. oxygen.   d. water.

12 Which of the following planets has the shortest period of revolution?
  a. Pluto   c. Mercury
  b. Earth   d. Jupiter

13 Which gas in Earth's atmosphere suggests that there is life on Earth?
  a. hydrogen   c. carbon dioxide
  b. oxygen   d. nitrogen

14 Which layer of the Earth has the lowest density?
  a. the core
  b. the mantle
  c. the crust
  d. None of the above

15 What is the measure of the average kinetic energy of particles in an object?
  a. temperature   c. gravity
  b. pressure   d. force

Understanding Key Ideas

7. c	10. d	13. b
8. b	11. c	14. c
9. c	12. c	15. a

16. Sample answer: A sunspot is a dark area of the photosphere of the sun that is cooler than the surrounding areas and that has a strong magnetic field. A solar flare is a region of extremely high temperature and brightness that develops on the sun's surface.

17. Sample answer: Scientists think that the oceans formed during Earth's second atmosphere, when Earth was cool enough for rain to fall and stay on the surface. After millions of years of rainfall, water began to cover Earth and formed a global ocean. The continents formed after the first few hundred million years and are made up of rocks that have melted and cooled many times.

18. Sample answer: Pressure and gravity may have become unbalanced in the solar nebula because of an external force, such as a collision of the solar nebula with another nebula or force from a nearby exploding star. This type of force was strong enough to overcome the pressure of the nebula and trigger its collapse.

Short Answer

16 Compare a sunspot with a solar flare.

17 Describe how the Earth's oceans and continents formed.

18 Explain how pressure and gravity may have become unbalanced in the solar nebula.

19 Define *nuclear fusion* in your own words. Describe how nuclear fusion generates the sun's energy.

CRITICAL THINKING

20 **Concept Mapping** Use the following terms to create a concept map: *solar nebula, solar system, planetesimals, sun, photosphere, core, nuclear fusion, planets,* and *Earth.*

21 **Making Comparisons** How did Newton's law of universal gravitation help explain the work of Johannes Kepler?

22 **Predicting Consequences** Using what you know about the relationship between living things and the development of Earth's atmosphere, explain how the formation of ozone holes in Earth's atmosphere could affect living things.

23 **Identifying Relationships** Describe Kepler's three laws of motion in your own words. Describe how each law relates to either the revolution, rotation, or orbit of a planetary body.

INTERPRETING GRAPHICS

Use the illustration below to answer the questions that follow.

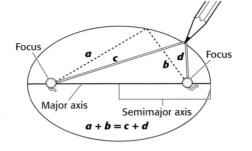

Focus   Focus

$a + b = c + d$

Major axis   Semimajor axis

24 Which of Kepler's laws of motion does the illustration represent?

25 How does the equation shown above support the law?

26 What is an ellipse's maximum length called?

$E = mc^2$

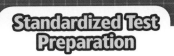

Standardized Test Preparation

Teacher's Note

To provide practice under more realistic testing conditions, give students 20 minutes to answer all of the questions in this Standardized Test Preparation.

MISCONCEPTION ALERT

Answers to the standardized test preparation can help you identify student misconceptions and misunderstandings.

READING

Passage 1

1. C
2. G
3. D

✚ TEST DOCTOR

Question 1: Answer C is correct because the passage states this fact in the last sentence. Students may incorrectly choose answers A, B, or D because these proper nouns also occur in the passage.

READING

Read each of the passages below. Then, answer the questions that follow each passage.

Passage 1 You know that you should not look at the sun, right? But how can we learn anything about the sun if we can't look at it? We can use a solar telescope! About 70 km southwest of Tucson, Arizona, is Kitt Peak National Observatory, where you will find three solar telescopes. In 1958, Kitt Peak was chosen from more than 150 mountain sites to be the site for a national observatory. Located in the Sonoran Desert, Kitt Peak is on land belonging to the Tohono O'odham Indian nation. On this site, the McMath-Pierce Facility houses the three largest solar telescopes in the world. Astronomers come from around the globe to use these telescopes. The largest of the three, the McMath-Pierce solar telescope, produces an image of the sun that is almost 1 m wide!

1. Which of the following is the largest telescope in the world?
   A Kitt Peak
   B Tohono O'odham
   C McMath-Pierce
   D Tucson

2. According to the passage, how can you learn about the sun?
   F You can look at it.
   G You can study it by using a solar telescope.
   H You can go to Kitt Peak National Observatory.
   I You can study to be an astronomer.

3. Which of the following is a fact in the passage?
   A One hundred fifty mountain sites contain solar telescopes.
   B Kitt Peak is the location of the smallest solar telescope in the world.
   C In 1958, Tucson, Arizona, was chosen for a national observatory.
   D Kitt Peak is the location of the largest solar telescope in the world.

Passage 2 Sunlight that has been focused can produce a great amount of thermal energy—enough to start a fire. Now, imagine focusing the sun's rays by using a magnifying glass that is 1.6 m in diameter. The resulting heat could melt metal. If a <u>conventional</u> telescope were pointed directly at the sun, it would melt. To avoid a meltdown, the McMath-Pierce solar telescope uses a mirror that produces a large image of the sun. This mirror directs the sun's rays down a diagonal shaft to another mirror, which is 50 m underground. This mirror is adjustable to focus the sunlight. The sunlight is then directed to a third mirror, which directs the light to an observing room and instrument shaft.

1. In this passage, what does the word *conventional* mean?
   A special
   B solar
   C unusual
   D ordinary

2. What can you infer from reading the passage?
   F Focused sunlight can avoid a meltdown.
   G Unfocused sunlight produces little energy.
   H A magnifying glass can focus sunlight to produce a great amount of thermal energy.
   I Mirrors increase the intensity of sunlight.

3. According to the passage, which of the following statements about solar telescopes is true?
   A Solar telescopes make it safe for scientists to observe the sun.
   B Solar telescopes don't need to use mirrors.
   C Solar telescopes are built 50 m underground.
   D Solar telescopes are 1.6 m in diameter.

Passage 2

1. D
2. H
3. A

✚ TEST DOCTOR

Question 1: Answer D is correct because the word *conventional* usually means "not unusual or extreme." The word *conventional* is used in the passage to compare an ordinary telescope with a solar telescope. Remind students that if they are asked to infer the meaning of a word in a passage, they should substitute each answer choice for the underlined word and reread the sentence to determine the correct answer choice.

The diagram below models the moon's orbit around the Earth. Use the diagram below to answer the questions that follow.

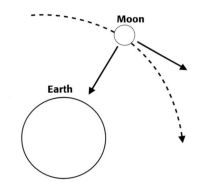

1. Which statement best describes the diagram?
   A Orbits are straight lines.
   B The force of gravity does not affect orbits.
   C Orbits result from a combination of gravitational attraction and inertia.
   D The moon moves in three different directions depending on its speed.

2. In which direction does gravity pull the moon?
   F toward the Earth
   G around the Earth
   H away from the Earth
   I toward and away from the Earth

3. If the moon stopped moving, what would happen?
   A It would fly off into space.
   B It would continue to orbit the Earth.
   C It would stay where it is in space.
   D It would move toward the Earth.

MATH

Read each question below, and choose the best answer.

1. An astronomer found 3 planetary systems in the nebula that she was studying. One system had 6 planets, another had 2 planets, and the third had 7 planets. What is the average number of planets in all 3 systems?
   A 3
   B 5
   C 8
   D 16

2. A newly discovered planet has a period of rotation of 270 Earth years. How many Earth days are in 270 Earth years?
   F 3,240
   G 8,100
   H 9,855
   I 98,550

3. A planet has seven rings. The first ring is 20,000 km from the center of the planet. Each ring is 50,000 km wide and 500 km apart. What is the total radius of the ring system from the planet's center?
   A 353,000 km
   B 373,000 km
   C 373,500 km
   D 370,000 km

4. If you bought a telescope for $87.75 and received a $10 bill, two $1 bills, and a quarter as change, how much money did you give the clerk?
   F $100
   G $99
   H $98
   I $90

Standardized Test Preparation

1. C
2. F
3. D

TEST DOCTOR

Question 2: Answer F is correct because the pull of gravity (shown by the arrow from the moon to Earth) causes the moon to be pulled toward Earth. Students may choose answer G because the orbit of the moon around Earth makes it appear as though the moon is being pulled around Earth. Answers H and I are incorrect because Earth's gravity does not pull the moon away from Earth.

MATH

1. B
2. I
3. B
4. F

TEST DOCTOR

Question 3: Answer B is the correct answer because 20,000 km + (500 km × 6) + (50,000 km × 7) = 373,000 km. Students may choose answer A if they did not include the 20,000 km distance from the center of the planet to the first ring. Students may choose answer C if they added an additional 500 km. Students may choose answer D if they did not include the distance between the rings.

CHAPTER RESOURCES

Chapter Resource File

• Standardized Test Preparation GENERAL

State Resources

For specific resources for your state, visit **go.hrw.com** and type in the keyword **HSMSTR**.

Science, Technology, and Society

Background

The McMath-Pierce solar telescope has a 91.5 m focal length. Such a long focal length allows scientists to see the details on a sunspot. The facility that houses the telescope looks like an upside-down *V*. The vertical side of this *V* is 30 m tall and contains the heliostat. The sun strikes the heliostat, which is a large, flat, rotating mirror. After sunlight strikes the heliostat, it travels 50 m underground to another mirror, which reflects it back to the observation room. In the observation room, the light is broken down into its different wavelengths by a spectrograph.

Scientific Discoveries

Homework ———— **GENERAL**

Giant Molecular Clouds Giant molecular clouds can send comets into orbit from the Oort cloud. Ask students to research giant molecular clouds and write a report about what they are, how often they interact with the Oort cloud, and what happens to the Oort cloud after it interacts with giant molecular clouds.

Science in Action

Science, Technology, and Society

Don't Look at the Sun!

How can we learn anything about the sun if we can't look at it? The answer is to use a special telescope called a *solar telescope*. The three largest solar telescopes in the world are located at Kitt Peak National Observatory near Tucson, Arizona. The largest of these telescopes, the McMath-Pierce solar telescope, creates an image of the sun that is almost 1 m wide! How is the image created? The McMath-Pierce solar telescope uses a mirror that is more than 2 m in diameter to direct the sun's rays down a diagonal shaft to another mirror, which is 152 m underground. This mirror is adjustable to focus the sunlight. The sunlight is then directed to a third mirror, which directs the light to an observing room and instrument shaft.

Math ACTIVITY

The outer skin of the McMath-Pierce solar telescope consists of 140 copper panels that measure 10.4 m × 2.4 m each. How many square meters of copper were used to construct the outer skin of the telescope?

Scientific Discoveries

The Oort Cloud

Have you ever wondered where comets come from? In 1950, Dutch astronomer Jan Oort decided to find out where comets originated. Oort studied 19 comets. He found that none of these comets had orbits indicating that the comets had come from outside the solar system. Oort thought that all of the comets had come from an area at the far edge of the solar system. In addition, he believed that the comets had entered the planetary system from different directions. These conclusions led Oort to theorize that the area from which comets come surrounds the solar system like a sphere and that comets can come from any point within the sphere. Today, this spherical zone at the edge of the solar system is called the *Oort Cloud*. Astronomers believe that billions or even trillions of comets may exist within the Oort Cloud.

Social Studies ACTIVITY

WRITING SKILL Before astronomers understood the nature of comets, comets were a source of much fear and misunderstanding among humans. Research some of the myths that humans have created about comets. Summarize your findings in a short essay.

Answer to Math Activity
10.4 m × 2.4 m × 140 copper panels = 3,494.4 m² of copper

Answer to Social Studies Activity
Suggest that students use library resources or the Internet to research myths about comets.

People in Science

Subrahmanyan Chandrasekhar

From White Dwarfs to Black Holes You may be familiar with the *Chandra X-Ray Observatory*. Launched by NASA in July 1999 to search for x-ray sources in space, the observatory is the most powerful x-ray telescope that has ever been built. However, you may not know how the observatory got its name. The *Chandra X-Ray Observatory* was named after the Indian American astrophysicist Subrahmanyan Chandrasekhar (SOOB ruh MAHN yuhn CHUHN druh SAY kuhr).

One of the most influential astrophysicists of the 20th century, Chandrasekhar was simply known as "Chandra" by his fellow scientists. Chandrasekhar made many contributions to physics and astrophysics. The contribution for which Chandrasekhar is best known was made in 1933, when he was a 23-year-old graduate student at Cambridge University in England. At the time, astrophysicists thought that all stars eventually became planet-sized stars known as *white dwarfs*. But from his calculations, Chandrasekhar believed that not all stars ended their lives as white dwarfs. He determined that the upper limit to the mass of a white dwarf was 1.4 times the mass of the sun. Stars that were more massive would collapse and would become very dense objects. These objects are now known as *black holes*. Chandrasekhar's ideas revolutionized astrophysics. In 1983, at the age of 73, Chandrasekhar was awarded the Nobel Prize in physics for his work on the evolution of stars.

Language Arts ACTIVITY

WRITING SKILL Using the Internet or another source, research the meaning of the word *chandra*. Write a paragraph describing your findings.

go.hrw.com
To learn more about these Science in Action topics, visit **go.hrw.com** and type in the keyword **HZ5SOLF.**

Current Science
Check out Current Science® articles related to this chapter by visiting go.hrw.com. Just type in the keyword **HZ5CS20.**

People in Science

 **GENERAL**

Chandrasekhar's Achievements
Most of the work that Chandrasekhar received a Nobel Prize for was completed in the 1930s. Have students research Chandrasekhar's other accomplishments in astrophysics. Have students research the other areas of astronomy and physics that Chandrasekhar studied. Have students share their findings with the class.

Answer to Language Arts Activity

The word *chandra* means "moon" in Sanskrit. Students should indicate in their paragraph how the word *chandra* relates to the feature.

A Family of Planets
Chapter Planning Guide

Compression guide:
To shorten instruction because of time limitations, omit Section 1.

OBJECTIVES	LABS, DEMONSTRATIONS, AND ACTIVITIES	TECHNOLOGY RESOURCES
PACING • 90 min pp. 92–97 **Chapter Opener**	SE **Start-up Activity**, p. 93 `GENERAL`	OSP **Parent Letter** ■ `GENERAL` CD **Student Edition on CD-ROM** CD **Guided Reading Audio CD** ■ TR **Chapter Starter Transparency*** VID **Brain Food Video Quiz**
Section 1 The Nine Planets • List the planets in the order in which they orbit the sun. • Explain how scientists measure distances in space. • Describe how the planets in our solar system were discovered. • Describe three ways in which the inner planets and outer planets differ.	TE **Connection Activity** Math, p. 95 `GENERAL` SE **Model-Making Lab** Why Do They Wander?, p. 168 `GENERAL` LB **Long-Term Projects & Research Ideas** What Did You See, Mr. Messier?* `ADVANCED`	CRF **Lesson Plans*** TR **Bellringer Transparency*** SE **Internet Activity**, p. 96 `GENERAL` CRF **SciLinks Activity*** `GENERAL`
PACING • 90 min pp. 98–103 **Section 2 The Inner Planets** • Explain the difference between a planet's period of rotation and period of revolution. • Describe the difference between prograde and retrograde rotation. • Describe the individual characteristics of Mercury, Venus, Earth, and Mars. • Identify the characteristics that make Earth suitable for life.	TE **Activity** Your Age in Venusian Years, p. 98 `GENERAL` TE **Connection Activity** Math, p. 101 `GENERAL` SE **Connection to Physics** Boiling Point on Mars, p. 102 `GENERAL` SE **Inquiry Lab** Create a Calendar, p. 124 `GENERAL` LB **Whiz-Bang Demonstrations** Crater Creator* ◆ `BASIC` LB **Whiz-Bang Demonstrations** Space Snowballs* ◆ `BASIC` LB **Labs You Can Eat** Meteorite Delight* ◆ `ADVANCED`	CRF **Lesson Plans*** TR **Bellringer Transparency*** TR The Inner Planets; The Outer Planets* VID **Lab Videos for Earth Science**
PACING • 45 min pp. 104–109 **Section 3 The Outer Planets** • Explain how gas giants are different from terrestrial planets. • Describe the individual characteristics of Jupiter, Saturn, Uranus, Neptune, and Pluto.	SE **School-to-Home Activity** Surviving Space, p. 109 `GENERAL` SE **Science in Action** Math, Social Studies, and Language Arts Activities, pp. 130–131 `GENERAL` TE **Connection Activity** Fine Arts, p. 108 `GENERAL`	CRF **Lesson Plans*** TR **Bellringer Transparency*** CD **Interactive Explorations CD-ROM** Space Case `GENERAL`
PACING • 45 min pp. 110–117 **Section 4 Moons** • Describe the current theory of the origin of Earth's moon. • Explain what causes the phases of Earth's moon. • Describe the difference between a solar eclipse and a lunar eclipse. • Describe the individual characteristics of the moons of other planets.	TE **Activity** Lunar Ice, p. 110 `GENERAL` TE **Activity** Modeling the Earth and Moon, p. 112 `GENERAL` TE **Connection Activity** Language Arts, p. 112 `BASIC` SE **Quick Lab** Clever Insight, p. 113 `GENERAL` TE **Connection Activity** Math, p. 113 `GENERAL` SE **Model-Making Lab** Eclipses, p. 170 `GENERAL` SE **Skills Practice Lab** Phases of the Moon, p. 171 `GENERAL`	CRF **Lesson Plans*** TR **Bellringer Transparency*** TR *LINK TO PHYSICAL SCIENCE* Two Motions Combine to Form Projectile Motion* TR Formation of the Moon* TR Phases of the Moon* TR Solar Eclipse; Lunar Eclipse*
PACING • 45 min pp. 118–123 **Section 5 Small Bodies in the Solar System** • Explain why comets, asteroids, and meteoroids are important to the study of the formation of the solar system. • Describe the similarities of and differences between asteroids and meteoroids. • Explain how cosmic impacts may affect life on Earth.	TE **Demonstration** Modeling Comets, p. 118 ◆ `GENERAL` SE **Connection to Language Arts** Interplanetary Journalist, p. 119 `GENERAL` TE **Activity** Collecting Micrometeorites, p. 121 ◆ `GENERAL` SE **Connection to Biology** Mass Extinctions, p. 122 `GENERAL`	CRF **Lesson Plans*** TR **Bellringer Transparency***

PACING • 90 min

CHAPTER REVIEW, ASSESSMENT, AND STANDARDIZED TEST PREPARATION

CRF **Vocabulary Activity*** `GENERAL`
SE **Chapter Review**, pp. 126–127 `GENERAL`
CRF **Chapter Review*** ■ `GENERAL`
CRF **Chapter Tests A*** `GENERAL`, **B*** `ADVANCED`, **C*** `SPECIAL NEEDS`
SE **Standardized Test Preparation**, pp. 128–129 `GENERAL`
CRF **Standardized Test Preparation*** `GENERAL`
CRF **Performance-Based Assessment*** `GENERAL`
OSP **Test Generator** `GENERAL`
CRF **Test Item Listing*** `GENERAL`

Online and Technology Resources

Visit **go.hrw.com** for a variety of free resources related to this textbook. Enter the keyword **HZ5FAM**.

Students can access interactive problem-solving help and active visual concept development with the *Holt Science and Technology* Online Edition available at **www.hrw.com**.

Guided Reading Audio CD

A direct reading of each chapter using instructional visuals as guideposts. For auditory learners, reluctant readers, and Spanish-speaking students. Available in English and Spanish.

KEY

SE	Student Edition	**CRF**	Chapter Resource File	**SS**	Science Skills Worksheets	***** Also on One-Stop Planner
TE	Teacher Edition	**OSP**	One-Stop Planner	**MS**	Math Skills for Science Worksheets	**♦** Requires advance prep
		LB	Lab Bank	**CD**	CD or CD-ROM	**■** Also available in Spanish
		TR	Transparencies	**VID**	Classroom Video/DVD	

SKILLS DEVELOPMENT RESOURCES	SECTION REVIEW AND ASSESSMENT	STANDARDS CORRELATIONS
SE Pre-Reading Activity, p. 92 `GENERAL` **OSP** Science Puzzlers, Twisters & Teasers `GENERAL`		National Science Education Standards SAI 1; HNS 1, 3; ES 3a
CRF Directed Reading A* ■ `BASIC`, B* `SPECIAL NEEDS` **CRF** Vocabulary and Section Summary* ■ `GENERAL` **SE** Reading Strategy Paired Summarizing, p. 94 `GENERAL` **SS** Science Skills Grasping Graphing* `GENERAL` **CRF** Critical Thinking Martian Holiday* `ADVANCED` **CRF** Reinforcement Worksheet The Planets of Our Solar System* `BASIC` **TE** Inclusion Strategies, p. 96	**SE** Reading Checks, pp. 95, 97 `GENERAL` **TE** Reteaching, p. 96 `BASIC` **TE** Quiz, p. 96 `GENERAL` **TE** Alternative Assessment, p. 96 `GENERAL` **SE** Section Review,* p. 97 ■ `GENERAL` **CRF** Section Quiz* ■ `GENERAL`	UCP 1, 3; SAI 1; ST 2; SPSP 5; HNS 1, 3; ES 1c, 3a, 3b, 3c; *LabBook:* UCP 2; SAI 1; ST 1
CRF Directed Reading A* ■ `BASIC`, B* `SPECIAL NEEDS` **CRF** Vocabulary and Section Summary* ■ `GENERAL` **SE** Reading Strategy Reading Organizer, p. 98 `GENERAL`	**SE** Reading Checks, pp. 99, 100, 102 `GENERAL` **TE** Homework, p. 99 `ADVANCED` **TE** Reteaching, p. 102 `BASIC` **TE** Quiz, p. 102 `GENERAL` **TE** Alternative Assessment, p. 102 `GENERAL` **SE** Section Review,* p. 103 ■ `GENERAL` **CRF** Section Quiz* ■ `GENERAL`	UCP 1, 3; SAI 1; ST 2; SPSP 5; HNS 1, 3; ES 1c, 3a, 3b; *Chapter Lab:* UCP 1, 2; SAI 1, 2; ST 1; ES 3b
CRF Directed Reading A* ■ `BASIC`, B* `SPECIAL NEEDS` **CRF** Vocabulary and Section Summary* ■ `GENERAL` **SE** Reading Strategy Prediction Guide, p. 104 `GENERAL` **TE** Reading Strategy Prediction Guide, p. 106 `GENERAL`	**SE** Reading Checks, pp. 105, 107 `GENERAL` **TE** Reteaching, p. 108 `BASIC` **TE** Quiz, p. 108 `GENERAL` **TE** Alternative Assessment, p. 108 `ADVANCED` **SE** Section Review,* p. 109 ■ `GENERAL` **CRF** Section Quiz* ■ `GENERAL`	UCP 1, 3; SAI 1; ST 2; SPSP 5; HNS 1, 3; ES 1c, 3a, 3b
CRF Directed Reading A* ■ `BASIC`, B* `SPECIAL NEEDS` **CRF** Vocabulary and Section Summary* ■ `GENERAL` **SE** Reading Strategy Reading Organizer, p. 110 `GENERAL` **TE** Reading Strategy Prediction Guide, p. 111 `GENERAL` **CRF** Reinforcement Worksheet Lunar and Solar Eclipses* `BASIC`	**SE** Reading Checks, pp. 111, 113, 114, 115, 116 `GENERAL` **TE** Homework, p. 114 `GENERAL` **TE** Reteaching, p. 116 `BASIC` **TE** Quiz, p. 116 `GENERAL` **TE** Alternative Assessment, p. 116 `GENERAL` **SE** Section Review,* p. 117 ■ `GENERAL` **CRF** Section Quiz* ■ `GENERAL`	UCP 1, 3; SAI 1; HNS 1, 3; ES 3a, 3b, 3c; *LabBook:* UCP 2; SAI 1; ST 1; ES 1a, 3b
CRF Directed Reading A* ■ `BASIC`, B* `SPECIAL NEEDS` **CRF** Vocabulary and Section Summary* ■ `GENERAL` **SE** Reading Strategy Discussion, p. 118 `GENERAL` **TE** Inclusion Strategies, p. 121	**SE** Reading Checks, pp. 119, 121, 122 `GENERAL` **TE** Homework, p. 119 `GENERAL` **TE** Reteaching, p. 122 `BASIC` **TE** Quiz, p. 122 `GENERAL` **TE** Alternative Assessment, p. 122 `GENERAL` **SE** Section Review,* p. 123 ■ `GENERAL` **CRF** Section Quiz* ■ `GENERAL`	UCP 1; ES 2a, 3a, 3b

One-Stop Planner® CD-ROM

This convenient CD-ROM includes:
- Lab Materials QuickList Software
- Holt Calendar Planner
- Customizable Lesson Plans
- Printable Worksheets
- ExamView® Test Generator

CNN student News™

cnnstudentnews.com

Find the latest news, lesson plans, and activities related to important scientific events.

SCiLINKS®

NSTA

www.scilinks.org

Maintained by the **National Science Teachers Association.** See Chapter Enrichment pages for a complete list of topics.

Current Science®

Check out *Current Science* articles and activities by visiting the HRW Web site at **go.hrw.com**. Just type in the keyword **HZ5CS21T.**

Classroom Videos

- **Lab Videos** demonstrate the chapter lab.
- **Brain Food Video Quizzes** help students review the chapter material.

Visual Resources

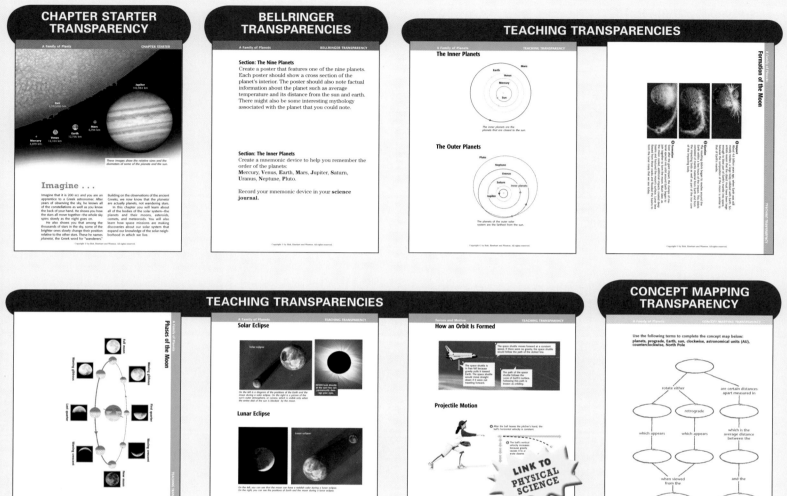

Planning Resources

LESSON PLANS

Lesson Plan SAMPLE

Section: Waves

Pacing
Regular Schedule: with lab(s):2 days   without lab(s):1 days
Block Schedule: with lab(s):1 1/2 days   without lab(s):1 day

Objectives
1. Relate the seven properties of life to a living organism.
2. Describe seven themes that can help you to organize what you learn about biology.
3. Identify the tiny structures that make up all living organisms.
4. Differentiate between reproduction and heredity and between metabolism and homeostasis.

National Science Education Standards Covered
LSInter6:Cells have particular structures that underlie their functions.
LSMat1:Most cell functions involve chemical reactions.
LSBeh1:Cells store and use information to guide their functions.
UCP1:Cell functions are regulated.
SI1: Cells can differentiate and form complete multicellular organisms.
PS1:Species evolve over time.
ESS1: The great diversity of organisms is the result of more than 3.5 billion years of evolution.
ESS2: Natural selection and its evolutionary consequences provide a scientific explanation for the fossil record of ancient life forms as well as for the striking molecular similarities observed among the diverse species of living organisms.
ST1: The millions of different species of plants, animals, and microorganisms that live on Earth today are related by descent from common ancestors.
ST2: The energy for life primarily comes from the sun.
SPSP1: The complexity and organization of organisms accommodates the need for obtaining, transforming, transporting, releasing, and eliminating the matter and energy used to sustain the organism.
SPSP6: As matter and energy flows through different levels of organization of living systems—cells, organs, communities—and between living systems and the physical environment, chemical elements are recombined in different ways.
HNS1: Organisms have behavioral responses to internal changes and to external stimuli.

PARENT LETTER

SAMPLE

Dear Parent,

Your son's or daughter's science class will soon begin exploring the chapter entitled "The World of Physical Science." In this chapter, students will learn about how the scientific method applies to the world of physical science and the role of physical science in the world. By the end of this chapter, students should demonstrate a clear understanding of the chapter's main ideas and be able to discuss the following topics:

1. physical science as the study of matter and energy (Section 1)
2. the role of physical science in the world around them (Section 1)
3. careers that rely on physical science (Section 1)
4. the steps used in the scientific method (Section 2)
5. examples of technology (Section 2)
6. how the scientific method is used to answer questions and solve problems (Section 2)
7. how our knowledge of science changes over time (Section 2)
8. how models represent real objects or systems (Section 3)
9. examples of different ways models are used in science (Section 3)
10. the importance of the International System of Units (Section 4)
11. the appropriate units to use for particular measurements (Section 4)
12. how area and density are derived quantities (Section 4)

Questions to Ask Along the Way

You can help your son or daughter learn about these topics by asking interesting questions such as the following:

• What are some surprising careers that use physical science?
• What is a characteristic of a good hypothesis?
• When is it a good idea to use a model?
• Why do Americans measure things in terms of inches and yards and meters?

ALSO IN SPANISH

TEST ITEM LISTING

TEST ITEM LISTING
The World of Earth Science SAMPLE

MULTIPLE CHOICE

1. A limitation of models is that
   a. they are large enough to see
   b. they do not act exactly like the things that they model.
   c. they are smaller than the things that they model.
   d. they model unfamiliar things.
   Answer: B   Difficulty: 1   Section: 3   Objective: 2

2. The length 10 m is equal to
   a. 100 cm.   c. 10,000 mm.
   b. 1,000 cm.   d. Both (b) and (c)
   Answer: B   Difficulty: 1   Section: 3   Objective: 2

3. To be valid, a hypothesis must be
   a. testable.   c. made into a law
   b. supported by evidence.   d. Both (a) and (b)
   Answer: B   Difficulty: 1   Section: 3   Objective: 2 1

4. The statement "Sheila has a stain on her shirt" is an example of a(n)
   a. law.   c. observation.
   b. hypothesis.   d. prediction.
   Answer: B   Difficulty: 1   Section: 3   Objective: 2

5. A hypothesis is often developed out of
   a. observations.   c. laws.
   b. experiments.   d. Both (a) and (b)
   Answer: B   Difficulty: 1   Section: 3   Objective: 2

6. How many milliliters are in 3.5 kL?
   a. 3,500 mL.   c. 3,500, 000 mL.
   b. 0.0035 mL.   d. 35,000 mL.
   Answer: B   Difficulty: 1   Section: 3   Objective: 2

7. A map of Seattle is an example of a
   a. law.   c. model
   b. theory.   d. unit
   Answer: B   Difficulty: 1   Section: 3   Objective: 2

8. A lab has the safety icons shown below. These icons mean that you should wear
   a. safety goggles.   c. safety goggles and a lab apron.
   b. only a lab apron.   d. safety goggles, a lab apron, and gloves.
   Answer: B   Difficulty: 1   Section: 3   Objective: 2

9. The law of conservation of mass says the test all states before a chemical change is
   a. more than the total mass after the change.
   b. less than the total mass after the change.
   c. the same as the total mass after the change.
   d. not the same as the total mass after the change.
   Answer: B   Difficulty: 1   Section: 3   Objective: 2

10. In which of the following areas might you find a geochemist at work?
    a. studying the chemistry of rocks   c. studying fishes
    b. studying forestry   d. studying the atmosphere
    Answer: B   Difficulty: 1   Section: 3   Objective: 2

One-Stop Planner® CD-ROM

This CD-ROM includes all of the resources shown here and the following time-saving tools:

• *Lab Materials QuickList Software*
• *Customizable lesson plans*
• *Holt Calendar Planner*
• *The powerful ExamView® Test Generator*

For a preview of available worksheets covering math and science skills, see pages T12–T19. All of these resources are also on the One-Stop Planner®.

Meeting Individual Needs

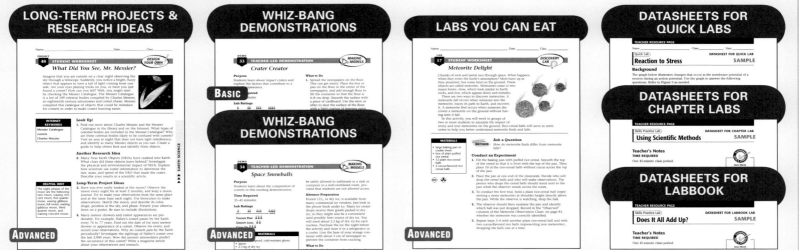

DIRECTED READING A

Skills Worksheet
Directed Reading A — SAMPLE

Section:
THAT'S SCIENCE!
1. How did James Czarnowski get his idea for the penguin
Explain.

ALSO IN SPANISH

that is unusual about the way that Proteus moves through

BASIC

DIRECTED READING B

Name / Class / Date
Skills Worksheet
Directed Reading B — SAMPLE

Section:
THAT'S SCIENCE!
1. How did James Czarnowski get his idea for the penguin boat, Proteus?
Explain.

2. What is unusual about the way that Proteus moves through the water?

SPECIAL NEEDS PHYSICAL SCIENCE
and a cheetah have in common?

VOCABULARY ACTIVITY

Activity
Vocabulary Activity — SAMPLE

Getting the Dirt on the Soil

After you finish reading Chapter: [Unique Title], try this puzzle! Use the clues below to unscramble the vocabulary words. Write your answer in the space provided.

breakdown of rock into
and smaller pieces:
NGNETH

the chemical breakdown of rocks
and minerals into new
substances: CAMILCHE
THEARIGWEN

GENERAL

VOCABULARY AND SECTION SUMMARY

Skills Worksheet
Vocabulary & Notes — SAMPLE

Section:
VOCABULARY
In your own words, write a definition of the following term in the space provided.

1. scientific method

2. technology

ALSO IN SPANISH

GENERAL

REINFORCEMENT

Skills Worksheet
Reinforcement — SAMPLE

The Plane Truth
Complete this worksheet after you finish reading the Section: [Unique Section Title]

You plan to enter a paper airplane contest sponsored by Talkin' Physical Science magazine. The person whose airplane flies the farthest wins a lifetime subscription to the magazine! The week before the contest, you watch an airplane landing at a nearby airport. You notice that the wings of the airplane have flaps, as shown in the illustration at right. The paper airplanes you've been testing do not have wing flaps.
What question would you ask yourself based on these observations? Write your

Flaps

BASIC

CRITICAL THINKING

Skills Worksheet
Critical Thinking — SAMPLE

A Solar Solution

Dear Mr. Burns,
I've got this great idea for a new product called the Solar Heater. It's a portable, solar-powered space heater. The heater's design includes these features:
• the heater will be so long
• the solar cell will use all as
  with it a solting box
• the handle will have a
  2 glass top set of as easy as
  to catch the sun's rays
• the inside of the heater
  it will be dark colored to
  absorb solar heat

Joseph D. Burns
Inventory Advisory Consultants
Portland, OR 97001

think my idea will work, I will make the Solar
fectively designed as heater using no tech-
tting models. Please write back soon with your

ADVANCED

SCILINKS ACTIVITY

Skills Worksheet
SciLinks Activity — SAMPLE

MARINE ECOSYSTEMS
Go to www.scilinks.com. To find links related to marine ecosystems, type in the keyword HL590. Then, use the links to answer the questions about marine ecosystems.

percentage of the Earth's surface is covered by water?

SciLinks
NSTA
Go to www.scilinks.org
Topic: Reproductive System
Irregularities
SciLinks code: HL5690

GENERAL

SCIENCE PUZZLERS, TWISTERS & TEASERS

CHAPTER
20 SCIENCE PUZZLERS, TWISTERS & TEASERS
A Family of Planets

Planetary Omission Puzzles
1. When combined and rearranged, the missing letters from the alphabets below will spell a word from the chapter. What are the words?

BCFGHJKLMNPQUVWXYZ

a. _____

ABDFGHIJKLNPQRSUVWXYZ

b. _____

Dr. Emma Nint
Dr. Emma Nint has a dream of renaming the planets of our solar system with a more rational, scientific system. She has devised a mnemonic system for naming the planets, but has not given you a mnemonic or an explanation of her naming system. Figure out what each set of numbers represents and complete her list. (You'll probably need to refer to your book for help.)

GENERAL

Labs and Activities

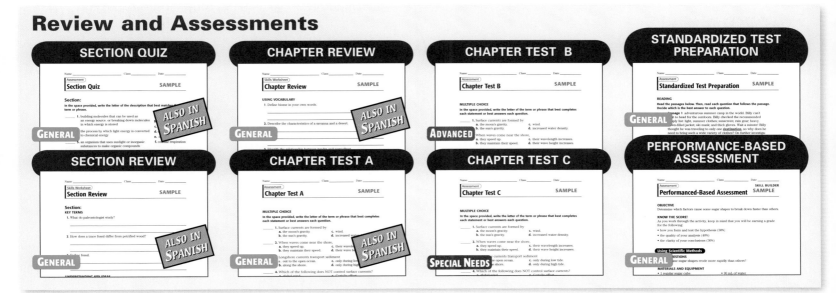

LONG-TERM PROJECTS & RESEARCH IDEAS

PROJECT
4K STUDENT WORKSHEET — DESIGN YOUR OWN
What Did You See, Mr. Messier?

Imagine that you are outside on a clear night observing the sky through a telescope. Suddenly, you notice a bright, fuzzy object that appears to have a tail of light coming from one side. Are your eyes playing tricks on you, or have you just found a comet? How can you tell? Well, you might start by checking the Messier Catalogue. The Messier Catalogue is a list of 109 celestial bodies compiled by Charles Messier, an eighteenth-century astronomer and comet chaser. Messier compiled this catalogue of objects that could be mistaken for comets in order to make comet hunting easier.

INTERNET KEYWORDS
Messier Catalogue
comets
Charles Messier

Look Up!
1. Find out more about Charles Messier and the Messier Catalogue in the library and on the Internet. What types of celestial bodies are included in the Messier Catalogue? Why are these celestial bodies likely to be confused with comets? Visit an area at night that does not have light interference, and identify as many Messier objects as you can. Create a guide to help others find and identify these objects.

Another Research Idea
2. Many Near Earth Objects (NEOs) have crashed into Earth. What clues did these objects leave behind? Investigate the physical and environmental impact of NEOs. Explain how scientists use crater information to determine the size, mass, and speed of the NEO that made the crater. Describe your results in a scientific article.

HELPFUL HINT
The eight phases of the moon are the following: new moon, waxing crescent moon, first quarter moon, waxing gibbous moon, full moon, waning gibbous moon, third quarter moon, and waning crescent moon.

Long-Term Project Ideas
3. Have you ever really looked at the moon? Observe the moon every night for at least 2 months, and keep a moon journal. Try to make your observations from the same place and at the same time each night. Use binoculars to make observations. Sketch the moon, and describe its color, shape, position in the sky, and phase. Present your observations on a poster. Be sure to include illustrations.

4. Many meteor showers and comet appearances are predictable. For example, Halley's comet can be seen from the Earth every 76 to 77 years. Find out the date of the next meteor shower or appearance of a comet. Observe the event, and record your observations. Why do comets pass by the Earth periodically? Investigate the sightings of Halley's comet over the last 2,000 years. How did ancient astronomers predict the occurrence of this comet? Write a magazine article about your observations and research.

EARTH SCIENCE

ADVANCED

WHIZ-BANG DEMONSTRATIONS

DEMO
33 TEACHER-LED DEMONSTRATION — MAKING MODELS
Crater Creator

Purpose
Students learn about impact craters and explore the factors that contribute to a appearance.

Lab Ratings

What to Do
1. Spread the newspapers on the floor. This can get messy. Place the box or pan on the floor in the center of the newspapers, and add enough flour to fill the container so that the flour is 6–8 cm deep. Smooth the surface with a piece of cardboard. Use the sieve or sifter to dust the surface of the flour with about 1 cm thick coating of tempera paint

BASIC

WHIZ-BANG DEMONSTRATIONS

DEMO
34 TEACHER-LED DEMONSTRATION — MAKING MODELS
Space Snowballs

Purpose
Students learn about the composition of comets in this exciting demonstration.

Time Required
35–45 minutes

Lab Ratings

TEACHER PREP
CONCEPT LEVEL

MATERIALS
• apron
• 2.5 kg of dry ice

be safely allowed to sublimate in a sink or container in a well-ventilated room, provided that students are not allowed access.

Advance Preparation
Frozen CO₂, or dry ice, is available from many commercial ice vendors. Just look in the phone book under Ice. Many ice cream shops receive their goods packed in dry ice, so they might also be a convenient (and possibly free) source of dry ice. You will need about 2.5 kg of dry ice to each nucleus. Purchase the ice the night before the activity and store it in a refrigerator or a cooler. Line the base of your storage container with about 3 cm of newspaper to prevent the container from cracking.

What to Do

proof, cold-resistant gloves

ADVANCED

LABS YOU CAN EAT

LAB
17 STUDENT WORKSHEET — DISCOVERY LAB
Meteorite Delight

Chunks of rock and metal race through space. What happens when they enter the Earth's atmosphere? Most burn up as they plummet, but some land on the ground. These objects are called meteorites. Meteorites come in two major forms: stony, which look similar to Earth rocks, and iron, which appear shiny and metallic.

There are two ways to discover meteorites. A meteorite fall occurs when someone sees the meteorite, traces its path to Earth, and recovers it. A meteorite find occurs when someone discovers a meteorite on the ground without having seen it fall.

In this activity, you will work in groups of two or more students to simulate the impact of stony and iron meteorites on the ground. Rice-cereal balls will serve as meteorites to help you better understand meteorite finds and falls.

MATERIALS
• large baking pan or cookie sheet
• box of plain puffed rice cereal
• 12 plain rice-cereal balls
• 2 cocoa-flavored rice cereal balls

METHOD
Ask a Question
How do meteorite finds differ from meteorite falls?

Conduct an Experiment
1. Fill the baking pan with puffed rice cereal. Smooth the top of the cereal so that it is level with the top of the pan. Then place 10 of the rice-cereal balls without cocoa across the top of the pan.

2. Place the pan at one end of the classroom. Decide who will drop the cereal balls and who will make observations. The person who drops the cereal balls should stand next to the pan while the observer stands across the room.

3. To conduct the first trial, hold a plain rice-cereal ball (representing a stony meteorite) at shoulder height directly above the pan. While the observer is watching, drop the ball.

4. The observer should then examine the pan and identify which ball was just dropped. Record in the "Observed" column of the Meteorite Observation Chart, on page 83, whether the meteorite was correctly identified.

5. Repeat steps 3–4 with another plain rice-cereal ball and with two cocoa-flavored rice balls (representing iron meteorites), dropping the balls one at a time.

ADVANCED

DATASHEETS FOR QUICK LABS

TEACHER RESOURCE PAGE
Quick Lab — DATASHEET FOR QUICK LAB
Reaction to Stress — SAMPLE

Background
The graph below illustrates changes that occur in the membrane potential of a neuron during an action potential. Use the graph to answer the following questions. Refer to Figure 3 as needed.

DATASHEETS FOR CHAPTER LABS

TEACHER RESOURCE PAGE
Skills Practice Lab — DATASHEET FOR CHAPTER LAB
Using Scientific Methods — SAMPLE

Teacher's Notes
TIME REQUIRED
One 45-minute class period.

DATASHEETS FOR LABBOOK

TEACHER RESOURCE PAGE
Skills Practice Lab — DATASHEET FOR LABBOOK LAB
Does It All Add Up? — SAMPLE

Teacher's Notes
TIME REQUIRED
One 45-minute class period.

Review and Assessments

SECTION QUIZ

Assessment
Section Quiz — SAMPLE

Section:
In the space provided, write the letter of the description that best matches term or phrase.

____ 1. building molecules that can be used as an energy source. or breaking down molecules in which energy is stored
____ that light energy is converted to chemical energy
____ an organism that uses sunlight or inorganic substances to make organic compounds

ALSO IN SPANISH

GENERAL

SECTION REVIEW

Name / Class / Date
Skills Worksheet
Section Review — SAMPLE

Section:
KEY TERMS
1. What do paleontologist study?

2. How does a trace fossil differ from petrified wood?

fossil.

ALSO IN SPANISH

GENERAL

CHAPTER REVIEW

Skills Worksheet
Chapter Review — SAMPLE

USING VOCABULARY
1. Define biome in your own words.

2. Describe the characteristics of a savanna and a desert.

ALSO IN SPANISH

GENERAL

CHAPTER TEST A

Assessment
Chapter Test A — SAMPLE

MULTIPLE CHOICE
In the space provided, write the letter of the term or phrase that best completes each statement or best answers each question.

____ 1. Surface currents are formed by
   a. the moon's gravity.   c. wind.
   b. the sun's gravity.   d. increased wa

____ 2. When waves come near the shore,
   a. they speed up.   c. their ware
   b. they maintain their speed. d. their ware b

____ 3. Longshore currents transport sediment
   a. out to the open ocean.   c. only during low
   b. along the shore.   d. only during high tide.

____ 4. Which of the following does NOT control surface currents?

ALSO IN SPANISH

GENERAL

CHAPTER TEST B

Assessment
Chapter Test B — SAMPLE

MULTIPLE CHOICE
In the space provided, write the letter of the term or phrase that best completes each statement or best answers each question.

____ 1. Surface currents are formed by
   a. the moon's gravity.   c. wind.
   b. the sun's gravity.   d. increased water density.

____ 2. When waves come near the shore,
   a. they speed up.   c. their wavelength increases.
   b. they maintain their speed. d. their wavelength increases.

Identify the relationship between speed and experiment

ADVANCED

CHAPTER TEST C

Assessment
Chapter Test C — SAMPLE

MULTIPLE CHOICE
In the space provided, write the letter of the term or phrase that best completes each statement or best answers each question.

____ 1. Surface currents are formed by
   a. the moon's gravity.   c. wind.
   b. the sun's gravity.   d. increased water density.

____ 2. When waves come near the shore,
   a. they speed up.   c. their wavelength increases.
   b. they maintain their speed. d. their wave height increases.

____ 3. Longshore currents transport sediment
   a. out to the open   c. only during low tide
   b. the shore.   d. only during high tide.

____ 4. Which of the following does NOT control surface currents?

SPECIAL NEEDS

STANDARDIZED TEST PREPARATION

Assessment
Standardized Test Preparation — SAMPLE

READING
Read the passages below. Then, read each question that follows the passage. Decide which is the best answer to each question.

Passage 1 Last summer, adventurous summer camp in the world. Billy can't wait to head for the outdoors. Billy checked the recommended packing list: light, summer clothes; sunscreen; rain gear; heavy, down-filled jacket; ski mask; and thick gloves. Wait a minute! Billy thought he was traveling to only one destination, so why does he need to bring such a wide variety of clothes? On further investiga-

GENERAL

PERFORMANCE-BASED ASSESSMENT

Assessment
Performanced-Based Assessment — SKILL BUILDER SAMPLE

OBJECTIVE
Determine which factors cause some sugar shapes to break down faster than others.

KNOW THE SCORE!
As you work through the activity, keep in mind that you will be earning a grade for the following:
• how you form and test your hypothesis (30%)
• the quality of your analysis (40%)
• the clarity of your conclusions (30%)

Using Scientific Methods
QUESTIONS
sugar shapes erode more rapidly than others?

MATERIALS AND EQUIPMENT
1 regular sugar cube   • 80 mL of water

GENERAL

This Chapter Enrichment provides relevant and interesting information to expand and enhance your presentation of the chapter material.

Section 1

The Nine Planets

Ptolemy

- In the second century CE, the astronomer Claudius Ptolemy formulated an elaborate scientific theory of a geocentric universe. He argued that everything in the universe revolves around Earth's center. In Ptolemy's model, the stars moved in a rotating sphere, and the motion of the planets, moons, and comets was explained by a series of large and small circles turning inside one another. Although Ptolemy's theories about the mechanism of planetary movement were later rejected, his basic model of the universe remained the predominant scientific theory until the work of Copernicus, Galileo, and Kepler from around 1500 to 1650.

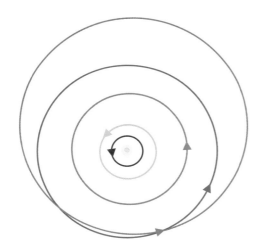

Section 2

The Inner Planets

A Day on Mercury

- Imagine waking up in the middle of winter just before dawn to find the outside temperature a frigid –173°C! As you watch the sun slowly rise over the next several days, you notice that it appears 3 times as big as it does from Earth. You also notice that the sky is black. The reason is that Mercury has an extremely thin atmosphere that doesn't scatter blue light as Earth's atmosphere does. Forty-four Earth days later, it would be noon on Mercury and the middle of summer. The temperature would be a toasty 427°C. The range of Mercury's surface temperatures is more extreme than that of any other planet in the solar system.

Is That a Fact!

- ◆ The planet Mercury has been known and studied for more than 2,000 years. Its wanderings may have been noted by Hypatia (415–370 BCE), an Egyptian mathematician and philosopher and the first known female astronomer. Hypatia was a student of Plato.

Section 3

The Outer Planets

Which Is Last?

- Pluto is not always the farthest planet from the sun. Pluto's orbit around the sun takes 248 Earth years to complete. Because its orbit is highly elliptical, Pluto spends about 20 years of its orbit closer to the sun than Neptune does. The last time that Pluto was closer to the sun than Neptune was occurred between 1979 and 1999. The next time will be in the 23rd century.

Section 4

Moons

Earth Tides

● As the moon revolves around the Earth, it causes tides—even on land! The distance from Earth's center to its surface increases by a few centimeters as the moon passes overhead. This change is not as noticeable as ocean tides, but it can be detected by very sensitive instruments.

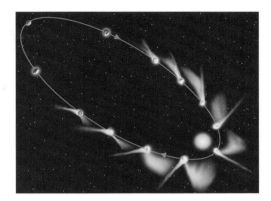

Is That a Fact!

◆ Four moons in the solar system are larger than Earth's moon: Jupiter's Ganymede, Callisto, and Io and Saturn's Titan. Earth's moon is special because it is very large relative to the planet it orbits. Pluto's moon, Charon, however, is more than half the size of Pluto.

The Kuiper Belt

● To explain the source of short-period comets (comets with a relatively short orbit), the Dutch American astronomer Gerard Kuiper proposed in 1949 that a belt of icy bodies must lie beyond the orbits of Pluto and Neptune. Kuiper argued that comets were icy planetesimals that formed during the condensation of our solar nebula. Because the icy bodies are so far from any large planet's gravitational field (30 AU to 100 AU), they can remain on the fringe of the solar system. Some theorists speculate that the large moons Triton and Charon were once members of the Kuiper belt before they were captured by Neptune and Pluto. These moons and short-period comets have similar physical and chemical properties.

Section 5

Small Bodies in the Solar System

The Oort Cloud

● To explain the origin of comets, a Dutch astronomer named Jan Oort suggested in the 1950s that a spherical cloud of comets surrounds the solar system. He estimated that the cloud is 40,000 to 100,000 astronomical units (AU) from the sun and that it may contain trillions of icy bodies.

SCILINKS®

NSTA
Developed and maintained by the
National Science Teachers Association

SciLinks is maintained by the National Science Teachers Association to provide you and your students with interesting, up-to-date links that will enrich your classroom presentation of the chapter.

Visit www.scilinks.org and enter the SciLinks code for more information about the topic listed.

Topic: The Nine Planets
SciLinks code: HSM1033

Topic: Moons of Other Planets
SciLinks code: HSM0993

Topic: The Inner Planets
SciLinks code: HSM0798

Topic: Comets, Asteroids, and Meteoroids
SciLinks code: HSM0317

Topic: The Outer Planets
SciLinks code: HSM1091

Overview

This chapter introduces the planets in the solar system. Students will learn about the differences between the inner planets and the outer planets. Students will also learn about moons and smaller bodies of the solar system, such, as comets, asteroids, and meteoroids.

Assessing Prior Knowledge

Students should be familiar with the following topics:

• methods of studying space

• stars

• the formation of the solar system

Identifying Misconceptions

Students may not understand that the mass and volume of planets varies greatly. Students may benefit from a comparison of the relative mass, volume, and location of the sun and planets. Point out that in terms of mass, the solar system has two main bodies—the sun and Jupiter. The sun makes up more than 99.5% of the mass in the solar system. Jupiter has one-thousandth the mass of the sun but is roughly 317 times more massive than Earth, and Jupiter's volume is 1,321 times the volume of Earth. As you teach this chapter, encourage students to make other such comparisons.

4

A Family of Planets

About the

These rich swirls of color may remind you of a painting you might see in an art museum. But this photograph is of the planet Jupiter. The red swirl, called the Great Red Spot, is actually a hurricane-like storm system that is 3 times the size of Earth!

PRE-READING ACTIVITY

FOLDNOTES **Booklet** Before you read the chapter, create the FoldNote entitled "Booklet" described in the **Study Skills** section of the Appendix. Label each page of the booklet with a name of a planet in our solar system. As you read the chapter, write what you learn about each planet on the appropriate page of the booklet.

Standards Correlations

National Science Education Standards

The following codes indicate the National Science Education Standards that correlate to this chapter. The full text of the standards is at the front of the book.

Chapter Opener
SAI 1; HNS 1, 3; ES 3a

Section 1 The Nine Planets
UCP 1, 3; SAI 1; ST 2; SPSP 5; HNS 1, 3; ES 1c, 3a, 3b, 3c; *LabBook:* UCP 2; SAI 1; ST 1

Section 2 The Inner Planets
UCP 1, 3; SAI 1; ST 2; SPSP 5; HNS 1, 3; ES 1c, 3a, 3b

Section 3 The Outer Planets
UCP 1, 3; SAI 1; ST 2; SPSP 5; HNS 1, 3; ES 1c, 3a, 3b

Section 4 Moons
UCP 1, 3; SAI 1; HNS 1, 3; ES 3a, 3b, 3c; *LabBook:* UCP 2; SAI 1; ST 1; ES 1a, 3b

Section 5 Small Bodies in the Solar System
UCP 1; ES 2a, 3a, 3b

START-UP ACTIVITY

MATERIALS

FOR EACH GROUP
- chalk
- chalkboard
- meterstick

Teacher's Notes: The solar and planetary data presented in this chapter are the most current data available at the time of publication. Because of the vast size of the solar system, instrument limitations, and differences in methods of gathering data, the values given have varying margins of error. As measuring precision increases, these values are updated. Therefore, other sources may show different values for the same statistics.

Answer

1. The inner four planets are closer together than the outer planets are.

START-UP ACTIVITY

Measuring Space

Do the following activity to get a better idea of your solar neighborhood.

Procedure

1. Use a **meterstick** and some **chalk** to draw a line 2 m long on a **chalkboard.** Draw a large dot at one end of the line. This dot represents the sun.

2. Draw smaller dots on the line to represent the relative distances of each of the planets from the sun, based on information in the table.

Analysis

1. What do you notice about how the planets are spaced?

Planet	Distance from sun	
	Millions of km	**Scaled to cm**
Mercury	57.9	
Venus	108.2	4
Earth	149.6	5
Mars	227.9	8
Jupiter	778.4	26
Saturn	1,424.0	48
Neptune	2,827.0	97
Uranus	4,499.0	151
Pluto	5,943.0	200

Chapter Lab
UCP 1, 2; SAI 1, 2; ES 3b

Chapter Review
UCP 1, 2, 3; SAI 1, 2; ES 1a, 3a, 3b, 3c

Science in Action
UCP 1, 2, SAI 2; ST 2; SPSP 5; HNS 1; ES 3a, 3b, 3c

Imagine . . .

Imagine that it is 200 BCE and you are an apprentice to a Greek astronomer. After years of observing the sky, he knows all

Building on the observations of the ancient Greeks, we now know that the planets are actually planets, not wandering stars.

Chapter Starter Transparency
Use this transparency to help students begin thinking about why planets were called "wanderers" by the Greeks.

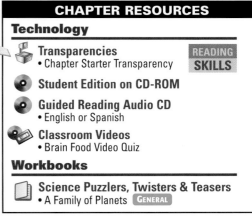

CHAPTER RESOURCES

Technology

Transparencies
- Chapter Starter Transparency

READING SKILLS

Student Edition on CD-ROM

Guided Reading Audio CD
- English or Spanish

Classroom Videos
- Brain Food Video Quiz

Workbooks

Science Puzzlers, Twisters & Teasers
- A Family of Planets **GENERAL**

Focus

Overview

This section describes the scale of the solar system, the discovery of the planets, and planetary motion. The section also introduces the differences between the inner and outer planets.

Bellringer

To introduce this chapter, assign each student a planet to research. Tell students to create a poster that features the planet and includes a cross section of the planet's interior. Students should provide factual information and mythology about the planet in their poster.

Motivate

Discussion ——— GENERAL

Solar System Questions

Display the posters that the students made for the Bellringer activity. Have students list questions they want to answer about the planets. Pool students' questions to form one list. Read the list aloud for the class, and have students propose possible answers to the questions. Then, list the questions on a bulletin board. Leave space for answers, and have students fill in the answers as they study this chapter. **LS** Visual/Verbal

READING WARM-UP

Objectives

- List the planets in the order in which they orbit the sun.
- Explain how scientists measure distances in space.
- Describe how the planets in our solar system were discovered.
- Describe three ways in which the inner planets and outer planets differ.

Terms to Learn

astronomical unit

READING STRATEGY

Paired Summarizing Read this section silently. In pairs, take turns summarizing the material. Stop to discuss ideas that seem confusing.

The Nine Planets

Did you know that planets, when viewed from Earth, look like stars to the naked eye? Ancient astronomers were intrigued by these "stars" which seemed to wander in the sky.

Ancient astronomers named these "stars" planets, which means "wanderers" in Greek. These astronomers knew planets were physical bodies and could predict their motions. But scientists did not begin to explore these worlds until the 17th century, when Galileo used the telescope to study planets and stars. Now, scientists have completed more than 150 successful missions to moons, planets, comets, and asteroids in our cosmic neighborhood.

Our Solar System

Our *solar system,* shown in **Figure 1,** includes the sun, the planets, and many smaller objects. In some cases, these bodies may be organized into smaller systems of their own. For example, the Saturn system is made of the planet Saturn and the several moons that orbit Saturn. In this way, our solar system is a combination of many smaller systems.

Figure 1 *These images show the relative sizes of the planets and the sun.*

Mercury
4,879 km

Venus
12,104 km

Earth
12,756 km

Mars
6,794 km

Sun
1,392,000 km

Jupiter
142,984 km

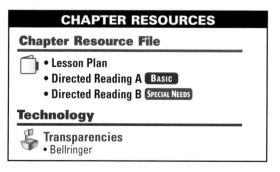

CHAPTER RESOURCES

Chapter Resource File

- Lesson Plan
- Directed Reading A **BASIC**
- Directed Reading B **SPECIAL NEEDS**

Technology

Transparencies
- Bellringer

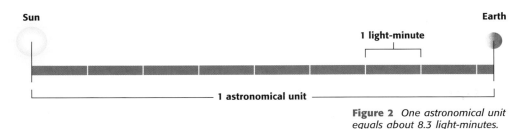

Figure 2 *One astronomical unit equals about 8.3 light-minutes.*

Measuring Interplanetary Distances

One way that scientists measure distances in space is by using the astronomical unit. One **astronomical unit** (AU) is the average distance between the sun and Earth. Another way to measure distances in space is by using the speed of light. Light travels at about 300,000 km/s in space. This means that in 1 s, light travels 300,000 km.

In 1 min, light travels nearly 18,000,000 km. This distance is also called a *light-minute*. Look at **Figure 2.** Light from the sun takes 8.3 min to reach Earth. So, the distance from Earth to the sun, or 1 AU, is 8.3 light-minutes. Distances in the solar system can be measured in light-minutes and light-hours.

Reading Check How far does light travel in 1 s? (*See the Appendix for answers to Reading Checks.*)

astronomical unit the average distance between the Earth and the sun; approximately 150 million kilometers (symbol, AU)

Saturn
120,536 km

Uranus
51,118 km

Neptune
49,528 km

Pluto
2,390 km

Answer to Reading Check
Light travels about 300,000 km/s.

Teach

Using the Figure—GENERAL

A Sense of Scale The images of the planets and the sun in **Figure 1** are shown to scale. Ask students to use a ruler to estimate how many Earths would fit side by side along the diameter of Jupiter. (Answers may vary, but students should determine that about 10 Earths would fit within the diameter of Jupiter.)
LS Visual/Kinesthetic

MISCONCEPTION
ALERT

Scale in Diagrams Tell students that the diameters of Earth and the sun are not to scale in **Figure 2.** If they were, the sun would be a little less than 2 mm across, and Earth would be too small to see. The lengths of the light-minute and the AU, however, are to scale. Ask students why the AU is the average distance between the Earth and sun. (The Earth's orbit is elliptical. Therefore, the distance between the Earth and the sun is continuously changing.) **LS** Logical

---

CONNECTION ACTIVITY
Math ——————— GENERAL

Calculating Distances If light travels 18 million kilometers in 1 min, how far away is Earth from the sun if sunlight takes 8.3 min to reach Earth?
(18,000,000 km/min × 8.3 min = 149,400,000 km)
How far is Mercury from the sun if sunlight takes 3.2 min to reach its surface?
(18,000,000 km/min × 3.2 min = 57,600,000 km)
Remind students that these numbers are rounded and are therefore approximations.
LS Logical

Reteaching — BASIC

Planetary Review On the board or an overhead transparency, draw a diagram of the solar system that shows the orbit of each planet. Ask students to help you indicate which planet occupies each orbital path. **LS** Visual

Quiz — GENERAL

1. Which planets are part of the inner solar system? (Mercury, Venus, Earth, and Mars)

2. If a rocket could travel at the speed of light, how far would it go in 15 min?
(18,000,000 km/min × 15 min = 270,000,000 km)

3. Why are the inner planets known as the *terrestrial planets*? (because their surfaces are dense and rocky)

Alternative Assessment — GENERAL

Build a Mobile Ask students to create a mobile or diorama of the planets. The mobile should have the planets in order, and their sizes should be in correct proportion to one another. As students learn more about the planets, they should add facts about each planet to their mobile. **LS** Visual/Kinesthetic

For another activity related to this chapter, go to **go.hrw.com** and type in the keyword **HZ5FAMW.**

The Discovery of the Solar System

Up until the 17th century, the universe was thought to have only eight bodies. These bodies included the planets Earth, Mercury, Venus, Mars, Jupiter, and Saturn, the sun, and the Earth's moon. These bodies are the only ones that can be seen from Earth without using a telescope.

After the telescope was invented in the 17th century, however, more discoveries were made. By the end of the 17th century, nine more large bodies were discovered. These bodies were moons of Jupiter and Saturn.

By the 18th century, the planet Uranus, along with two of its moons and two more of Saturn's moons, was discovered. In the 19th century, Neptune, as well as moons of several other planets, was discovered. Finally, in the 20th century, the ninth planet, Pluto, was discovered.

The Inner and Outer Solar Systems

The solar system is divided into two main parts: the inner solar system and the outer solar system. The inner solar system contains the four planets that are closest to the sun. The outer solar system contains the planets that are farthest from the sun.

The Inner Planets

The planets of the inner solar system, shown in **Figure 3,** are more closely spaced than the planets of the outer solar system. The inner planets are also known as the *terrestrial planets* because their surfaces are dense and rocky. However, each of the inner planets is unique.

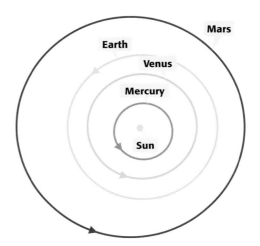

Figure 3 *The inner planets are the planets that are closest to the sun.*

INCLUSION Strategies

- *Learning Disabled*
- *Developmentally Delayed*
- *Visually Impaired*

Organize students into small teams to play a planetary quiz game. Each team should choose a category that relates to a heading in the section. Have students write five questions and answers for the category on separate index cards. The difficulty and point value of the questions should increase incrementally.

Review each team's questions and answers. Play the game by allowing teams to ask questions to another team. If a team cannot answer a question, the team should work with another team to find the answer. If teams cooperate, they should share the points earned. When the game is over, hand out a review sheet that contains all the questions with corresponding answers. **LS** Verbal   English Language Learners

The Outer Planets

The planets of the outer solar system include Jupiter, Saturn, Uranus, Neptune, and Pluto. The outer planets are very different from the inner planets, as you will soon find out.

Unlike the inner planets, the outer planets, except for Pluto, are large and are composed mostly of gases. Because of this, Jupiter, Saturn, Uranus, and Neptune are known as gas giants. The atmospheres of these planets blend smoothly into the denser layers of their interiors. The icy planet Pluto is the only planet of the outer solar system that is small, dense, and rocky. You can see a diagram of the outer solar system in **Figure 4.**

✓ Reading Check Which planets are in the outer solar system?

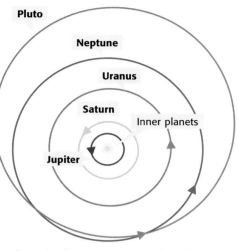

Figure 4 The planets of the outer solar system are the farthest from the sun.

Pluto
Neptune
Uranus
Saturn
Inner planets
Jupiter

SECTION Review

Summary

- In the order in which they orbit the sun, the nine planets are Mercury, Venus, Earth, Mars, Jupiter, Saturn, Uranus, Neptune, and Pluto.
- Two ways in which scientists measure distances in space are to use astronomical units and to use light-years.
- The inner planets are spaced more closely together, are smaller, and are rockier than the outer planets.

Using Key Terms

1. In your own words, write a definition for the term *astronomical unit.*

Understanding Key Ideas

2. When was the planet Uranus discovered?
   a. before the 17th century
   b. in the 18th century
   c. in the 19th century
   d. in the 20th century

3. The invention of what instrument helped early scientists discover more bodies in the solar system?

4. Which of the nine planets are included in the outer solar system?

5. Describe how the inner planets are different from the outer planets.

Math Skills

6. If Venus is 6.0 light-minutes from the sun, what is Venus's distance from the sun in astronomical units?

Critical Thinking

7. **Analyzing Methods** The distance between Earth and the sun is measured in light-minutes, but the distance between Pluto and the sun is measured in light-hours. Explain why.

SciLINKS.

Developed and maintained by the National Science Teachers Association

For a variety of links related to this chapter, go to www.scilinks.org

Topic: The Nine Planets
SciLinks code: HSM1033

Answer to Reading Check
Jupiter, Saturn, Uranus, Neptune, and Pluto are in the outer solar system.

CHAPTER RESOURCES

Chapter Resource File
- Section Quiz **GENERAL**
- Section Review **GENERAL**
- Vocabulary and Section Summary **GENERAL**
- SciLinks Activity **GENERAL**

Technology
Transparencies
- The Inner Planets; The Outer Planets

Workbooks
Math Skills for Science
- A Shortcut for Multiplying Large Numbers **GENERAL**

Answers to Section Review

1. Sample answer: An astronomical unit is the average distance from the Earth to the sun.

2. b

3. the telescope

4. Jupiter, Saturn, Uranus, Neptune, and Pluto

5. Sample answer: The surfaces of the inner planets are dense and rocky. Except for Pluto, the outer planets are extremely large and composed mostly of gases. Pluto is the only outer planet that is small, dense, and rocky.

6. Use a ratio to calculate the relationship between the distance from Venus to the sun in light-minutes and in astronomical units.

   6 light-minutes $\div X$ AU = 8.3 light-minutes $\div$ 1 AU

   This fraction reduces to X = 6 light-minutes $\div$ 8.3 light-minutes

   X = 0.7 AU

7. Sample answer: Pluto is so far from the sun that expressing this distance in light-minutes would require a very large number. The distance is measured in light-hours, which is a smaller number than light-minutes.

Focus

Overview

This section teaches students about the four inner planets of the solar system: Mercury, Venus, Earth, and Mars.

Bellringer

Have students create a mnemonic device to help them remember the order of the planets:

Mercury, **V**enus, **E**arth, **M**ars, **J**upiter, **S**aturn, **U**ranus, **N**eptune, **P**luto

Example: My very eccentric mother just sent us nine pigs.

Motivate

ACTIVITY ———— GENERAL

Your Age in Venusian Years
Tell students that Venus is the second-brightest object in the night sky. Venus is often called the *morning star* or the *evening star* because it is visible only at dawn or dusk. Also, the planet rotates much more slowly than Earth does. A Venusian day is longer than a Venusian year! A Venusian year is 224 Earth days long, but a Venusian day is 243 Earth days long. Have students calculate their age in Venusian years. (A 10-year-old student is 3,656 days old: 3,656 Earth days ÷ 224 Earth days/Venusian year = 16.3 Venusian years old.)
LS Logical

The Inner Planets

In the inner solar system, you will find one of the hottest places in our solar system as well as the only planet known to support life.

The inner planets are also called **terrestrial planets** because, like Earth, they are very dense and rocky. The inner planets are smaller, denser, and rockier than the outer planets. In this section, you will learn more about the individual characteristics of Mercury, Venus, Earth, and Mars.

Mercury: Closest to the Sun

If you visited the planet Mercury, shown in **Figure 1,** you would find a very strange world. For one thing, on Mercury you would weigh only 38% of what you weigh on Earth. The weight you have on Earth is due to surface gravity, which is less on less massive planets. Also, because of Mercury's slow rotation, a day on Mercury is almost 59 Earth days long! The amount of time that an object takes to rotate once is called its *period of rotation*. So, Mercury's period of rotation is almost 59 Earth days long.

A Year on Mercury

Another curious thing about Mercury is that its year is only 88 Earth days long. As you know, a *year* is the time that a planet takes to go around the sun once. The motion of a body orbiting another body in space is called *revolution*. The time an object takes to revolve around the sun once is called its *period of revolution*. Every 88 Earth days, or 1.5 Mercurian days, Mercury revolves once around the sun.

READING WARM-UP

Objectives

● Explain the difference between a planet's period of rotation and period of revolution.

● Describe the difference between prograde and retrograde rotation.

● Describe the individual characteristics of Mercury, Venus, Earth, and Mars.

● Identify the characteristics that make Earth suitable for life.

Terms to Learn

terrestrial planet
prograde rotation
retrograde rotation

READING STRATEGY

Reading Organizer As you read this section, create an outline of the section. Use the headings from the section in your outline.

terrestrial planet one of the highly dense planets nearest to the sun; Mercury, Venus, Mars, and Earth

Figure 1 *This image of Mercury was taken by the* Mariner 10 *spacecraft on March 24, 1974, from a distance of 5,380,000 km.*

Mercury Statistics	
Distance from sun	3.2 light-minutes
Period of rotation	58 days, 19 h
Period of revolution	88 days
Diameter	4,879 km
Density	5.43 g/cm³
Surface temperature	−173°C to 427°C
Surface gravity	38% of Earth's

CHAPTER RESOURCES

Chapter Resource File

- Lesson Plan
- Directed Reading A **BASIC**
- Directed Reading B **SPECIAL NEEDS**

Technology

Transparencies
- Bellringer

Venus Statistics	
Distance from sun	6.0 light-minutes
Period of rotation	243 days, 16 h (R)*
Period of revolution	224 days, 17 h
Diameter	12,104 km
Density	5.24 g/cm^3
Surface temperature	464°C
Surface gravity	91% of Earth's

*R = retrograde rotation

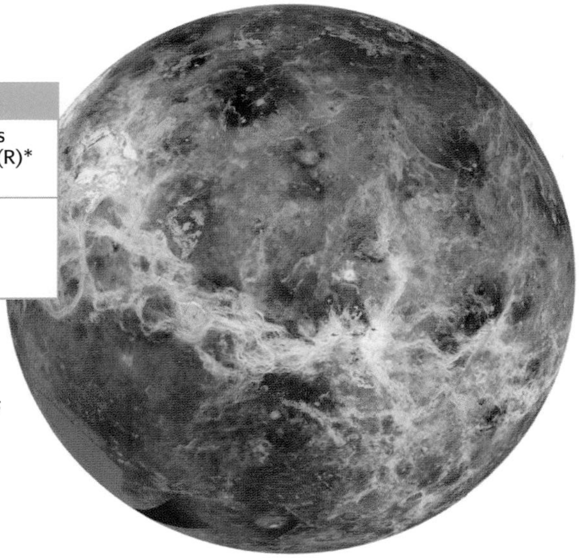

Figure 2 *This image of Venus was taken by Mariner 10 on February 5, 1974. The uppermost layer of clouds contains sulfuric acid.*

Venus: Earth's Twin?

Look at **Figure 2.** In many ways, Venus is more like Earth than any other planet. Venus is only slightly smaller, less massive, and less dense than Earth. But in other ways, Venus is very different from Earth. On Venus, the sun rises in the west and sets in the east. The reason is that Venus and Earth rotate in opposite directions. Earth is said to have **prograde rotation** because it appears to spin in a *counterclockwise* direction when it is viewed from above its North Pole. If a planet spins in a *clockwise* direction, the planet is said to have **retrograde rotation.**

The Atmosphere of Venus

Of the terrestrial planets, Venus has the densest atmosphere. Venus's atmosphere has 90 times the pressure of Earth's atmosphere! The air on Venus is mostly carbon dioxide, but the air is also made of some of the most destructive acids known. The carbon dioxide traps thermal energy from sunlight in a process called the *greenhouse effect.* The greenhouse effect causes Venus's surface temperature to be very high. At 464°C, Venus has the hottest surface of any planet in the solar system.

Mapping Venus's Surface

Between 1990 and 1992, the *Magellan* spacecraft mapped the surface of Venus by using radar waves. The radar waves traveled through the clouds and bounced off the planet's surface. Data gathered from the radar waves showed that Venus, like Earth, has volcanoes.

Reading Check What technology was used to map the surface of Venus? (*See the Appendix for answers to Reading Checks.*)

prograde rotation the counterclockwise spin of a planet or moon as seen from above the planet's North Pole; rotation in the same direction as the sun's rotation

retrograde rotation the clockwise spin of a planet or moon as seen from above the planet's North Pole

Teach

CONNECTION to Environmental Science ——— GENERAL

Writing **Atmospheric Change** The concentration of carbon dioxide in Venus's atmosphere causes a severe greenhouse effect. As a result, surface temperatures on Venus are hot enough to melt lead. Could an increased greenhouse effect cause Earth's atmosphere to become more like Venus's? Earth's current atmosphere is primarily nitrogen and oxygen. CO_2 levels in our atmosphere have risen steadily since the Industrial Revolution. Have students write a short story describing how Earth's atmosphere could become more like Venus's atmosphere. Ask students to describe how life on Earth would change. **LS Verbal**

MISCONCEPTION ALERT

Variations in Brightness Students may believe that the planets always have the same brightness. Actually, planets appear brighter when they are closer to Earth. For example, as Mars and Earth orbit the sun, the distance between the two planets varies from about 75 million kilometers to about 375 million kilometers. This difference in distance causes the apparent brightness of Mars to vary by a factor of 25.

Homework ——— ADVANCED

Using Maps Most of the features of Venus are named after female scientists, female historical figures, and goddesses. Many of Mercury's craters are named after artists and musicians, and most craters of the moon are named after famous scientists. Have students use maps of the inner planets to learn more about their features and the origin of their names. **LS Visual**

Answer to Reading Check

Radar technology was used to map the surface of Venus.

Discussion — **BASIC**

Defining Terms Ask students what an oasis is. (An oasis is a hospitable place in an otherwise inhospitable area.) Discuss with students why Earth is referred to in the text as an *oasis*. Tell them to focus on the importance of Earth's distance from the sun and the presence of large amounts of liquid water on Earth's surface. Earth's mass also plays an important role because the mass "holds" the gases that constitute Earth's life-sustaining atmosphere around the Earth.
LS Verbal

MISCONCEPTION ALERT

The World Is Not Round
Earth is not a perfect sphere. The diameter of Earth as measured from the North Pole to the South Pole is 44 km less than the diameter as measured at the equator. None of the other planets or stars is perfectly spherical either. Therefore, all of the planetary diameters given in this chapter are equatorial diameters.

Figure 3 *Earth is the only planet known to support life.*

Earth: An Oasis in Space

As viewed from space, Earth is like a sparkling blue oasis in a black sea of stars. Constantly changing weather patterns create the swirls of clouds that blanket the blue and brown sphere we call home. Look at **Figure 3.** Why did Earth have such good fortune, while its two nearest neighbors, Venus and Mars, are unsuitable for life as we know it?

Water on Earth

Earth formed at just the right distance from the sun. Earth is warm enough to keep most of its water from freezing. But unlike Venus, Earth is cool enough to keep its water from boiling away. Liquid water is a vital part of the chemical processes that living things depend on for survival.

The Earth from Space

The picture of Earth shown in **Figure 4** was taken from space. You might think that the only goal of space exploration is to make discoveries beyond Earth. But the National Aeronautics and Space Administration (NASA) has a program to study Earth by using satellites in the same way that scientists study other planets. This program is called the Earth Science Enterprise. Its goal is to study the Earth as a global system that is made of smaller systems. These smaller systems include the atmosphere, land, ice, the oceans, and life. The program will also help us understand how humans affect the global environment. By studying Earth from space, scientists hope to understand how different parts of the global system interact.

✓ Reading Check What is the Earth Science Enterprise?

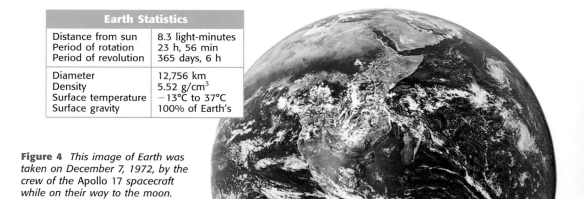

Earth Statistics	
Distance from sun	8.3 light-minutes
Period of rotation	23 h, 56 min
Period of revolution	365 days, 6 h
Diameter	12,756 km
Density	5.52 g/cm^3
Surface temperature	−13°C to 37°C
Surface gravity	100% of Earth's

Figure 4 *This image of Earth was taken on December 7, 1972, by the crew of the* Apollo 17 *spacecraft while on their way to the moon.*

WEIRD SCIENCE

The Earth and its moon revolve around the sun like a "double planet." They can be thought of as the unequal ends of a barbell. The center of gravity for the Earth-moon system, called the *barycenter*, is actually 1,700 km below the Earth's surface. The barycenter is what follows the curved line of Earth's orbit. The moon and the Earth wobble around this center of gravity as they circle the sun.

Answer to Reading Check

Earth's global system includes the atmosphere, the oceans, and the biosphere.

Mars Statistics	
Distance from sun	12.7 light-minutes
Period of rotation	24 h, 40 min
Period of revolution	1 year, 322 days
Diameter	6,794 km
Density	3.93 g/cm^3
Surface temperature	−123°C to 37°C
Surface gravity	38% of Earth's

Mars: Our Intriguing Neighbor

Mars, shown in **Figure 5,** is perhaps the most studied planet in the solar system other than Earth. Much of our knowledge of Mars has come from information gathered by spacecraft. *Viking 1* and *Viking 2* landed on Mars in 1976, and *Mars Pathfinder* landed on Mars in 1997.

The Atmosphere of Mars

Because of its thinner atmosphere and greater distance from the sun, Mars is a cold planet. Midsummer temperatures recorded by the *Mars Pathfinder* range from −13°C to −77°C. Martian air is so thin that the air pressure on the surface of Mars is about the same as it is 30 km above Earth's surface. This distance is about 3 times higher than most planes fly! The air pressure is so low that any liquid water would quickly boil away. The only water found on the surface of Mars is in the form of ice.

Water on Mars

Even though liquid water cannot exist on Mars's surface today, there is strong evidence that it existed there in the past. **Figure 6** shows an area on Mars with features that might have resulted from deposition of sediment in a lake. This finding means that in the past Mars might have been a warmer place and had a thicker atmosphere.

Figure 5 *This Viking orbiter image shows the eastern hemisphere of Mars. The large circular feature in the center is the impact crater Schiaparelli, which has a diameter of 450 km.*

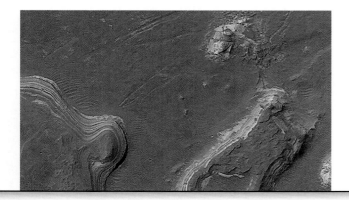

Figure 6 *The origin of the features shown in this image is unknown. The features might have resulted from deposition of sediment in a lake.*

Science Bloopers

Giovanni Schiaparelli (1835–1910), an Italian astronomer, studied Mars in the late 1800s. He thought that he saw straight lines crisscrossing the surface of the red planet. He called these lines *canali*. In Italian, *canali* means "channels," but the word was erroneously translated to English as "canals." Partly because of this misconception, for nearly 100 years, many people believed that Mars had supported at some time in its past intelligent beings that had built canals. This belief was disproved in the 1960s when a spacecraft sent to Mars did not find any canals.

Reteaching ——— BASIC

Planetary Data Table Have students create a table summarizing the data in this section. (Answers may vary. Answers could include a row for each planet and columns for distance from the sun, period of rotation, period of revolution, diameter, density, surface temperature, and surface gravity.)
LS Logical

Quiz ——— GENERAL

Have students complete the following sentences:

1. The counterclockwise spin of a planet or moon as seen from above the planet's North Pole is _____. (prograde rotation)
2. The space probes that have been sent to Mars are _____, _____, _____, and _____. (*Viking 1, Viking 2, Mars Pathfinder,* and *Mars Express Orbiter*)

Alternative Assessment ——— GENERAL

Adaptations for Space Ask students to think about the conditions to which hypothetical life-forms on Mercury, Venus, or Mars would need to adapt. Have students draw a poster that show an imaginary organism for each planet. Students should include a description of adaptations that the life-forms have developed to live on each planet. **LS** Verbal

CONNECTION TO Physics

WRITING SKILL **Boiling Point on Mars** At sea level on Earth's surface, water boils at 100°C. But if you try to boil water on top of a high mountain, you will find that the boiling point is lower than 100°C. Do some research to find out why. Then, in your own words, explain why liquid water cannot exist on Mars, based on what you learned.

Where Is the Water Now?

Mars has two polar icecaps made of both frozen water and frozen carbon dioxide. But the polar icecaps do not have enough water to create a thick atmosphere or rivers. Looking closely at the walls of some Martian craters, scientists have found that the debris around the craters looks as if it were made by the flow of mud rather than by dry soil. In this case, where might some of the "lost" Martian water have gone? Many scientists think that it is frozen beneath the Martian soil.

Martian Volcanoes

Mars has a rich volcanic history. Unlike Earth, where volcanoes exist in many places, Mars has only two large volcanic systems. The largest, the Tharsis region, stretches 8,000 km across the planet. The largest mountain in the solar system, Olympus Mons, is an extinct shield volcano similar to Mauna Kea on the island of Hawaii. Mars not only is smaller and cooler than Earth but also has a slightly different chemical makeup. This makeup may have kept the Martian crust from moving around as Earth's crust does. As a result, the volcanoes kept building up in the same spots on Mars. Images and data sent back by probes such as the *Sojourner* rover, shown in **Figure 7**, are helping to explain Mars's mysterious past.

Reading Check What characteristics of Mars may explain why Mars has only two large volcanic systems?

Figure 7 The *Sojourner rover*, part of the Mars Pathfinder mission, is shown here creeping up to a rock named Yogi to measure its composition. The solar panel on the rover's back collected the solar energy used to power the rover's motor.

Answer to Reading Check

Mars' crust is chemically different from Earth's crust, so the Martian crust does not move. As a result, volcanoes build up in the same spots on Mars.

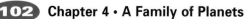

Missions to Mars

Scientists are still intrigued by the mysteries of Mars. Several recent missions to Mars were launched to gain a better understanding of the Martian world. **Figure 8** shows the *Mars Express Orbiter,* which was launched by the European Space Agency (ESA) in 2003, and was designed to help scientists determine the composition of the Martian atmosphere and Martian climate. Also, in 2003, NASA launched the Twin Rover mission to Mars. These exploration rovers are designed to gather information that may help scientists determine if life ever existed on Mars. In addition, information collected by these rovers may help scientists prepare for human exploration on Mars.

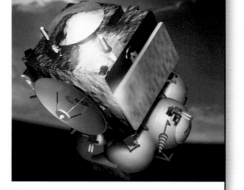

Figure 8 *The* Mars Express Orbiter *will help scientists study Mars's atmosphere.*

SECTION Review

Summary

- A period of rotation is the length of time that an object takes to rotate once on its axis.
- A period of revolution is the length of time that an object takes to revolve around the sun.
- Mercury is the planet closest to the sun. Of all the terrestrial planets, Venus has the densest atmosphere. Earth is the only planet known to support life. Mars has a rich volcanic history and shows evidence of once having had water.

Using Key Terms

1. In your own words, write a definition for the term *terrestrial planet.*

For the pair of terms below, explain how the meanings of the terms differ.

2. *prograde rotation* and *retrograde rotation*

Understanding Key Ideas

3. Scientists believe that the water on Mars now exists as
   a. polar icecaps.
   b. dry riverbeds.
   c. ice beneath the Martian soil.
   d. Both (a) and (c)

4. List three differences between and three similarities of Venus and Earth.

5. What is the difference between a planet's period of rotation and its period of revolution?

6. What are some of the characteristics of Earth that make it suitable for life?

7. Explain why the surface temperature of Venus is higher than the surface temperatures of the other planets in our solar system.

Math Skills

8. Mercury has a period of rotation equal to 58.67 Earth days. Mercury's period of revolution is equal to 88 Earth days. How many times does Mercury rotate during one revolution around the sun?

Critical Thinking

9. **Making Inferences** What type of information can we get by studying Earth from space?

10. **Analyzing Ideas** What type of evidence found on Mars suggests that Mars may have been a warmer place and had a thicker atmosphere?

SCLINKS®

NSTA

Developed and maintained by the
National Science Teachers Association

For a variety of links related to this chapter, go to www.scilinks.org

Topic: The Inner Planets
SciLinks code: HSM0798

Answers to Section Review

1. Sample answer: A terrestrial planet is a planet that has a solid, rocky surface.

2. Answers may vary. Answers should indicate that prograde rotation is counterclockwise and retrograde rotation is clockwise when viewed from above the planet's North Pole.

3. d

4. Sample answer: Unlike Earth, Venus has retrograde rotation, Venus's surface temperature is very hot, and Venus's atmosphere is very dense. Also, Venus's atmosphere contains destructive acids. Similarities between Earth and Venus include size, density, mass, and surface gravity.

5. Answers may vary. Answers should indicate that the period of rotation is the amount of time a planet takes to spin on its axis. The period of revolution is amount of time the planet takes to make one trip around the sun.

6. Sample answer: Earth is warm enough to keep most of its water from freezing and cool enough to keep its water from boiling away. Living things depend on liquid water for survival.

7. The surface temperature of Venus is high because of a severe greenhouse effect.

8. 1.5 rotations per revolution (88 days/revolution ÷ 58.67 days/rotation = 1.5 rotations/revolution)

9. Answers may vary. Sample answer: We can learn about Earth's systems and how they interact.

10. Answers may vary. Sample answer: Some features on Mars might have resulted from the deposition of sediment in a lake. For this reason, some scientists think that Mars might have been a warmer place and had a thicker atmosphere in the past. If Mars had not been warmer and had a thicker atmosphere, a lake would not have formed.

Focus

Overview

This section teaches students about the atmosphere and characteristics of the outer planets: Jupiter, Saturn, Uranus, Neptune, and Pluto.

 Bellringer

All planets with atmospheres have weather. Jupiter's Great Red Spot appears to be very similar to a hurricane system on Earth, but it has lasted for centuries, driven by the planet's internal thermal energy. Have students write and tape-record a humorous but accurate weather forecast for one of the planets with an atmosphere.

Motivate

Discussion ———— GENERAL

Almost a Star Some astronomers think of Jupiter as a small star that never reached maturity. The *Galileo* probe found that the relative amounts of hydrogen and helium in Jupiter's atmosphere are very similar to those in the sun. However, Jupiter's 30,000°C core temperature is not high enough to initiate the fusion reactions that occur in the sun's core. **LS Verbal**

READING WARM-UP

Objectives

● Explain how gas giants are different from terrestrial planets.

● Describe the individual characteristics of Jupiter, Saturn, Uranus, Neptune, and Pluto.

Terms to Learn

gas giant

READING STRATEGY

Prediction Guide Before reading this section, write the title of each heading in this section. Next, under each heading write what you think you will learn.

---

gas giant a planet that has a deep, massive atmosphere, such as Jupiter, Saturn, Uranus, or Neptune

---

Figure 1 *This* Voyager 2 *image of Jupiter was taken at a distance of 28.4 million kilometers. Io, one of Jupiter's largest moons, can also be seen in this image.*

The Outer Planets

What do all the outer planets except for Pluto have in common?

Except for Pluto, the outer planets are very large planets that are made mostly of gases. These planets are called gas giants. **Gas giants** are planets that have deep, massive atmospheres rather than hard and rocky surfaces like those of the inner planets.

Jupiter: A Giant Among Giants

Jupiter is the largest planet in our solar system. Like the sun, Jupiter is made mostly of hydrogen and helium. The outer part of Jupiter's atmosphere is made of layered clouds of water, methane, and ammonia. The beautiful colors you see in **Figure 1** are probably due to small amounts of organic compounds. At a depth of about 10,000 km into Jupiter's atmosphere, the pressure is high enough to change hydrogen gas into a liquid. Deeper still, the pressure changes the liquid hydrogen into a liquid, metallic state. Unlike most planets, Jupiter radiates much more energy into space than it receives from the sun. The reason is that Jupiter's interior is very hot. Another striking feature of Jupiter is the Great Red Spot, a storm system that is more than 400 years old and is about 1.5 times as large as the Earth!

NASA Missions to Jupiter

NASA has sent five missions to Jupiter. These include two Pioneer missions, two Voyager missions, and the recent Galileo mission. The *Voyager 1* and *Voyager 2* spacecraft sent back images that revealed a thin, faint ring around Jupiter. The Voyager missions also gave us the first detailed images of Jupiter's moons. The *Galileo* spacecraft reached Jupiter in 1995 and sent a probe into Jupiter's atmosphere. The probe sent back data on Jupiter's composition, temperature, and pressure.

Jupiter Statistics	
Distance from sun	43.3 light-minutes
Period of rotation	9 h, 54 min
Period of revolution	11 years, 313 days
Diameter	142,984 km
Density	1.33 g/cm³
Temperature	−110°C
Gravity	236% of Earth's

CHAPTER RESOURCES

Chapter Resource File

- Lesson Plan
- Directed Reading A **BASIC**
- Directed Reading B **SPECIAL NEEDS**

Technology

- Transparencies
  - Bellringer

CONNECTION to Meteorology ———— GENERAL

Wind Advisory Saturn has the most violent winds of any planet in our solar system. At Saturn's equator, the wind blows at nearly 1,700 km/h (approximately 1,000 m/h)—not exactly good weather for playing outside.

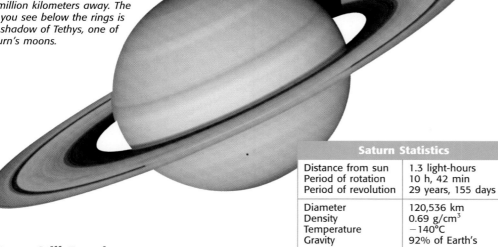

Figure 2 *This Voyager 2 image of Saturn was taken from 21 million kilometers away. The dot you see below the rings is the shadow of Tethys, one of Saturn's moons.*

Saturn Statistics	
Distance from sun	1.3 light-hours
Period of rotation	10 h, 42 min
Period of revolution	29 years, 155 days
Diameter	120,536 km
Density	0.69 g/cm³
Temperature	−140°C
Gravity	92% of Earth's

Saturn: Still Forming

Saturn, shown in **Figure 2,** is the second-largest planet in the solar system. Saturn has roughly 764 times the volume of Earth and is 95 times more massive than Earth. Its overall composition, like Jupiter's, is mostly hydrogen and helium. But methane, ammonia, and ethane are found in the upper atmosphere. Saturn's interior is probably much like Jupiter's. Also, like Jupiter, Saturn gives off much more energy than it receives from the sun. Scientists think that Saturn's extra energy comes from helium falling out of the atmosphere and sinking to the core. In other words, Saturn is still forming!

The Rings of Saturn

Although all of the gas giants have rings, Saturn's rings are the largest. Saturn's rings have a total diameter of 272,000 km. Yet, Saturn's rings are only a few hundred meters thick. The rings are made of icy particles that range in size from a few centimeters to several meters wide. **Figure 3** shows a close-up view of Saturn's rings.

 **Reading Check** What are Saturn's rings made of? (*See the Appendix for answers to Reading Checks.*)

NASA's Exploration of Saturn

Launched in 1997, the *Cassini* spacecraft is designed to study Saturn's rings, moons, and atmosphere. The spacecraft is also designed to return more than 300,000 color images of Saturn.

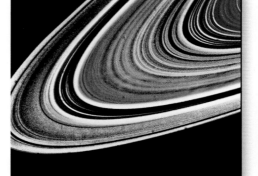

Figure 3 *The different colors in this Voyager 2 image of Saturn's rings show differences in the rings' chemical composition.*

Answer to Reading Check

Saturn's rings are made of icy particles ranging in size from a few centimeters to several meters wide.

 **WEIRD SCIENCE**

Although Saturn's composition is similar to Jupiter's, Saturn appears less colorful. The reason is that Saturn's colder atmosphere causes thick, white ammonia clouds to condense and block our view.

Prediction Guide Before students read this page, ask them whether each of the following statements is true or false.

- Uranus was discovered in the 18th century. (true)
- Uranus's axis of rotation is almost parallel to the planet's orbit around the sun. (true)

LS Verbal

Using the Figure— GENERAL

Axial Tilt Using **Figure 5,** point out that because Uranus has an axial tilt of 82°, its poles point toward the sun during part of its year. (For simplicity, the figure shows the angle between the pole and the plane of orbit.) In contrast, Earth's 23° tilt means that like the poles of most planets, Earth's poles never point directly toward the sun. Students can simulate these axial tilts using a globe and an object to represent the sun. As students revolve around the sun, have them tilt the globe to represent the axial tilts of Venus (3°), Earth, and Uranus. Point out the times that Uranus's poles point toward the sun.

LS Kinesthetic Co-op Learning

Figure 4 *This image of Uranus was taken by* Voyager 2 *at a distance of 9.1 million kilometers.*

Uranus Statistics	
Distance from sun	2.7 light-hours
Period of rotation	17 h, 12 min (R)*
Period of revolution	83 years, 273 days
Diameter	51,118 km
Density	1.27 g/cm³
Temperature	−195°C
Gravity	89% of Earth's

*R = retrograde rotation

Uranus: A Small Giant

Uranus (YOOR uh nuhs) was discovered by the English amateur astronomer William Herschel in 1781. The atmosphere of Uranus is mainly hydrogen and methane. Because these gases absorb the red part of sunlight very strongly, Uranus appears blue-green in color, as shown in **Figure 4.** Uranus and Neptune have much less mass than Jupiter, but their densities are similar. This suggests that their compositions are different from Jupiter's. They may have lower percentages of light elements and a greater percentage of water.

A Tilted Planet

Unlike most other planets, Uranus is tipped over on its side. So, its axis of rotation is tilted by almost 90° and lies almost in the plane of its orbit, as shown in **Figure 5.** For part of a Uranus year, one pole points toward the sun while the other pole is in darkness. At the other end of Uranus's orbit, the poles are reversed. Some scientists think that early in its history, Uranus may have been hit by a massive object that tipped the planet over.

Figure 5 *Uranus's axis of rotation is tilted so that the axis is nearly parallel to the plane of Uranus's orbit. In contrast, the axes of most other planets are closer to being perpendicular to the plane of the planets' orbits.*

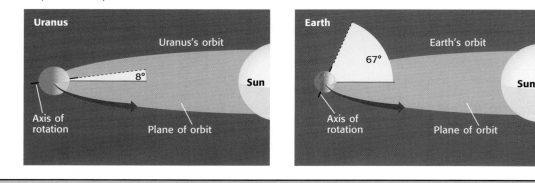

Science Bloopers

Uranus was discovered in 1781 by an English music teacher named William Herschel. Herschel originally named the planet *Georgium Sidus,* Latin for "George's Star," after England's King George III. No one outside England liked the name. A few years later, an astronomer named J. E. Bode suggested the name *Uranus* to continue the tradition of naming the planets after Greek or Roman gods.

Is That a Fact!

Measuring Tilt The *obliquity* values for planets, which also measure the tilt of a planet's axis, are greater than 90° for planets with retrograde rotation. These planets are thought to have tipped over after colliding with other massive bodies shortly after they formed. In essence, their "rotational north poles" now point "south" in space. For this reason, the obliquity values for Venus, Uranus, and Pluto are 177°, 98°, and 123°, respectively.

Neptune: The Blue World

Irregularities in the orbit of Uranus suggested to early astronomers that there must be another planet beyond it. They thought that the gravity of this new planet pulled Uranus off its predicted path. By using the predictions of the new planet's orbit, astronomers discovered the planet Neptune in 1846. Neptune is shown in **Figure 6.**

The Atmosphere of Neptune

The *Voyager 2* spacecraft sent back images that provided much new information about Neptune's atmosphere. Although the composition of Neptune's atmosphere is similar to that of Uranus's atmosphere, Neptune's atmosphere has belts of clouds that are much more visible. At the time of *Voyager 2*'s visit, Neptune had a Great Dark Spot like the Great Red Spot on Jupiter. And like the interiors of Jupiter and Saturn, Neptune's interior releases thermal energy to its outer layers. This release of energy helps the warm gases rise and the cool gases sink, which sets up the wind patterns in the atmosphere that create the belts of clouds. *Voyager 2* images also revealed that Neptune has a set of very narrow rings.

Reading Check What characteristic of Neptune's interior accounts for the belts of clouds in Neptune's atmosphere?

Figure 6 *This* Voyager 2 *image of Neptune, taken at a distance of more than 7 million kilometers, shows the Great Dark Spot as well as some bright cloud bands.*

Neptune Statistics	
Distance from sun	4.2 light-hours
Period of rotation	16 h, 6 min
Period of revolution	163 years, 263 days
Diameter	49,528 km
Density	1.64 g/cm^3
Temperature	−200°C
Gravity	112% of Earth's

CONNECTION to Environmental Science ——— GENERAL

Global Warming Since the *Voyager* spacecraft passed Neptune's moon Triton in 1989, scientists have noticed an interesting trend in Triton's atmosphere. Images from the *Hubble Space Telescope* taken in 1998 indicate that Triton is going through a rapid period of global warming. As Triton warms, frozen nitrogen on its surface melts and contributes nitrogen gas to its thin atmosphere. This process has happened so rapidly that the atmospheric pressure of Triton has doubled in less than 10 years! Scientists hope to use the global-warming trends on Triton to understand warming patterns on Earth. Ask students to discuss what they would expect to happen to Triton's atmosphere in the next 10 years. (Answers may vary. Sample answer: Triton's surface nitrogen will melt and contribute to the atmosphere, causing it to become even thicker and trap more heat. The atmospheric pressure will double again.) **LS Verbal**

Reteaching ─── **BASIC**

Planetary Quiz Have students write at least two characteristics of each planet on the back of separate index cards. Then, have them exchange index cards with a partner and try to guess the identity of each planet. **LS Verbal**

Quiz ────────── **GENERAL**

1. Which of the outer planets have retrograde rotation? (Uranus and Pluto)

2. Which of the outer planets has the shortest period of rotation? (Jupiter)

Alternative Assessment ─── **GENERAL**

Planetary Postcards
Ask students to write a postcard as if they were a tourist on an outer planet of their choice. Students should describe the weather, the view, the gravity, and other features, but they should not name the planet that they are visiting. Have students "mail" the postcards to a partner and then ask the partner to try to determine which planet the tourist is visiting. If the partner guesses incorrectly, the tourist should send another postcard describing more details about the planet. If the partner guesses correctly, the tourist can move to another planet. Have partners continue this game until they have both visited all five of the outer planets. **LS Kinesthetic/Visual**

CONNECTION ACTIVITY
Fine Arts ─── **GENERAL**

Alien Horizons Have students choose a planet and imagine what the sky would look like from their planet. Have students make a poster showing the view from their planet. Students could include moons, rings, or the sun in their illustration. Display the artwork in the class. **LS Visual**

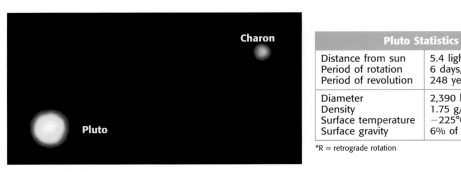

Pluto Statistics	
Distance from sun	5.4 light-hours
Period of rotation	6 days, 10 h (R)*
Period of revolution	248 years, 4 days
Diameter	2,390 km
Density	1.75 g/cm³
Surface temperature	−225°C
Surface gravity	6% of Earth's

*R = retrograde rotation

Figure 7 *This* Hubble Space Telescope *image is one of the clearest ever taken of Pluto* (left) *and its moon, Charon* (right).

Pluto: The Mystery Planet

Further study of Neptune showed some irregularities in Neptune's orbit. This finding led many scientists to believe there was yet another planet beyond Neptune. The mystery planet was finally discovered in 1930.

A Small World

The mystery planet, now called Pluto, is the farthest planet from the sun. Less than half the size of Mercury, Pluto is also the smallest planet. Pluto's moon, Charon (KER uhn), is more than half its size! In fact, Charon is the largest satellite relative to its planet in the solar system. **Figure 7** shows Pluto and Charon together. From Earth, it is hard to separate the images of Pluto and Charon because the bodies are so far away. **Figure 8** shows how far from the sun Pluto and Charon really are. From Pluto, the sun looks like a very distant bright star.

From calculations of Pluto's density, scientists know that Pluto must be made of rock and ice. Pluto is covered by frozen nitrogen, but Charon is covered by frozen water. Scientists believe Pluto has a thin atmosphere of methane.

Figure 8 *An artist's view of the sun and Charon from Pluto shows just how little light and heat Pluto receives from the sun.*

A True Planet?

Because Pluto is so small and is so unusual, some scientists think that it should not be classified as a planet. In fact, some scientists agree that Pluto could be considered a large asteroid or comet—large enough to have its own satellite. However, because Pluto was historically classified as a planet, it most likely will remain so.

Pluto is the only planet that has not been visited by a NASA mission. However, plans are underway to visit Pluto and Charon in 2006. During this mission, scientists hope to learn more about this unusual planet and map the surface of both Pluto and Charon.

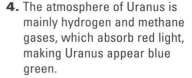

SCHOOL to HOME

Surviving Space

WRITING SKILL Imagine it is the year 2150 and you are flying a spacecraft to Pluto. Suddenly, your systems fail, giving you only one chance to land safely. You can't head back to Earth. With a parent, write a paragraph explaining which planet you would choose to land on.

ACTIVITY

SECTION Review

Summary

- Jupiter is the largest planet in our solar system. Energy from the interior of Jupiter is transferred to its exterior.
- Saturn is the second-largest planet and, in some ways, is still forming as a planet.
- Uranus's axis of rotation is tilted by almost 90°.
- Neptune has a faint ring, and its atmosphere contains belts of clouds.
- Pluto is the smallest planet, and its moon, Charon, is more than half its size.

Using Key Terms

1. In your own words, write a definition for the term *gas giant*.

Understanding Key Ideas

2. The many colors of Jupiter's atmosphere are probably caused by _____ in the atmosphere.
   a. clouds of water
   b. methane
   c. ammonia
   d. organic compounds

3. Why do scientists claim that Saturn, in a way, is still forming?

4. Why does Uranus have a blue green color?

5. What is unusual about Pluto's moon, Charon?

6. What is the Great Red Spot?

7. Explain why Jupiter radiates more energy into space than it receives from the sun.

8. How do the gas giants differ from the terrestrial planets?

9. What is so unusual about Uranus's axis of rotation?

Math Skills

10. Pluto is 5.5 light-hours from the sun. How far is Pluto from the sun in astronomical units? (Hint: 1 AU = 8.3 light-minutes)

11. If Jupiter is 43.3 light-minutes from the sun and Neptune is 4.2 light-hours from the sun, how far from Jupiter is Neptune?

Critical Thinking

12. **Evaluating Data** What conclusions can your draw about the properties of a planet just by knowing how far it is from the sun?

13. **Applying Concepts** Why isn't the word *surface* included in the statistics for the gas giants?

SCILINKS.

Developed and maintained by the National Science Teachers Association

For a variety of links related to this chapter, go to www.scilinks.org

Topic: The Outer Planets
SciLinks code: HSM1091

---

4. The atmosphere of Uranus is mainly hydrogen and methane gases, which absorb red light, making Uranus appear blue green.

5. Charon is the largest satellite relative to its planet in the solar system.

6. Sample answer: The Great Red Spot is a storm system on Jupiter that is more than 400 years old and is about 1.5 times as large as the Earth.

7. Jupiter radiates energy because Jupiter's interior is very hot.

8. Sample answer: Gas giants are much larger and more massive than terrestrial planets, and gas giants have deep, massive atmospheres rather than hard, rocky surfaces.

9. Sample answer: Uranus is tipped over on its side. Uranus's axis of rotation is tilted so that it lies almost in plane with its orbit. Each pole points toward the sun for part of Uranus's year.

10. about 40 AU (5.5 light-hours × 60 min/h = 330 light-minutes; 330 light-minutes ÷ 8.3 light-minutes/AU = 39.8 AU)

11. 208.7 light-minutes (4.2 light-hours × 60 min/h = 252 light-minutes; 252 light-minutes − 43.3 light-minutes = 208.7 light-minutes)

12. Answers may vary. Planets farther from the sun tend to have lower surface temperatures, they are spaced farther apart, their period of revolution is much longer than that of the inner planets, and they are more likely to be gas giants. (This pattern does not necessarily apply to other solar systems.)

13. Sample answer: Gas giants have no definite surface. Their atmosphere blends smoothly into the dense layers of their interior.

Answers to Section Review

1. Sample answer: A gas giant is a planet that has a deep, massive atmosphere.

2. d

3. Sample answer: Helium is currently falling out of Saturn's atmosphere and sinking to its core. So, scientists think that Saturn is still forming.

CHAPTER RESOURCES

Chapter Resource File
- Section Quiz GENERAL
- Section Review GENERAL
- Vocabulary and Section Summary GENERAL

Technology
- Interactive Explorations CD-ROM
- Space Case GENERAL

Workbooks
- Science Skills
- Grasping Graphing GENERAL

Focus

Overview

In this section, students learn about Earth's moon and the major moons of the other planets in the solar system.

🔔 Bellringer

The first astronauts to land on the moon were quarantined after their mission. NASA wanted to make sure that the astronauts didn't bring back any disease-causing organisms from the moon. Ask students to discuss whether or not they think this would be possible.

Motivate

ACTIVITY ——— GENERAL

Lunar Ice The 1998 Lunar Prospector mission showed that there could be water on our moon. The craters near the poles may contain as much as 300 million metric tons of ice! The ice can exist only in the permanently shadowed regions of the moon's poles; elsewhere, daytime temperatures can reach 134°C. Have students illustrate a lunar base, including a description of how lunar pioneers could obtain liquid water. **LS Visual**

READING WARM-UP

Objectives

● Describe the current theory of the origin of Earth's moon.
● Explain what causes the phases of Earth's moon.
● Describe the difference between a solar eclipse and a lunar eclipse.
● Describe the individual characteristics of the moons of other planets.

Terms to Learn

satellite
phase
eclipse

READING STRATEGY

Reading Organizer As you read this section, make a table comparing solar eclipses and lunar eclipses.

Moons

If you could, which moon would you visit? With volcanoes, craters, and possible underground oceans, the moons in our solar system would be interesting places to visit.

Natural or artificial bodies that revolve around larger bodies such as planets are called **satellites.** Except for Mercury and Venus, all of the planets have natural satellites called *moons*.

Luna: The Moon of Earth

Scientists have learned a lot from studying Earth's moon, which is also called *Luna*. The lunar rocks brought back during the Apollo missions were found to be about 4.6 billion years old. Because these rocks have hardly changed since they formed, scientists know the solar system itself is about 4.6 billion years old.

The Surface of the Moon

As you can see in **Figure 1,** the moon's history is written on its face. The surfaces of bodies that have no atmospheres preserve a record of almost all of the impacts that the bodies have had. Because scientists now know the age of the moon, they can count the number of impact craters to find the rate of cratering since the birth of our solar system. By knowing the rate of cratering, scientists are able to use the number of craters on any body to estimate how old the body's surface is. That way, scientists don't need to bring back rock samples.

Figure 1 *This image of the moon was taken by the Galileo spacecraft while on its way to Jupiter. The large, dark areas are lava plains called* maria.

Moon Statistics	
Period of rotation	27 days, 9 hours
Period of revolution	27 days, 7 hours
Diameter	3,475 km
Density	3.34 g/cm³
Surface temperature	−170 to 134°C
Surface gravity	16% of Earth's

CHAPTER RESOURCES

Chapter Resource File

📁 • Lesson Plan
• Directed Reading A **BASIC**
• Directed Reading B **SPECIAL NEEDS**

Technology

💠 **Transparencies**
• Bellringer
• **LINK TO PHYSICAL SCIENCE** How an Orbit Is Formed; Projectile Motion
• Formation of the Moon

MISCONCEPTION ///ALERT\\\

An Optical Illusion Students may think the moon is larger when it is close to the horizon. The moon appears larger because the observer's reference point is the skyline. The same phenomenon makes the sun appear larger. Challenge students to devise a way to verify this for themselves. Caution them not to look directly at the sun.

Lunar Origins

Before scientists had rock samples from the moon, there were three popular explanations for the moon's formation: (1) The moon was a separate body captured by Earth's gravity, (2) the moon formed at the same time and from the same materials as the Earth, and (3) the newly formed Earth was spinning so fast that a piece flew off and became the moon.

When rock samples of the moon were brought back from the Apollo mission, the mystery was solved. Scientists found that the composition of the moon was similar to that of Earth's mantle. This evidence from the lunar rock samples supported the third explanation for the moon's formation.

The current theory is that a large, Mars-sized object collided with Earth while the Earth was still forming, as shown in **Figure 2.** The collision was so violent that part of the Earth's mantle was blasted into orbit around Earth to form the moon.

Reading Check What is the current explanation for the formation of the moon? (*See the Appendix for answers to Reading Checks.*)

satellite a natural or artificial body that revolves around a planet

Figure 2 Formation of the Moon

❶ Impact
About 4.6 billion years ago, when Earth was still mostly molten, a large body collided with Earth. Scientists reason that the object must have been large enough to blast part of Earth's mantle into space, because the composition of the moon is similar to that of Earth's mantle.

❷ Ejection
The resulting debris began to revolve around the Earth within a few hours of the impact. This debris consisted of mantle material from Earth and from the impacting body as well as part of the iron core of the impacting body.

❸ Formation
Soon after the giant impact, the clumps of material ejected into orbit around Earth began to join together to form the moon. Much later, as the moon cooled, additional impacts created deep basins and fractured the moon's surface. Lunar lava flowed from those cracks and flooded the basins to form the lunar maria that we see today.

Answer to Reading Check
The moon formed from a piece of Earth's mantle, which broke off during a collision between Earth and a large object.

Teach

READING STRATEGY — GENERAL

Prediction Guide Before students read this page, ask the following questions: "Where did the moon come from?" and "What evidence is there to support this theory?" As students read the rest of this section, have them construct a chart that describes some of the major moons in our solar system. **LS Logical**

CONNECTION to Physics — ADVANCED

Orbital Motion Every second, the moon travels 1 km in its orbit, but during that second, it also falls about 14 mm toward the Earth. Because of the moon's velocity and the pull of gravity, the moon travels along a path that follows the curved surface of the Earth. This condition, known as *free fall,* keeps the moon in orbit around the Earth. Explain to students that the condition of free fall, or weightlessness, does not mean that there is no gravity. The Earth's gravity acts on the moon in the same way gravity acts on an apple that falls from a tree. The difference is that the moon, unlike the apple, is moving forward much, much faster than it is falling. Use the transparency entitled "Projectile Motion" to help students understand the moon's orbit. **LS Logical**

Modeling the Earth and Moon

Pick three volunteers. One will be the moon, another will be Earth, and the third will be the sun. Have the sun stand 5 m from Earth and hold a bright flashlight. Instruct the moon to stand 1 m away from Earth. Tell the moon to slowly orbit Earth, keeping his or her face turned toward Earth. Have the sun turn on the flashlight and point the light toward Earth and the moon. Darken the room. Ask students:

- How much of the moon is lit by the flashlight? (half)

- When the moon is between Earth and the sun, what phase is the moon in? (new)

As the moon moves around Earth, ask the students which phase is being demonstrated by the moon's motion and the pattern of light on the moon.

English Language Learners

LS Visual/ Kinesthetic

CONNECTION ACTIVITY

Language Arts ——— BASIC

Which Is Waxing? Students may have a difficult time remembering how to tell whether the moon is *waxing* or *waning*. To help students remember, have them develop a mnemonic device such as the following: "Light on the left is leaving, light on the right is returning." (Note: This statement is true only when you are studying a diagram of the moon's orbit.) **LS** Verbal

Phases of the Moon

From Earth, one of the most noticeable aspects of the moon is its continually changing appearance. Within a month, the moon's Earthward face changes from a fully lit circle to a thin crescent and then back to a circle. These different appearances of the moon result from its changing position relative to Earth and the sun. As the moon revolves around Earth, the amount of sunlight on the side of the moon that faces Earth changes. The different appearances of the moon due to its changing position are called **phases.** The phases of the moon are shown in **Figure 3.**

phase the change in the sunlit area of one celestial body as seen from another celestial body

Waxing and Waning

When the moon is *waxing*, the sunlit fraction that we can see from Earth is getting larger. When the moon is *waning*, the sunlit fraction is getting smaller. Notice in **Figure 3** that even as the phases of the moon change, the total amount of sunlight that the moon gets remains the same. Half the moon is always in sunlight, just as half the Earth is always in sunlight. But because the moon's period of rotation is the same as its period of revolution, on Earth you always see the same side of the moon. If you lived on the far side of the moon, you would see the sun for half of each lunar day, but you would never see the Earth!

Figure 3 *The positions of the moon, sun, and Earth determine which phase the moon is in. The photo insets show how the moon looks from Earth at each phase.*

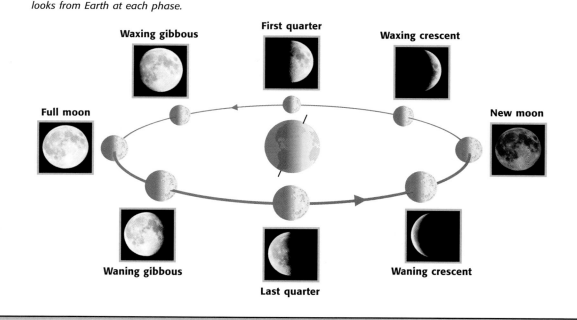

Waxing gibbous

First quarter

Waxing crescent

Full moon

New moon

Waning gibbous

Last quarter

Waning crescent

Technology

🗄 **Transparencies**
• Phases of the Moon

Is That a Fact!

Waxing means "growing," and *waning* means "shrinking." In the first quarter of a lunar phase, the moon is one-quarter of the way through its cycle of phases. At this point, sunlight is shining on the right half of the moon. During the last quarter, sunlight is shining on the left half of the moon.

Solar eclipse

NEVER look directly at the sun! You can permanently damage your eyes.

Figure 4 *On the left is a diagram of the positions of the Earth and the moon during a solar eclipse. On the right is a picture of the sun's outer atmosphere, or corona, which is visible only when the entire disk of the sun is blocked by the moon.*

Eclipses

When the shadow of one celestial body falls on another, an **eclipse** occurs. A *solar eclipse* happens when the moon comes between Earth and the sun and the shadow of the moon falls on part of Earth. A *lunar eclipse* happens when Earth comes between the sun and the moon and the shadow of Earth falls on the moon.

Solar Eclipses

Because the moon's orbit is elliptical, the distance between the moon and the Earth changes. During an *annular eclipse*, the moon is farther from the Earth. The disk of the moon does not completely cover the disk of the sun. A thin ring of the sun shows around the moon's outer edge. When the moon is closer to the Earth, the moon appears to be the same size as the sun. During a *total solar eclipse*, the disk of the moon completely covers the disk of the sun, as shown in **Figure 4**.

eclipse an event in which the shadow of one celestial body falls on another

✓ Reading Check Describe what happens during a solar eclipse.

Clever Insight

1. Cut out a circle of **heavy, white paper**. This circle will represent Earth.
2. Find **two spherical objects** and **several other objects** of different shapes.
3. Hold up each object in front of a **lamp** (which represents the sun) so that the object's shadow falls on the white paper circle.
4. Rotate your objects in all directions, and record the shapes of the shadows that the objects make.
5. Which objects always cast a curved shadow?

Answer to Reading Check

During a solar eclipse, the moon blocks out the sun and casts a shadow on Earth.

BRAIN FOOD

Why Are Moons Round?
Like planets, most moons are spherical in shape. Why aren't some moons square, tube shaped, or pyramidal? The force of gravity and the origin of celestial bodies have something to do with the answer. As the mass of an object increases, the gravitational force that it exerts also increases. When a rocky object reaches a diameter of about 350 km, the gravitational force becomes greater than the strength of the material, and the moon starts to become spherical.

Homework ——— GENERAL

NASA in the News NASA is planning several missions to explore the moons of different planets. Have students select one of NASA's projects and find out what was discovered or what NASA hopes to find. Have students imagine that they are reporters assigned to the project. Have them write a newspaper article with the details of the project. **LS Verbal**

Lunar eclipse

Figure 5 *On the left, you can see that the moon can have a reddish color during a lunar eclipse. On the right, you can see the positions of Earth and the moon during a lunar eclipse.*

Lunar Eclipses

As shown in **Figure 5,** the view during a lunar eclipse is spectacular. Earth's atmosphere acts like a lens and bends some of the sunlight into the Earth's shadow. When sunlight hits the particles in the atmosphere, blue light is filtered out. As a result, most of the remaining light that lights the moon is red.

The Tilted Orbit of the Moon

You may be wondering why you don't see solar and lunar eclipses every month. The reason is that the moon's orbit around Earth is tilted—by about 5°—relative to the orbit of Earth around the sun. This tilt is enough to place the moon out of Earth's shadow for most full moons and Earth out of the moon's shadow for most new moons.

✓ Reading Check Explain why you don't see solar and lunar eclipses every month.

The Moons of Other Planets

The moons of the other planets range in size from very small to as large as terrestrial planets. All of the gas giants have multiple moons, and scientists are still discovering new moons. Some moons have very elongated, or elliptical, orbits, and some moons even orbit their planet backward! Many of the very small moons may be captured asteroids. As scientists are learning from recent space missions, moons may be some of the most bizarre and interesting places in the solar system!

CHAPTER RESOURCES

Technology

📦 **Transparencies**
• Solar Eclipse; Lunar Eclipse

Answer to Reading Check

We don't see solar and lunar eclipses every month because the moon's orbit around Earth is tilted.

The Moons of Mars

Mars's two moons, Phobos and Deimos, are small, oddly shaped satellites. Both moons are very dark. Their surface materials are much like those of some asteroids—large, rocky bodies in space. Scientists think that these two moons are asteroids caught by Mars's gravity.

The Moons of Jupiter

Jupiter has dozens of moons. The four largest moons—Ganymede, Callisto, Io, and Europa—were discovered in 1610 by Galileo. They are known as the *Galilean satellites*. The largest moon, Ganymede, is even larger than the planet Mercury! Many of the smaller moons probably are captured asteroids.

The Galilean satellite closest to Jupiter is Io, a truly bizarre world. Io is caught in a gravitational tug of war between Jupiter and Io's nearest neighbor, the moon Europa. This constant tugging stretches Io a little and causes it to heat up. As a result, Io is the most volcanically active body in the solar system!

Recent pictures of the moon Europa, shown in **Figure 6,** support the idea that liquid water may lie beneath the moon's icy surface. This idea makes many scientists wonder if life could have evolved in the underground oceans of Europa.

The Moons of Saturn

Like Jupiter, Saturn has dozens of moons. Most of these moons are small bodies that are made mostly of frozen water but contain some rocky material. The largest satellite, Titan, was discovered in 1655 by Christiaan Huygens. In 1980, the *Voyager 1* spacecraft flew past Titan and discovered a hazy orange atmosphere, as shown in **Figure 7.** Earth's early atmosphere may have been much like Titan's is now. In 1997, NASA launched the *Cassini* spacecraft to study Saturn and its moons, including Titan. By studying Titan, scientists hope to learn more about how life began on Earth.

✔ **Reading Check** How can scientists learn more about how life began on Earth by studying Titan?

Figure 6 *Europa, Jupiter's fourth largest moon, might have liquid water beneath the moon's icy surface.*

Figure 7 *Titan is Saturn's largest moon.*

Answer to Reading Check

Because Titan's atmosphere is similar to the atmosphere on Earth before life evolved, scientists can study Titan's atmosphere to learn how life began.

CONNECTION to Life Science —— ADVANCED

Is There Life Out There? Life as we know it requires liquid water. If liquid water exists on other planets or their moons, it is possible that there are also living organisms there. Scientists have been studying a group of organisms called *extremophiles* for clues about what extraterrestrial life might be like. Extremophiles are organisms that live in extreme environments, such as deep-ocean volcanic vents, hot springs, or highly acidic or basic environments. Scientists hope that studying these organisms will help in their search for life elsewhere in the solar system. Some of the most likely places to search for evidence of life are Mars and some of the moons of Jupiter. Have interested students find out about the status of NASA projects that are searching for extraterrestrial life and about organisms classified as extremophiles. **LS Verbal**

Is That a Fact!

Io, one of Jupiter's moons, is well known for the volcanoes on its surface. These volcanoes, which are the hottest in the solar system, regularly erupt yellow and red clouds of sulfur up to 300 km above the surface!

Modeling Eclipses Have students describe the location of the moon in relation to Earth and the sun during solar and lunar eclipses. Have them demonstrate the positions using paper cutouts to represent the sun, the moon, and Earth.
LS Verbal/Visual

Quiz ———————— GENERAL

1. What characteristic of Earth's moon supports the current theory of its formation? (The moon has a composition similar to that of Earth's mantle.)

2. Why doesn't a solar eclipse occur every month? (because the moon's orbit around Earth is titled, which places Earth out of the moon's shadow for most new moons)

3. What is the difference between an annular eclipse and a total solar eclipse? (During an annular eclipse, the moon is farther from Earth and doesn't completely cover the disk of the sun. During a total solar eclipse, the moon completely covers the sun.)

Alternative Assessment ———— GENERAL

Writing **Moon Mission** Have students research the moons of a planet other than Earth and write a short paper about them. **LS** Verbal

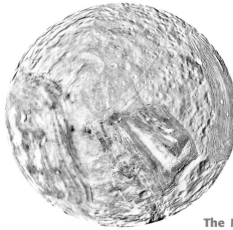

Figure 8 *This* Voyager 2 *image shows Miranda, the most unusual moon of Uranus. Its patchwork terrain indicates that it has had a violent history.*

The Moons of Uranus

Uranus has several moons. Like the moons of Saturn, Uranus's largest moons are made of ice and rock and are heavily cratered. The small moon Miranda, shown in **Figure 8,** has some of the strangest features in the solar system. Miranda's surface has smooth, cratered plains as well as regions that have grooves and cliffs. Scientists think that Miranda may have been hit and broken apart in the past. Gravity pulled the pieces together again, leaving a patchwork surface.

The Moons of Neptune

Neptune has several known moons, only one of which is large. This large moon, Triton, is shown in **Figure 9.** It revolves around the planet in a *retrograde,* or "backward," orbit. This orbit suggests that Triton may have been captured by Neptune's gravity. Triton has a very thin atmosphere made mostly of nitrogen gas. Triton's surface is mostly frozen nitrogen and methane. *Voyager 2* images reveal that Triton is geologically active. "Ice volcanoes," or geysers, eject nitrogen gas high into the atmosphere. The other moons of Neptune are small, rocky worlds much like the smaller moons of Saturn and Jupiter.

The Moon of Pluto

Pluto's only known moon, Charon, was discovered in 1978. Charon's period of revolution is the same as Pluto's period of rotation—about 6.4 days. So, one side of Pluto always faces Charon. In other words, if you stood on the surface of Pluto, Charon would always occupy the same place in the sky. Charon's orbit around Pluto is tilted relative to Pluto's orbit around the sun. As a result, Pluto, as seen from Earth, is sometimes eclipsed by Charon. But don't hold your breath; this eclipse happens only once every 120 years!

✓ **Reading Check** How often is Pluto eclipsed by Charon?

Figure 9 *This* Voyager 2 *image shows Neptune's largest moon, Triton. The polar icecap currently facing the sun may have a slowly evaporating layer of nitrogen ice, adding to Triton's thin atmosphere.*

Answer to Reading Check
Pluto is eclipsed by Charon every 120 years.

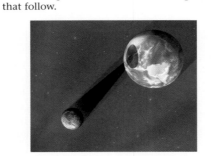

SECTION Review

Summary

- Scientists reason that the moon formed from the debris that was created after a large body collided with Earth.
- As the moon revolves around Earth, the amount of sunlight on the side of the moon changes. Because the amount of sunlight on the side of the moon changes, the moon's appearance from Earth changes. These changes in appearance are the phases of the moon.
- A solar eclipse happens when the shadow of the moon falls on Earth.

- A lunar eclipse happens when the shadow of Earth falls on the moon.
- Mars has 2 moons: Phobos and Deimos.
- Jupiter has dozens of moons. Ganymede, Io, Callisto, and Europa are the largest.
- Saturn has dozens of moons. Titan is the largest.
- Uranus has several moons.
- Neptune has several moons. Triton is the largest.
- Pluto has 1 known moon, Charon.

Using Key Terms

Complete each of the following sentences by choosing the correct term from the word bank.

    satellite   eclipse

1. A(n) _____, or a body that revolves around a larger body, can be either artificial or natural.

2. A(n) _____ occurs when the shadow of one body in space falls on another body.

Understanding Key Ideas

3. Which of the following is a Galilean satellite?
   a. Phobos
   b. Deimos
   c. Ganymede
   d. Charon

4. Describe the current theory for the origin of Earth's moon.

5. What is the difference between a solar eclipse and a lunar eclipse?

6. What causes the phases of Earth's moon?

Critical Thinking

7. **Analyzing Methods** How can astronomers use the age of a lunar rock to estimate the age of the surface of a planet such as Mercury?

8. **Identifying Relationships** Charon stays in the same place in Pluto's sky, but the moon moves across Earth's sky. What causes this difference?

Interpreting Graphics

Use the diagram below to answer the questions that follow.

9. What type of eclipse is shown in the diagram?

10. Describe what is happening in the diagram.

11. Make a sketch of the type of eclipse that is not shown in the diagram.

Developed and maintained by the National Science Teachers Association

For a variety of links related to this chapter, go to www.scilinks.org

Topic: The Moons of Other Planets
SciLinks code: HSM5495

CHAPTER RESOURCES

Chapter Resource File

- Section Quiz **GENERAL**
- Section Review **GENERAL**
- Vocabulary and Section Summary **GENERAL**
- Reinforcement Worksheet **BASIC**
- Datasheet for Quick Lab

Answers to Section Review

1. satellite
2. eclipse
3. c
4. The current theory for the origin of the moon is that a large, Mars-sized body collided with Earth during Earth's formation. The collision ejected part of Earth's mantle into orbit around Earth. This material became the moon.
5. Sample answer: During a lunar eclipse, Earth comes between the sun and the moon, and Earth's shadow falls on the moon. During a solar eclipse, the moon comes between Earth and the sun, and the moon's shadow falls on Earth.
6. Lunar phases result from the moon's changing position relative to Earth and the sun. As the moon orbits Earth, the amount of sunlight on the side of the moon that faces Earth changes.
7. Answers may vary. Sample answer: Dating the lunar rock will determine the age of the moon. Then, scientists can compare the amount of cratering on the moon's surface with the amount of cratering on Mercury's surface to estimate the age of Mercury.
8. Sample answer: Charon's period of revolution is about the same as Pluto's period of rotation. So, Charon remains in a fixed position in Pluto's sky. The moon's period of revolution is not the same as Earth's period of rotation, so the moon moves across Earth's sky.
9. a solar eclipse
10. The moon is between Earth and the sun. The moon is casting a shadow on Earth. As viewed from Earth, the sun would be obscured by the moon.
11. Sketches should illustrate a lunar eclipse.

SECTION

5

Focus

Overview

This section explores the minor bodies of the solar system, including comets, asteroids, and meteoroids.

🔊 Bellringer

Ask students if scientists have ever brought extraterrestrial material to Earth. (The only samples came from the moon missions.) Then, point out that scientists have studied rocks from Mars and other parts of the solar system. Ask students how scientists obtained these rocks. (The rocks are meteorites.)

Motivate

Demonstration — GENERAL

Modeling Comets To simulate a comet and its tail, mix 2 cups of water, 2 tbsp of dirt, and a few pebbles in a container. While wearing protective gloves, crush 2 cups of dry ice in a plastic bag. Slowly pour the liquid mixture into the bag, mixing constantly. Mold this mixture to produce a model comet. Spread plastic over your work area, and place the comet on top of an inverted foam cup. Use a hair dryer to simulate the solar wind that produces a comet's tail when a comet approaches the sun.
LS Visual/Kinesthetic

READING WARM-UP

Objectives

● Explain why comets, asteroids, and meteoroids are important to the study of the formation of the solar system.

● Describe the similarities of and differences between asteroids and meteoroids.

● Explain how cosmic impacts may affect life on Earth.

Terms to Learn

comet	meteoroid
asteroid	meteorite
asteroid belt	meteor

READING STRATEGY

Discussion Read this section silently. Write down questions that you have about this section. Discuss your questions in a small group.

Small Bodies in the Solar System

Imagine you are traveling in a spacecraft to explore the edge of our solar system. You see several small bodies, as well as the planets and their satellites, moving through space.

The solar system contains not only planets and moons but other small bodies, including comets, asteroids, and meteoroids. Scientists study these objects to learn about the composition of the solar system.

Comets

A small body of ice, rock, and cosmic dust loosely packed together is called a **comet**. Some scientists refer to comets as "dirty snowballs" because of their composition. Comets formed in the cold, outer solar system. Nothing much has happened to comets since the birth of the solar system 4.6 billion years ago. Comets are probably left over from the time when the planets formed. As a result, each comet is a sample of the early solar system. Scientists want to learn more about comets to piece together the history of our solar system.

Comet Tails

When a comet passes close enough to the sun, solar radiation heats the ice so that the comet gives off gas and dust in the form of a long tail, as shown in **Figure 1**. Sometimes, a comet has two tails—an *ion tail* and a *dust tail*. The ion tail is made of electrically charged particles called *ions*. The solid center of a comet is called its *nucleus*. Comet nuclei can range in size from less than half a kilometer to more than 100 km in diameter.

Figure 1 *This image shows the physical features of a comet when it is close to the sun. The nucleus of a comet is hidden by brightly lit gases and dust.*

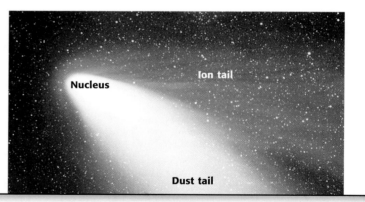

Nucleus

Ion tail

Dust tail

CHAPTER RESOURCES

Chapter Resource File

📂 • Lesson Plan
• Directed Reading A BASIC
• Directed Reading B SPECIAL NEEDS

Technology

💻 Transparencies
• Bellringer

Is That a Fact!

Comets are fairly fragile; some break apart on their own, or the gravity of a planet can pull them apart. For example, comet Shoemaker-Levy 9 broke apart when it passed too close to Jupiter. When Shoemaker-Levy 9 returned in 1994, fragments of the comet crashed into Jupiter's atmosphere. Some of the fragments generated explosions that produced fireballs larger than Earth.

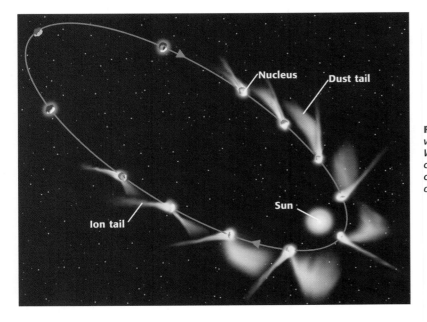

Nucleus

Dust tail

Ion tail

Sun

Figure 2 *Comets have very elongated orbits. When a comet gets close to the sun, the comet can develop one or two tails.*

Comet Orbits

The orbits of all bodies that move around the sun are ellipses. *Ellipses* are circles that are somewhat stretched out of shape. The orbits of most planets are close to perfect circles, but the orbits of comets are very elongated.

Notice in **Figure 2** that a comet's ion tail always points away from the sun. The reason is that the ion tail is blown away from the sun by *solar wind*, which is also made of ions. The dust tail tends to follow the comet's orbit around the sun. Dust tails do not always point away from the sun. When a comet is close to the sun, its tail can extend millions of kilometers through space!

Comet Origins

Where do comets come from? Many scientists think that comets come from the Oort (AWRT) cloud, a spherical region that surrounds the solar system. When the gravity of a passing planet or star disturbs part of this cloud, comets can be pulled toward the sun. Another recently discovered region where comets exist is the Kuiper (KIE puhr) belt, which is the region outside the orbit of Neptune.

Reading Check From which two regions do comets come? *(See the Appendix for answers to Reading Checks.)*

comet a small body of ice, rock, and cosmic dust that follows an elliptical orbit around the sun and that gives off gas and dust in the form of a tail as it passes close to the sun

CONNECTION TO Language Arts

WRITING SKILL **Interplanetary Journalist** In 1994, the world watched in awe as parts of the comet Shoemaker-Levy 9 collided with Jupiter, which caused enormous explosions. Imagine you were an interplanetary journalist who traveled through space to observe the comet during this time. Write an article describing your adventure.

Answer to Reading Check
Comets come from the Oort cloud and the Kuiper belt.

MISCONCEPTION ALERT

Comet Tails Students may be surprised to learn that comets don't have a tail during most of their orbit. Only when they near the sun do they warm up and release a tail made of gas and dust. The comet nucleus has an irregular shape. Sometimes, gas leaves the comet's surface unevenly in "jets." These jets can act like miniature rocket engines, pushing a comet off course and making it difficult to find during its next orbit.

BRAIN FOOD

Tracking Asteroids The orbits of some asteroids cross Earth's orbit. Every few million years, one of these asteroids hits the Earth. If an asteroid is larger than 10 km across, its impact can have catastrophic global effects. In the first few seconds of an impact event, both the impactor and part of the target become liquid, and an impact crater forms. Shock waves spread out from the site, and debris is ejected high into the atmosphere. About 65 million years ago, a large asteroid struck Earth on the Yucatán Peninsula. This event may have led to the mass extinction of the dinosaurs. The collision of comet Shoemaker-Levy 9 with Jupiter in 1994 led NASA to devote more of its resources to finding and tracking asteroids whose orbits cross Earth's. Have students find out more about NASA's asteroid-tracking program.

asteroid a small, rocky object that orbits the sun, usually in a band between the orbits of Mars and Jupiter

asteroid belt the region of the solar system that is between the orbits of Mars and Jupiter and in which most asteroids orbit

Asteroids

Small, rocky bodies that revolve around the sun are called **asteroids.** They range in size from a few meters to more than 900 km in diameter. Asteroids have irregular shapes, although some of the larger ones are spherical. Most asteroids orbit the sun in the asteroid belt. The **asteroid belt** is a wide region between the orbits of Mars and Jupiter. Like comets, asteroids are thought to be material left over from the formation of the solar system.

Types of Asteroids

The composition of asteroids varies depending on where they are located within the asteroid belt. In the outermost region of the asteroid belt, asteroids have dark reddish brown to black surfaces. This coloring may indicate that the asteroids are rich in organic material. Asteroids that have dark gray surfaces are rich in carbon. In the innermost part of the asteroid belt are light gray asteroids that have either a stony or metallic composition. **Figure 3** shows three asteroids: Hektor, Ceres, and Vesta.

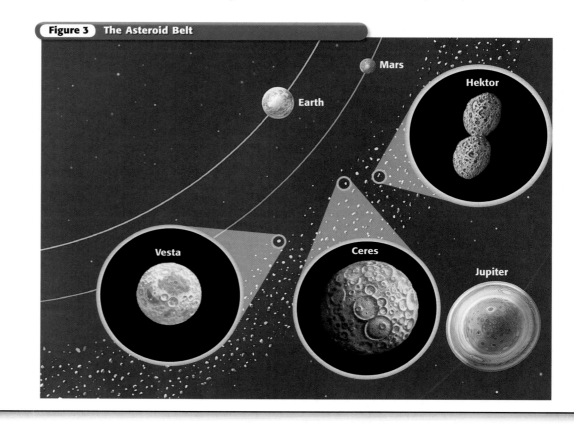

Figure 3 The Asteroid Belt

WEIRD SCIENCE

In 1908, an object thought to be a comet about 60 m in diameter exploded less than 10 km above a remote part of Siberia. The blast flattened trees in an area greater than 2,000 km². The crater has never been found.

Meteoroids

Meteoroids are similar to but much smaller than asteroids. A **meteoroid** is a small, rocky body that revolves around the sun. Most meteoroids are probably pieces of asteroids. A meteoroid that enters Earth's atmosphere and strikes the ground is called a **meteorite.** As a meteoroid falls into Earth's atmosphere, the meteoroid moves so fast that its surface melts. As the meteoroid burns up, it gives off an enormous amount of light and thermal energy. From the ground, you see a spectacular streak of light, or a shooting star. A **meteor** is the bright streak of light caused by a meteoroid or comet dust burning up in the atmosphere.

Meteor Showers

Many of the meteors that we see come from very small (dust-sized to pebble-sized) rocks. Even so, meteors can be seen on almost any night if you are far enough away from a city to avoid the glare of its lights. At certain times of the year, you can see large numbers of meteors, as shown in **Figure 4.** These events are called *meteor showers*. Meteor showers happen when Earth passes through the dusty debris that comets leave behind.

Types of Meteorites

Like their asteroid relatives, meteorites have different compositions. The three major types of meteorites—stony, metallic, and stony-iron meteorites—are shown in **Figure 5.** Many of the stony meteorites probably come from carbon-rich asteroids. Stony meteorites may contain organic materials and water. Scientists use meteorites to study the early solar system. Like comets and asteroids, meteorites are some of the building blocks of planets.

✓ Reading Check What are the major types of meteorites?

meteoroid a relatively small, rocky body that travels through space

meteorite a meteoroid that reaches the Earth's surface without burning up completely

meteor a bright streak of light that results when a meteoroid burns up in the Earth's atmosphere

Figure 4 *Meteors are the streaks of light caused by meteoroids as they burn up in Earth's atmosphere.*

Figure 5 **Three Major Types of Meteorites**

| **Stony meteorite** rocky material | **Metallic meteorite** iron and nickel | **Stony-iron meteorite** rocky material, iron, and nickel |

Answer to Reading Check

The major types of meteorites are stony, metallic, and stony-iron meteorites.

SCIENTISTS AT ODDS

What Are Meteorites? As late as the 1800s, scientists were skeptical that meteorites originate in space—despite records from the Chinese, Romans, and Greeks describing stones falling from the sky. In 1803, meteorites fell in France. A physicist documented the event, finally convincing scientists that meteorites fall from the sky.

Close

Reteaching ——— BASIC

Draw a diagram of a comet on the board. Ask volunteers to label the parts and to describe the composition of each part.

LS Visual/Kinesthetic   *English Language Learners*

Quiz ——— GENERAL

Have students complete the following sentences:

1. _____ are small bodies of ice and cosmic dust. (Comets)

2. Most asteroids in our solar system are found between _____ and _____. (Mars, Jupiter)

3. _____ are meteoroids that fall to Earth. (Meteorites)

Alternative Assessment ——— GENERAL

PORTFOLIO **Field Guide to the Solar System**   Have students create an illustrated field guide to small bodies in our solar system. The guide should incorporate all of the vocabulary used in this section as well as drawings, diagrams, and explanations for each object they include. **LS Visual**   *English Language Learners*

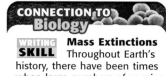

CONNECTION TO Biology

WRITING SKILL **Mass Extinctions**   Throughout Earth's history, there have been times when large numbers of species suddenly became extinct. Many scientists think that these mass extinctions may have been caused by impacts of large objects on Earth. However, other scientists are not so sure. Use the Internet or another source to research this idea. In your **science journal**, write a paragraph describing the different theories scientists have for past mass extinctions.

The Role of Impacts in the Solar System

An impact happens when an object in space collides with another object in space. Often, the result of such a collision is an impact crater. Many planets and moons have visible impact craters. In fact, several planets and moons have many more impact craters than Earth does. Planets and moons that do not have atmospheres have more impact craters than do planets and moons that have atmospheres.

Look at **Figure 6.** Earth's moon has many more impact craters than the Earth does because the moon has no atmosphere to slow objects down. Fewer objects strike Earth because Earth's atmosphere acts as a shield. Smaller objects burn up before they ever reach the surface. Also, most craters left on Earth are no longer visible because of weathering, erosion, and tectonic activity.

Future Impacts on Earth?

Most objects that come close to Earth are small and usually burn up in the atmosphere. However, larger objects are more likely to strike Earth's surface. Scientists estimate that impacts that are powerful enough to cause a natural disaster might happen once every few thousand years. An impact that is large enough to cause a global catastrophe is estimated to happen once every few hundred thousand years, on average.

✓ Reading Check   How often do large objects strike Earth?

Figure 6 *The surface of the moon preserves a record of billions of years of cosmic impacts.*

WEIRD SCIENCE

In 1954, Mrs. E. Hulitt Hodge, of Alabama, was struck by a meteorite as she was taking her afternoon nap. Bruised, but not badly injured, she is one of only two people known to have been struck by a meteorite.

Answer to Reading Check
Large objects strike Earth every few thousand years.

The Torino Scale

The Torino scale is a system that allows scientists to rate the hazard level of an object moving toward Earth. The object is carefully observed and then assigned a number from the scale. The scale ranges from 0 to 10. Zero indicates that the object has a very small chance of striking Earth. Ten indicates that the object will definitely strike Earth and cause a global disaster. The Torino scale is also color coded. White represents 0, and green represents 1. White and green objects rarely strike Earth. Yellow represents 2, 3, and 4 and indicates a higher chance that objects will hit Earth. Orange, which represents 5, 6, and 7, refers to objects highly likely to hit Earth. Red refers to objects that will definitely hit Earth.

SECTION Review

Summary

- Studying comets, asteroids, and meteoroids can help scientists understand more about the formation of the solar system.

- Asteroids are small bodies that orbit the sun. Meteoroids are similar to but smaller than asteroids. Most meteoroids come from asteroids.

- Most objects that collide with Earth burn up in the atmosphere. Large impacts, however, may cause a global catastrophe.

Using Key Terms

For each pair of terms, explain how the meanings of the terms differ.

1. *comet* and *asteroid*

2. *meteor* and *meteorite*

Understanding Key Ideas

3. Which of the following is NOT a type of meteorite?
   a. stony meteorite
   b. rocky-iron meteorite
   c. stony-iron meteorite
   d. metallic meteorite

4. Why is the study of comets, asteroids, and meteoroids important in understanding the formation of the solar system?

5. Why do a comet's two tails often point in different directions?

6. How can a cosmic impact affect life on Earth?

7. What is the difference between an asteroid and a meteoroid?

8. Where is the asteroid belt located?

9. What is the Torino scale?

10. Describe why we see several impact craters on the moon but few on Earth.

Math Skills

11. The diameter of comet A's nucleus is 55 km. If the diameter of comet B's nucleus is 30% larger than comet A's nucleus, what is the diameter of comet B's nucleus?

Critical Thinking

12. **Expressing Opinions** Do you think the government should spend money on programs to search for asteroids and comets that have Earth-crossing orbits? Explain.

13. **Making Inferences** What is the likelihood that scientists will discover an object belonging in the red category of the Torino scale in the next 500 years? Explain your answer.

SCILINKS®

NSTA

Developed and maintained by the National Science Teachers Association

For a variety of links related to this chapter, go to www.scilinks.org

Topic: Comets, Asteroids, and Meteoroids
SciLinks code: HSM0317

Answers to Section Review

1. Sample answer: A comet is a small body of ice, rock, and dust that forms a tail when it passes close to the sun. An asteroid is a small, rocky body that does not have much ice and does not form a tail.

2. Sample answer. A meteor is a streak of light that results when a meteoroid burns up in Earth's atmosphere. A meteorite is a meteoroid that reaches Earth's surface without burning up completely.

3. b

CHAPTER RESOURCES

Chapter Resource File

- **Section Quiz** GENERAL
- **Section Review** GENERAL
- **Vocabulary and Section Summary** GENERAL

---

4. Sample answer: Comets, asteroids, and meteoroids represent the leftover building blocks of the solar system. Studying these bodies will help scientists learn about the composition of the solar system.

5. A comet's ion tail is blown away from the sun by the solar wind, but its dust tail is not. So, a comet's two tails may point in different directions.

6. Answers may vary. Sample answer: A cosmic impact can change the global climate, causing plants and animals not suited to the new climate to die.

7. Sample answer: Meteoroids are similar to asteroids but are much smaller.

8. The asteroid belt is located between Mars and Jupiter.

9. The Torino scale is a system that enables scientists to rate the hazard level of an object moving toward Earth.

10. Answers may vary. Answers should include that most objects burn up in Earth's atmosphere before striking Earth's surface. The moon does not have an atmosphere to slow objects down. In addition, erosion and plate tectonics cause the surface features of Earth to change.

11. 71.5 km
    (55 km × 0.30 = 16.5 km;
    55 km + 16.5 km = 71.5 km)

12. Sample answer: yes; Tracking asteroids and comets with Earth-crossing orbits is important because doing so could help people prepare for a possible disaster.

13. Answers may vary. Sample answer: Because large impacts occur every few thousand years, it is somewhat likely that scientists will discover a red-category object in the next 500 years.

Create a Calendar

Teacher's Notes

Time Required
One 45-minute class period

Lab Ratings

EASY ——————————— HARD

Teacher Prep 🧪🧪
Student Set-Up 🧪
Concept Level 🧪🧪🧪
Clean Up 🧪

MATERIALS

The materials listed on the student page are enough for a group of 2 or 3 students.

Preparation Notes

This activity will require math skills. As a class, you may need to review how to multiply fractions.

As an extension activity, students may research the rotation and revolution of other planets. Students can then create a calendar for one of the other planets.

Create a Calendar

Imagine that you live in the first colony on Mars. You have been trying to follow the Earth calendar, but it just isn't working anymore. Mars takes almost 2 Earth years to revolve around the sun—almost 687 Earth days to be exact! That means that there are only two Martian seasons for every Earth calendar year. One year, you get winter and spring, but the next year, you get only summer and fall! And Martian days are longer than Earth days. Mars takes 24.6 Earth hours to rotate on its axis. Even though they are similar, Earth days and Martian days just don't match. Using the Earth calendar won't work!

OBJECTIVES

Create a calendar based on the Martian cycles of rotation and revolution.

Describe why it is useful to have a calendar that matches the cycles of the planet on which you live.

MATERIALS

- calculator (optional)
- marker
- pencils, assorted colors
- poster board
- ruler, metric

Ask a Question

❶ How can I create a calendar based on the Martian cycles of rotation and revolution that includes months, weeks, and days?

Form a Hypothesis

❷ Write a few sentences that answer your question.

Test the Hypothesis

❸ Use the following formulas to determine the number of Martian days in a Martian year:

$$\frac{687 \text{ Earth days}}{1 \text{ Martian year}} \times \frac{24 \text{ Earth hours}}{1 \text{ Earth day}} = \text{Earth hours per Martian year}$$

$$\text{Earth hours per Martian year} \times \frac{1 \text{ Martian day}}{24.6 \text{ Earth hours}} = \text{Martian days per Martian year}$$

CLASSROOM
TESTED & APPROVED

Michael E. Kral
West Hardin Middle School
Cecilia, Kentucky

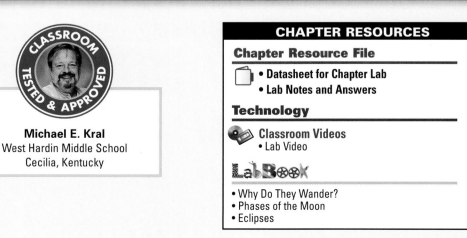

CHAPTER RESOURCES

Chapter Resource File
- • Datasheet for Chapter Lab
- • Lab Notes and Answers

Technology
💿 **Classroom Videos**
- • Lab Video

LabBook

- • Why Do They Wander?
- • Phases of the Moon
- • Eclipses

4 Decide how to divide your calendar into a system of Martian months, weeks, and days. Will you have a leap day, a leap week, a leap month, or a leap year? How often will it occur?

5 Choose names for the months and days of your calendar. Explain why you chose each name. If you have time, explain how you would number the Martian years. For instance, would the first year correspond to a certain Earth year?

6 Follow your design to create your own calendar for Mars. Construct your calendar by using a computer to help organize your data. Draw the calendar on your piece of poster board. Make sure it is brightly colored and easy to follow.

7 Present your calendar to the class. Explain how you chose your months, weeks, and days.

Analyze the Results

1 **Analyzing Results** What advantages does your calendar design have? Are there any disadvantages to your design?

2 **Classifying** Which student or group created the most original calendar? Which design was the most useful? Explain.

3 **Analyzing Results** What might you do to improve your calendar?

Draw Conclusions

4 **Evaluating Models** Take a class vote to decide which design should be chosen as the new calendar for Mars. Why was this calendar chosen? How did it differ from the other designs?

5 **Drawing Conclusions** Why is it useful to have a calendar that matches the cycles of the planet on which you live?

CHAPTER RESOURCES

Workbooks

Whiz-Bang Demonstrations
• Crater Creator `BASIC`
• Space Snowballs `BASIC`

Labs You Can Eat
• Meteorite Delight `ADVANCED`

Long-Term Projects & Research Ideas
• What Did You See, Mr. Messier? `ADVANCED`

Test the Hypothesis

7. Accept all reasonable responses. Students should have used the formula in the procedure to calculate the number of days in a Martian year. There are 670.25 Martian days in a Martian year. Students should explain the system they used for grouping Martian days into weeks and months.

Analyze the Results

1. Accept all reasonable responses. Sample answer: A calendar with a leap year that has one extra day every four Martian years would be less confusing than having leap days with extra hours every year.

2. Accept all reasonable responses.

3. Accept all reasonable responses. Sample answer: An improvement might be to divide the Martian year into 10 months with 67 days per month. Students may suggest simplifying their design

Draw Conclusions

4. Accept all reasonable responses.

5. Accept all reasonable responses. Sample answer: It would be difficult to keep track of the changing seasons of a planet using a calendar that doesn't match the period of revolution of the planet.

Chapter Review

Assignment Guide

Section	Questions
1	1, 4, 12–13
2	6, 11, 16–18
3	26
4	7, 19-20, 22, 24–25
5	2–3, 5, 15
2 and 3	9–10, 14, 27–28
2, 3, and 4	8, 23
2, 3, 4, and 5	21

ANSWERS

Using Key Terms

1. Sample answer: The terrestrial planets are the small, rocky planets of the inner solar system. The gas giants are the large, gaseous planets of the outer solar system.

2. Sample answer: Asteroids are small bodies made of rocky material, and comets are small bodies made of ice, rock, and cosmic dust.

3. Sample answer: A meteor is a streak of light that results from a meteoroid burning up in the atmosphere. A meteorite is a meteoroid that has passed through the atmosphere and struck the ground.

4. astronomical unit

5. meteoroid

6. prograde

7. satellite

USING KEY TERMS

For each pair of terms, explain how the meanings of the terms differ.

1. *terrestrial planet* and *gas giant*

2. *asteroid* and *comet*

3. *meteor* and *meteorite*

Complete each of the following sentences by choosing the correct term from the word bank.

astronomical unit   meteorite
meteoroid   prograde
retrograde   satellite

4. The average distance between the sun and Earth is 1 ___.

5. A small rock in space is called a(n) ___.

6. When viewed from above its north pole, a body that moves in a counter-clockwise direction is said to have ___ rotation.

7. A(n) ___ is a natural or artificial body that revolves around a planet.

UNDERSTANDING KEY IDEAS

Multiple Choice

8. Of the following, which is the largest body?
   a. the moon
   b. Pluto
   c. Mercury
   d. Ganymede

9. Which of the following planets have retrograde rotation?
   a. the terrestrial planets
   b. the gas giants
   c. Mercury, Venus, and Uranus
   d. Venus, Uranus, and Pluto

10. Which of the following planets does NOT have any moons?
    a. Mercury
    b. Mars
    c. Uranus
    d. None of the above

11. Why can liquid water NOT exist on the surface of Mars?
    a. The temperature is too high.
    b. Liquid water once existed there.
    c. The gravity of Mars is too weak.
    d. The atmospheric pressure is too low.

Short Answer

12. List the names of the planets in the order the planets orbit the sun.

13. Describe three ways in which the inner planets are different from the outer planets.

14. What are the gas giants? How are the gas giants different from the terrestrial planets?

15. What is the difference between asteroids and meteoroids?

16. What is the difference between a planet's period of rotation and period of revolution?

Understanding Key Ideas

8. d
9. d
10. a
11. d
12. Mercury, Venus, Earth, Mars, Jupiter, Saturn, Uranus, Neptune, and Pluto
13. Sample answer: The inner planets are small, rocky, and closely spaced. The outer planets are large, gaseous, and far apart.

14. Sample answer: The gas giants are the large, gaseous planets of the outer solar system. In contrast, the terrestrial planets are the small, rocky planets of the inner solar system.

15. Asteroids are small bodies made of rocky material. Meteoroids are similar to asteroids, only much smaller.

16. The period of rotation is how long a planet takes to complete a turn on its axis. The period of revolution is how long a planet takes to orbit the sun.

17 Explain the difference between prograde rotation and retrograde rotation.

18 Which characteristics of Earth make it suitable for life?

19 Describe the current theory for the origin of Earth's moon.

20 What causes the phases of the moon?

CRITICAL THINKING

21 **Concept Mapping** Use the following terms to create a concept map: *solar system, terrestrial planets, gas giants, moons, comets, asteroids,* and *meteoroids.*

22 **Applying Concepts** Even though we haven't yet retrieved any rock samples from Mercury's surface for radiometric dating, scientists know that the surface of Mercury is much older than that of Earth. How do scientists know this?

23 **Making Inferences** Where in the solar system might scientists search for life, and why?

24 **Analyzing Ideas** Is the far side of the moon always dark? Explain your answer.

25 **Predicting Consequences** If scientists could somehow bring Europa as close to the sun as the Earth is, 1 AU, how do you think Europa would be affected?

26 **Identifying Relationships** How did variations in the orbit of Uranus help scientists discover Neptune?

INTERPRETING GRAPHICS

The graph below shows density versus mass for Earth, Uranus, and Neptune. Mass is given in Earth masses—the mass of Earth is equal to 1 Earth mass. The relative volumes for the planets are shown by the size of each circle. Use the graph below to answer the questions that follow.

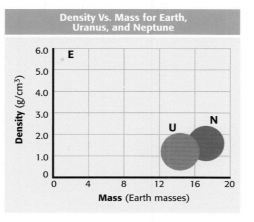

Density Vs. Mass for Earth, Uranus, and Neptune

27 Which planet is denser, Uranus or Neptune? How can you tell?

28 You can see that although Earth has the smallest mass, it has the highest density of the three planets. How can Earth be the densest of the three when Uranus and Neptune have so much more mass than Earth does?

20. Answers may vary. Sample answer: As Earth and the moon revolve together around the sun, the side of the moon facing Earth gets varying amounts of sunlight.

Critical Thinking

21. An answer to this exercise can be found at the end of this book.

22. Sample answer: Mercury's surface is covered with impact craters that record the planet's history. Earth's surface has only a few craters, indicating that the rocks on Earth's surface are continually recycled.

23. Sample answer: The search for life should include areas where liquid water is present because life as we know it depends on liquid water for survival.

24. Sample answer: no; The far side of the moon gets as much sunlight as the near side. As the moon revolves around Earth, the moon also rotates.

25. Answers will vary. If Europa were closer to the sun, it would heat up considerably. Europa is made mostly of ice, so much of its surface might melt to form oceans and an atmosphere.

26. Sample answer: Neptune's gravitational field prevented Uranus from following its predicted orbit. The irregular orbit of Uranus indicated to scientists that there was something, probably a planet, beyond Uranus.

Interpreting Graphics

27. Neptune is denser. It has a higher density value on the chart. (Neptune has a smaller volume and more mass, giving it a greater density than Uranus.)

28. The masses of Neptune and Uranus occupy a much larger volume than the mass of Earth does. (Density is the amount of mass that exists within a given volume of space.)

17. Prograde rotation is counterclockwise when viewed from a planet's North Pole. Retrograde rotation is clockwise when viewed from a planet's North Pole.

18. Sample answer: The presence of liquid water makes Earth suitable for life. Earth's atmosphere also regulates temperature on Earth.

19. The current theory is that a collision with a Mars-sized object blasted part of Earth's mantle into space. This material formed the moon.

CHAPTER RESOURCES

Chapter Resource File

- Chapter Review **GENERAL**
- Chapter Test A **GENERAL**
- Chapter Test B **ADVANCED**
- Chapter Test C **SPECIAL NEEDS**
- Vocabulary Activity **GENERAL**

Workbooks

Study Guide
- Assessment resources are also available in Spanish.

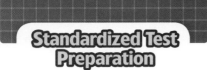

Teacher's Note

To provide practice under more realistic testing conditions, give students 20 minutes to answer all of the questions in this Standardized Test Preparation.

Answers to the standardized test preparation can help you identify student misconceptions and misunderstandings.

READING

Passage 1

1. D

2. F

3. B

+ TEST DOCTOR

Question 2: None of the answers provided have been specifically stated in the passage. However, students should infer from the statement "After years of observing the sky" that Greek astronomers were both patient and observant.

READING

Read each of the passages below. Then, answer the questions that follow each passage.

Passage 1   Imagine that it is 200 BCE and you are an apprentice to a Greek astronomer. After years of observing the sky, the astronomer knows all of the constellations as well as the back of his hand. He shows you how the stars all move together—the whole sky spins slowly as the night goes on. He also shows you that among the thousands of stars in the sky, some of the brighter ones slowly change their position relative to the other stars. He names these stars *planetai*, the Greek word for "wanderers." Building on the observations of the ancient Greeks, we now know that the *planetai* are actually planets, not wandering stars.

1. Which of the following did the ancient Greeks know to be true?

   A All planets have at least one moon.

   B The planets revolve around the sun.

   C The planets are much smaller than the stars.

   D The planets appear to move relative to the stars.

2. What can you infer from the passage about the ancient Greek astronomers?

   F They were patient and observant.

   G They knew much more about astronomy than we do.

   H They spent all their time counting stars.

   I They invented astrology.

3. What does the word *planetai* mean in Greek?

   A planets

   B wanderers

   C stars

   D moons

Passage 2   To explain the source of short-period comets (comets that have a relatively short orbit), the Dutch-American astronomer Gerard Kuiper proposed in 1949 that a belt of icy bodies must lie beyond the orbits of Pluto and Neptune. Kuiper argued that comets were icy <u>planetesimals</u> that formed from the condensation that happened during the formation of our galaxy. Because the icy bodies are so far from any large planet's gravitational field (30 to 100 AU), they can remain on the fringe of the solar system. Some theorists speculate that the large moons Triton and Charon were once members of the Kuiper belt before they were captured by Neptune and Pluto. These moons and short-period comets have similar physical and chemical properties.

1. According to the passage, why can icy bodies remain at the edge of the solar system?

   A The icy bodies are so small that they naturally float to the edge of the solar system.

   B The icy bodies have weak gravitational fields and therefore do not orbit individual planets.

   C The icy bodies are short-period comets, which can reside only at the edge of the solar system.

   D The icy bodies are so far away from any large planet's gravitational field that they can remain at the edge of the solar system.

2. According to the passage, which of the following best describes the meaning of the word *planetesimal*?

   F a small object that existed during the early development of the solar system

   G an extremely tiny object in space

   H a particle that was once part of a planet

   I an extremely large satellite that was the result of a collision of two objects

Passage 2

1. D

2. F

 TEST DOCTOR

Question 2: Several of the choices may seem as if they could be correct. However, only one of the phrases most completely describes the underlined term. Answer F most fully defines planetesimals.

Use the diagrams below to answer the questions that follow.

Planet A 115 craters/km²

Planet B 75 craters/km²

Planet C 121 craters/km²

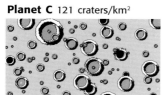

Planet D 97 craters/km²

1. According to the information above, which planet has the oldest surface?

A planet A

B planet B

C planet C

D planet D

2. How many more craters per square kilometer are there on planet C than on planet B?

F 46 craters per square kilometer

G 24 craters per square kilometer

H 22 craters per square kilometer

I 6 craters per square kilometer

MATH

Read each question below, and choose the best answer.

1. Venus's surface gravity is 91% of Earth's. If an object weighs 12 N on Earth, how much would it weigh on Venus?

A 53 N

B 13 N

C 11 N

D 8 N

2. Earth's overall density is 5.52 g/cm³, while Saturn's density is 0.69 g/cm³. How many times denser is Earth than Saturn?

F 8 times

G 9 times

H 11 times

I 12 times

3. If Earth's history spans 4.6 billion years and the Phanerozoic eon was 543 million years, what percentage of Earth's history does the Phanerozoic eon represent?

A about 6%

B about 12%

C about 18%

D about 24%

4. The diameter of Venus is 12,104 km. The diameter of Mars is 6,794 km. What is the difference between the diameter of Venus and the diameter of Mars?

F 5,400 km

G 5,310 km

H 4,890 km

I 890 km

Standardized Test Preparation

---

1. C

2. G

 **TEST DOCTOR**

Question 2: Remind students to use the data provided above the graphic rather than try to count craters themselves.

MATH

1. B

2. I

3. B

4. H

TEST DOCTOR

Question 1: Remind students that to find the percentage of a number, they should multiply the number by the percentage, written as a decimal. Therefore, students should multiply 12 by 0.91 to find weight on Venus.

CHAPTER RESOURCES

Chapter Resource File

• Standardized Test Preparation GENERAL

State Resources

 For specific resources for your state, visit **go.hrw.com** and type in the keyword **HSMSTR**.

Science Fiction

Background

In 1934, Stanley Weinbaum published his first science fiction story, "A Martian Odyssey." Before becoming a writer, Weinbaum studied chemical engineering. During the Great Depression, Weinbaum gave up his science career to be a writer. Sadly, less than two years after his first story was published, Stanley Weinbaum died of cancer. Although his list of works is short, many people consider him among the best science fiction writers.

Discuss the author's description of Io. Ask students, "How does the author's description compare with what we now know about Io?" Have students use the story as inspiration to write their own description of a human colony on one of Jupiter's moons.

Scientific Debate

Background

In many ways, Pluto resembles Neptune's moon Triton. Pluto and Triton are similar in size, and both rotate in a direction counter to that of the other planets. Some scientists believe that there was a collision between Pluto and Triton and that the force of the collision ejected Pluto from the Neptune system.

Science in Action

Science Fiction

"The Mad Moon" by Stanley Weinbaum

The third largest moon of Jupiter, called Io, can be a hard place to live. Grant Calthorpe is finding this out the hard way. Although living comfortably is possible in the small cities at the polar regions of Io, Grant has to spend most of his time in the moon's hot and humid jungles. Grant treks into the jungles of Io to gather ferva leaves so that they can be converted into useful medications for humans. During Grant's quest, he encounters loonies and slinkers, and he has to avoid blancha, a kind of tropical fever that causes hallucinations, weakness, and vicious headaches. Without proper medication a person with blancha can go mad or even die. In "The Mad Moon," you'll discover a dozen adventures with Grant Calthorpe as he struggles to stay alive—and sane.

Language Arts ACTIVITY

WRITING SKILL Read "The Mad Moon" by Stanley Weinbaum. Write a short story describing the adventures that you would have on Io if you were chosen as Grant Calthorpe's assistant.

HOLT ANTHOLOGY OF
Science Fiction

HOLT, RINEHART AND WINSTON

Scientific Debate

Is Pluto a Planet?

Is it possible that Pluto isn't a planet? Some scientists think so! Since 1930, Pluto has been included as one of the nine planets in our solar system. But observations in the 1990s led many astronomers to refer to Pluto as an object, not a planet. Other astronomers disagree with this change. Astronomers that refer to Pluto as an object do not think that it fits well with the other outer planets. Unlike the other outer planets, which are large and gaseous, Pluto is small and made of rock and ice. Pluto also has a very elliptical orbit that is unlike its neighboring planets. Astronomers that think Pluto is a planet point out that Pluto, like all other planets, has its own atmosphere and its own moon, called Charon. These and other factors have fueled a debate as to whether Pluto should be classified as a planet.

Math ACTIVITY

How many more kilometers is Earth's diameter compared to Pluto's diameter if Earth's diameter is 12,756 km and Pluto's diameter is 2,390 km?

Answer to Language Arts Activity

Accept any reasonable answer that demonstrates that students have completed their assigned reading.

Answer to Math Activity

10,366 km (12,756 km − 2,390 km = 10,366 km)

Adriana C. Ocampo

Planetary Geologist Sixty-five million years ago, in what is now Mexico, a giant meteor at least six miles wide struck Earth. The meteor made a hole nine miles deep and over 100 miles wide. The meteor sent billions of tons of dust into Earth's atmosphere. This dust formed thick clouds. After forming, these clouds may have left the planet in total darkness for six months, and the temperature near freezing for ten years. Some scientists think that this meteor crash and its effect on the Earth's climate led to the extinction of the dinosaurs. Adriana Ocampo studies the site in Mexico made by the crater known as the Chicxulub (cheeks OO loob) impact crater. Ocampo is a planetary geologist and has been interested in space exploration since she was young. Ocampo's specialty is studying "impact craters." "Impact craters are formed when an asteroid or a comet collides with the Earth or any other terrestrial planet," explains Ocampo. Ocampo visits crater sites around the world to collect data. She also uses computers to create models of how the impact affected the planet. Ocampo has worked for NASA and has helped plan space exploration missions to Mars, Jupiter, Saturn, and Mercury. Ocampo currently works for the European Space Agency (ESA) and is part of the team getting ready to launch the next spacecraft that will go to Mars.

Social Studies ACTIVITY

Research information about impact craters. Find the different locations around the world where impact craters have been found. Make a world map that highlights these locations.

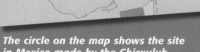

The circle on the map shows the site in Mexico made by the Chicxulub impact crater.

To learn more about these Science in Action topics, visit **go.hrw.com** and type in the keyword **HZ5FAMF.**

Current Science
Check out Current Science® articles related to this chapter by visiting go.hrw.com. Just type in the keyword **HZ5CS21.**

5

Exploring Space
Chapter Planning Guide

Compression guide:
To shorten instruction because of time limitations, omit the Chapter Lab.

OBJECTIVES	LABS, DEMONSTRATIONS, AND ACTIVITIES	TECHNOLOGY RESOURCES
PACING • 120 min pp. 132–137 **Chapter Opener**	SE **Start-up Activity,** p. 133 ◆ GENERAL	OSP **Parent Letter** ▣ GENERAL CD **Student Edition on CD-ROM** CD **Guided Reading Audio CD** ▣ TR **Chapter Starter Transparency*** VID **Brain Food Video Quiz**
Section 1 Rocket Science • Outline the development of rocket technology. • Describe how a rocket accelerates. • Explain the difference between orbital velocity and escape velocity.	TE **Demonstration** Expanding Gas, p. 134 ◆ GENERAL TE **Connection Activity** History, p. 135 GENERAL TE **Connection Activity** Language Arts, p. 135 GENERAL SE **Skills Practice Lab** Water Rockets Save the Day!, p. 156 ◆ GENERAL CRF **Datasheet for Chapter Lab*** LB **Whiz-Bang Demonstrations** Rocket Science* ◆ GENERAL	CRF **Lesson Plans*** TR **Bellringer Transparency*** TR **40 Years of NASA*** TR **How a Rocket Works*** VID **Lab Videos for Earth Science**
PACING • 45 min pp. 138–143 **Section 2 Artificial Satellites** • Identify the first satellites. • Compare low Earth orbits with geostationary orbits. • Explain the functions of military, communications, and weather satellites. • Explain how remote sensing from satellites has helped us study Earth as a global system.	TE **Connection Activity** Real World, p. 138 ADVANCED SE **Quick Lab** Modeling LEO and GEO, p. 139 GENERAL CRF **Datasheet for Quick Lab*** SE **School-to-Home Activity** Tracking Satellites, p. 141 GENERAL SE **Connection to Environmental Science** Space Junk, p. 142 GENERAL LB **Inquiry Labs** Crash Landing* ◆ BASIC	CRF **Lesson Plans*** TR **Bellringer Transparency*** TR **GEO and LEO*** TR **Landsat Data***
PACING • 45 min pp. 144–149 **Section 3 Space Probes** • Describe five discoveries made by space probes. • Explain how space-probe missions help us better understand the Earth. • Describe how NASA's new strategy of "faster, cheaper, and better" relates to space probes.	TE **Activity** Design Your Own Space Mission, p. 144 ADVANCED TE **Activity** Designing a Mission Patch, p. 145 GENERAL TE **Connection Activity** History, p. 146 GENERAL TE **Connection Activity** History, p. 147 GENERAL SE **Connection to Social Studies** Cosmic Message in a Bottle, p. 148 GENERAL LB **Long-Term Projects & Research Ideas** Space Voyage* ADVANCED	CRF **Lesson Plans*** TR **Bellringer Transparency*** TR **Space Probes in the Outer Solar System*** TR *LINK TO PHYSICAL SCIENCE* Forming Positive and Negative Ions*
PACING • 45 min pp. 150–155 **Section 4 People in Space** • Summarize the history and future of human spaceflight. • Explain the benefits of crewed space programs. • Identify five "space-age spinoffs" that are used in everyday life.	TE **Demonstration** O-Ring Failure, p. 151 ◆ BASIC SE **Connection to Biology** Effects of Weightlessness, p. 152 GENERAL TE **Group Activity** Skylab Results, p. 152 GENERAL SE **Connection to Social Studies** Oral Histories, p. 153 GENERAL SE **Model-Making Lab** Reach for the Stars, p. 172 GENERAL CRF **Datasheet for Lab Book*** LB **Inquiry Labs** Space Fitness* ◆ ADVANCED LB **EcoLabs & Field Activities** There's a Space for Us* ◆ GENERAL	CRF **Lesson Plans*** TR **Bellringer Transparency*** CRF **SciLinks Activity*** GENERAL

PACING • 90 min

CHAPTER REVIEW, ASSESSMENT, AND STANDARDIZED TEST PREPARATION

CRF **Vocabulary Activity*** GENERAL
SE **Chapter Review,** pp. 158–159 GENERAL
CRF **Chapter Review*** ▣ GENERAL
CRF **Chapter Tests A*** ▣ GENERAL, **B*** ADVANCED, **C*** SPECIAL NEEDS
SE **Standardized Test Preparation,** pp. 160–161 GENERAL
CRF **Standardized Test Preparation*** GENERAL
CRF **Performance-Based Assessment*** GENERAL
OSP **Test Generator** GENERAL
CRF **Test Item Listing*** GENERAL

Online and Technology Resources

Visit **go.hrw.com** for a variety of free resources related to this textbook. Enter the keyword **HZ5EXP.**

Holt Online Learning

Students can access interactive problem-solving help and active visual concept development with the *Holt Science and Technology* Online Edition available at **www.hrw.com.**

Guided Reading Audio CD

A direct reading of each chapter using instructional visuals as guideposts. For auditory learners, reluctant readers, and Spanish-speaking students. Available in English and Spanish.

SKILLS DEVELOPMENT RESOURCES	SECTION REVIEW AND ASSESSMENT	STANDARDS CORRELATIONS
SE Pre-Reading Activity, p. 132 `GENERAL` **OSP** Science Puzzlers, Twisters & Teasers `GENERAL`		National Science Education Standards UCP 2, 3; SAI 1; ST 1
CRF Directed Reading A* ■ `BASIC`, B* `SPECIAL NEEDS` **CRF** Vocabulary and Section Summary* ■ `GENERAL` **SE** Reading Strategy Discussion, p. 134 `GENERAL` **TE** Inclusion Strategies, p. 137 **CRF** Reinforcement Worksheet Ronnie Rocket* `BASIC`	**SE** Reading Checks, pp. 134, 136 `GENERAL` **TE** Homework, p. 135 `GENERAL` **TE** Reteaching, p. 136 `BASIC` **TE** Quiz, p. 136 `GENERAL` **TE** Alternative Assessment, p. 136 `GENERAL` **SE** Section Review,* p. 137 ■ `GENERAL` **CRF** Section Quiz* ■ `GENERAL`	UCP 2, 3; HNS 1, 3; *Chapter Lab:* SAI 1; ST 1
CRF Directed Reading A* ■ `BASIC`, B* `SPECIAL NEEDS` **CRF** Vocabulary and Section Summary* ■ `GENERAL` **SE** Reading Strategy Reading Organizer, p. 138 `GENERAL` **TE** Inclusion Strategies, p. 139 **SE** Math Practice Triangulation, p. 140 `GENERAL`	**SE** Reading Checks, pp. 139, 141, 143 `GENERAL` **TE** Reteaching, p. 142 `BASIC` **TE** Quiz, p. 142 `GENERAL` **TE** Alternative Assessment, p. 142 `GENERAL` **SE** Section Review,* p. 143 ■ `GENERAL` **CRF** Section Quiz* ■ `GENERAL`	UCP 2, 3; SAI 1; ST 2; SPSP 5
CRF Directed Reading A* ■ `BASIC`, B* `SPECIAL NEEDS` **CRF** Vocabulary and Section Summary* ■ `GENERAL` **SE** Reading Strategy Reading Organizer, p. 144 `GENERAL` **TE** Reading Strategy Paired Summarizing, p. 145 `GENERAL` **CRF** Critical Thinking Spacecraft R' Us* `ADVANCED` **CRF** Reinforcement Worksheet Probing Space* `BASIC`	**SE** Reading Checks, pp. 145, 146, 148 `GENERAL` **TE** Reteaching, p. 148 `BASIC` **TE** Quiz, p. 148 `GENERAL` **TE** Alternative Assessment, p. 148 `GENERAL` **SE** Section Review,* p. 149 ■ `GENERAL` **CRF** Section Quiz* ■ `GENERAL`	UCP 5; ST 2
CRF Directed Reading A* ■ `BASIC`, B* `SPECIAL NEEDS` **CRF** Vocabulary and Section Summary* ■ `GENERAL` **SE** Reading Strategy Reading Organizer, p. 150 `GENERAL` **TE** Reading Strategy Prediction Guide, p. 151 `GENERAL`	**SE** Reading Checks, pp. 151, 153, 155 `GENERAL` **TE** Homework, p. 152 `GENERAL` **TE** Reteaching, p. 154 `BASIC` **TE** Quiz, p. 154 `GENERAL` **TE** Alternative Assessment, p. 154 `GENERAL` **TE** Homework, p. 154 `GENERAL` **SE** Section Review,* p. 155 ■ `GENERAL` **CRF** Section Quiz* ■ `GENERAL`	UCP 5; ST 2; SPSP 5; HNS 1, 3; *LabBook:* UCP 2; SAI 1; ST 1, 2

One-Stop Planner® CD-ROM

This convenient CD-ROM includes:
- Lab Materials QuickList Software
- Holt Calendar Planner
- Customizable Lesson Plans
- Printable Worksheets
- ExamView® Test Generator

cnnstudentnews.com

Find the latest news, lesson plans, and activities related to important scientific events.

www.scilinks.org

Maintained by the **National Science Teachers Association.** See Chapter Enrichment pages for a complete list of topics.

 Current Science®

Check out *Current Science* articles and activities by visiting the HRW Web site at **go.hrw.com.** Just type in the keyword **HZ5CS22T.**

Classroom Videos
- **Lab Videos** demonstrate the chapter lab.
- **Brain Food Video Quizzes** help students review the chapter material.

Visual Resources

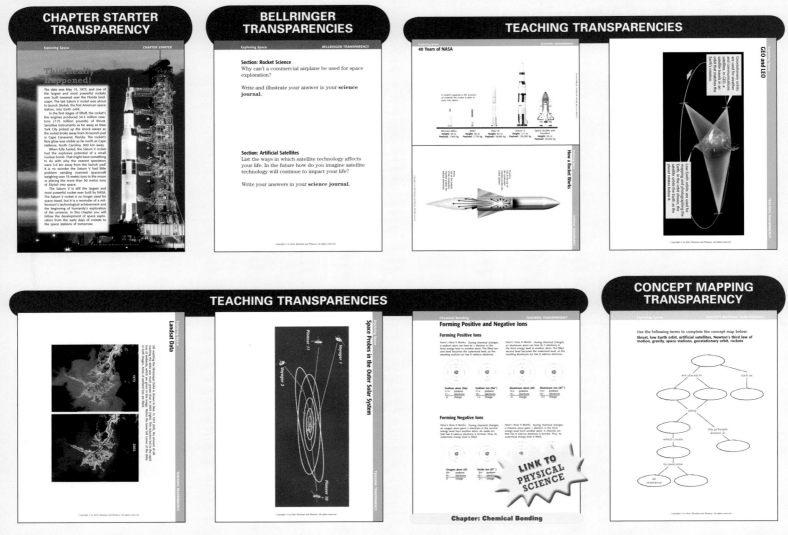

CHAPTER STARTER TRANSPARENCY

BELLRINGER TRANSPARENCIES

TEACHING TRANSPARENCIES

TEACHING TRANSPARENCIES

CONCEPT MAPPING TRANSPARENCY

Planning Resources

LESSON PLANS

PARENT LETTER

TEST ITEM LISTING

One-Stop Planner® CD-ROM

This CD-ROM includes all of the resources shown here and the following time-saving tools:

- *Lab Materials QuickList Software*
- *Customizable lesson plans*
- *Holt Calendar Planner*
- *The powerful ExamView® Test Generator*

Meeting Individual Needs

DIRECTED READING A

Skills Worksheet
Directed Reading A   SAMPLE

Section:
THAT'S SCIENCE!
1. How did James Czarnowski get his idea for the penguin
Explain.

ALSO IN SPANISH

BASIC

DIRECTED READING B

Skills Worksheet
Directed Reading B   SAMPLE

Section:
THAT'S SCIENCE!
1. How did James Czarnowski get his idea for the penguin boat, Proteus?
Explain.

2. What is unusual about the way that Proteus moves through the water?

SPECIAL NEEDS

VOCABULARY ACTIVITY

Activity
Vocabulary Activity   SAMPLE

Getting the Dirt on the Soil
After you finish reading Chapter: [Unique Title], try this puzzle! Use the clues below
to unscramble the vocabulary words. Write your answer in the space provided.
9. the chemical breakdown of rocks
and minerals into new
substances CAMILCHE
THEAIRGWEN

GENERAL

VOCABULARY AND SECTION SUMMARY

Skills Worksheet
Vocabulary & Notes   SAMPLE

Section:
VOCABULARY
In your own words, write a definition of the following term in the space provided.
1. scientific method

2. technology

ALSO IN SPANISH

GENERAL

REINFORCEMENT

Skills Worksheet
Reinforcement   SAMPLE

The Plane Truth
Complete this worksheet after you finish reading the Section: [Unique Section Title]

BASIC

CRITICAL THINKING

Skills Worksheet
Critical Thinking   SAMPLE

A Solar Solution

ADVANCED

SCILINKS ACTIVITY

Activity
SciLinks Activity   SAMPLE

MARINE ECOSYSTEMS
Go to www.scilinks.com. To find links related
to marine ecosystems, type in the keyword

GENERAL

SCIENCE PUZZLERS, TWISTERS & TEASERS

CHAPTER
22   SCIENCE PUZZLERS, TWISTERS & TEASERS
Exploring Space

Up There in the Sky

GENERAL

Labs and Activities

ECOLABS & FIELD ACTIVITIES

ECOLAB
16   TEACHER'S PREPARATORY GUIDE   DESIGN YOUR OWN
There's a Space for Us

Cooperative Learning Activity
Group size: 4–8 students
Group goal:

ADVANCED

LONG-TERM PROJECTS & RESEARCH IDEAS

PROJECT
50   STUDENT WORKSHEET   DESIGN YOUR OWN
Space Voyage

The Lead Scientist Speaks

Research Ideas

ADVANCED

WHIZ-BANG DEMONSTRATIONS

DEMO
36   TEACHER-LED DEMONSTRATION   MAKING MODELS
Rocket Science

Purpose
Students learn about Newton's third law of
motion and the basics of rocketry.

Time Required
20–25 minutes

Lab Ratings

MATERIALS

Safety Information

What to Do

GENERAL

INQUIRY LABS

LAB
14   TEACHER'S PREPARATORY GUIDE   DESIGN YOUR OWN
Crash Landing

Purpose
Students design and test a model of an
early Soviet descent module to investigate
the hazards of landing a space capsule.

Time Required
Two 45-minute class periods

Lab Ratings

BASIC

INQUIRY LABS

15   TEACHER'S PREPARATORY GUIDE   DESIGN YOUR OWN
Space Fitness

Purpose
Students create and test exercise equipment for use in the microgravity environment of space.

Time Required
One to two 45-minute class periods

Lab Ratings

ADVANCED

DATASHEETS FOR QUICK LABS

TEACHER RESOURCE PAGE
Quick Lab   DATASHEET FOR QUICK LAB
Reaction to Stress   SAMPLE

Background

DATASHEETS FOR CHAPTER LABS

TEACHER RESOURCE PAGE
Using Scientific Methods   DATASHEET FOR CHAPTER LAB   SAMPLE

Teacher's Notes
TIME REQUIRED
One 45-minute class period.

DATASHEETS FOR LABBOOK

TEACHER RESOURCE PAGE
Skills Practice Lab   DATASHEET FOR LABBOOK LAB
Does It All Add Up?   SAMPLE

Teacher's Notes
TIME REQUIRED
One 45-minute class period.

Review and Assessments

SECTION QUIZ

Assessment
Section Quiz   SAMPLE

Section:
In the space provided, write the letter of the description that best matches
term or phrase.

ALSO IN SPANISH

GENERAL

SECTION REVIEW

Skills Worksheet
Section Review   SAMPLE

Section:
KEY TERMS
1. What do paleontologist study?

2. How does a trace fossil differ from petrified wood?

ALSO IN SPANISH

GENERAL

CHAPTER REVIEW

Skills Worksheet
Chapter Review   SAMPLE

USING VOCABULARY
1. Define biome in your own words.

2. Describe the characteristics of a savanna and a desert.

ALSO IN SPANISH

GENERAL

CHAPTER TEST A

Assessment
Chapter Test A   SAMPLE

MULTIPLE CHOICE
In the space provided, write the letter of the term or phrase that best completes
each statement or best answers each question.
1. Surface currents are formed by
a. the moon's gravity.   c. wind.
b. the sun's gravity.   d. increased
2. When waves come near the shore,
a. they speed up.   c. their wave
b. they maintain their speed.   d. their wave

ALSO IN SPANISH

GENERAL

CHAPTER TEST B

Assessment
Chapter Test B   SAMPLE

MULTIPLE CHOICE
In the space provided, write the letter of the term or phrase that best completes
each statement or best answers each question.
1. Surface currents are formed by
a. the moon's gravity.   c. wind.
b. the sun's gravity.   d. increased water density.
When waves come near the shore,
a. they speed up.   c. their wavelength increases.
b. they maintain their speed.   d. their wave height increases.

ADVANCED

CHAPTER TEST C

Assessment
Chapter Test C   SAMPLE

MULTIPLE CHOICE
In the space provided, write the letter of the term or phrase that best completes
each statement or best answers each question.
1. Surface currents are formed by
a. the moon's gravity.   c. wind.
b. the sun's gravity.   d. increased water density.
2. When waves come near the shore,
a. they speed up.   c. their wavelength increases.
b. they maintain their speed.   d. their wave height increases.

SPECIAL NEEDS

STANDARDIZED TEST PREPARATION

Assessment
Standardized Test Preparation   SAMPLE

READING
Read the passages below. Then, read each question that follows the passage.
Decide which is the best answer to each question.

GENERAL

PERFORMANCE-BASED ASSESSMENT

Performanced-Based Assessment   SKILL BUILDER   SAMPLE

OBJECTIVE
Determine which factors cause some sugar shapes to break down faster than others.

KNOW THE SCORE!
As you work through the activity, keep in mind that you will be earning a grade
for the following:
• how you form and test the hypothesis (30%)
• the quality of your analysis (40%)
• the clarity of your conclusions (30%)

Using Scientific Methods

MATERIALS AND SOURCES
• 1 regular sugar cube   • 90 mL of water

GENERAL

This Chapter Enrichment provides relevant and interesting information to expand and enhance your presentation of the chapter material.

Section 1

Rocket Science

Konstantin Tsiolkovsky (1857–1935)

- As a youth, Konstantin Tsiolkovsky, the father of rocket theory, demonstrated a keen interest in science and mathematics. At age 9, a bout of scarlet fever left him partially deaf, and he spent much of his time studying on his own. After studying chemistry, astronomy, mathematics, and mechanics in Moscow, Tsiolkovsky got a job in 1876 as a mathematics teacher in a community north of Moscow. There, he continued his scientific pursuits. In 1903, Tsiolkovsky published the article "Exploration of Cosmic Space by Means of Reaction Devices," the culmination of years of theorization about the use of rocket engines for space travel.

- In later years, Tsiolkovsky elaborated on his earlier theories, developing a theory of rocket propulsion and anticipating a number of technologies used in contemporary space exploration, including multistage boosters and the use of chemical propellants to achieve enough thrust to overcome Earth's gravity.

Robert Goddard (1882–1945)

- In his youth, Robert Goddard was an enthusiastic reader of science fiction tales of space travel, and at an early age he wrote a paper titled "The Navigation of Space." In 1912, Goddard developed a mathematical theory of rocket propulsion. He achieved a major breakthrough in 1915, when he proved that rocket engines would work in a vacuum and thus could be used for space travel.

- In 1919, Goddard published his research in the landmark paper "A Method of Reaching Extreme Altitudes," in which he argued that rockets could be used to escape Earth's gravity. Some people found Goddard's theories ludicrous. *The New York Times,* for example, scoffed at Goddard and questioned his scientific qualifications. Undeterred, Goddard continued to design and experiment with rockets. In 1926, using a liquid fuel mixture of gasoline and oxygen, Goddard launched his first liquid-fueled rocket, which ascended to a height of nearly 13 m in 2.5 sec.

- In 1929, Goddard launched his first rocket to carry scientific instruments. The rocket rose about 30 m and then crashed to Earth, where it caught fire. People living nearby called the state fire marshal, who banned Goddard from doing any further rocket tests in Massachusetts. With a Guggenheim grant of $50,000, Goddard set up a test site in an unpopulated area outside of Roswell, New Mexico. He launched increasingly complex rockets that featured innovations such as steering systems, fuel pumps, and cooling mechanisms.

Is That a Fact!

- ◆ When Goddard died in 1945, his immense contributions to rocket technology were still relatively unknown. By 1960, however, the U.S. Department of Defense and NASA had fully recognized Goddard's achievements and paid his estate $1 million for the use of his 214 patented rocket-componentry designs. A year later, NASA named the Goddard Space Flight Center in Greenbelt, Maryland, in his honor.

Section 2

Artificial Satellites

The Echo Satellites

- The United States launched its first communications satellite, *Echo I,* into orbit on August 12, 1960. Surprisingly simple in its design, *Echo I* consisted of an aluminum-coated plastic balloon that inflated to a diameter of 30 m when it reached orbit. From a low Earth orbit, *Echo I* reflected radio signals back to Earth until 1968.

Is That a Fact!

◆ The *Echo II* satellite was part of the first cooperative space effort between the United States and the Soviet Union. A radio signal from an observatory in England was reflected off *Echo II* and was received in the Soviet Union.

Section 3

Space Probes

Soviet Lunar Probes

● Although Soviet cosmonauts never landed on the moon, their Luna space probes gathered a remarkable amount of lunar data using robotics and remotely controlled devices. In 1966, *Luna 9* became the first space probe to make a soft landing on the moon (previous probes crash-landed, and one shot past the moon into space). On impact, *Luna 9*'s egg-shaped instrument capsule rolled itself upright and automatically stabilized itself with four spring-loaded mechanisms. *Luna 9* sent the first television images of the lunar landscape back to Earth.

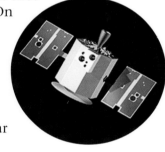

● Perhaps the most impressive of the Soviets' lunar space probes were *Luna 17* and *Luna 18*, which carried the eight-wheeled, heavy-duty lunar rovers *Lunokhod 1* and *2*. From Earth, the vehicles were directed around treacherous craters to cover vast expanses of the moon's surface. The rovers took photos, collected soil samples, and carried out other tests. *Lunokhod 1* traveled over the moon's surface for 11 months.

Section 4

People in Space

The Daily Routine Aboard Skylab

● Measuring 36 m long and 6.4 m high, *Skylab* was luxuriously large in comparison with previous space stations and had both working quarters and a living space. The living area included private sleeping quarters, a galley, a shower, and a suction toilet. Crew members carried out hundreds of astronomical and medical experiments.

● Astronauts were required to document everything they ate; measure the girth of their limbs, waists, and necks to check for muscle-tone loss; and wear electrodes while exercising so that their vital signs could be monitored. Astronauts did enjoy diversions such as "astrobatics"; in fact, *Skylab* astronaut Charles "Pete" Conrad commented, "We never went anywhere straight. We always did a somersault or a flip on the way."

Is That a Fact!

◆ By the time of the *Skylab* missions, the infamous spacebars and tubes of gooey "spacefood" had been replaced with more-palatable frozen, canned, and dehydrated foods. With more than 80 food items to choose from, a crew might whip up a breakfast of scrambled eggs, sausage, strawberries, bread and jam, orange juice, and coffee and finish out the day with a dinner of filet mignon, potato salad, and ice cream.

SciLINKS®

Developed and maintained by the National Science Teachers Association

SciLinks is maintained by the National Science Teachers Association to provide you and your students with interesting, up-to-date links that will enrich your classroom presentation of the chapter.

Visit www.scilinks.org and enter the SciLinks code for more information about the topic listed.

Topic: Rocket Technology
SciLinks code: HSM1323

Topic: History of NASA
SciLinks code: HSM0745

Topic: Artificial Satellites
SciLinks code: HSM0101

Topic: Space Probes
SciLinks code: HSM1432

Topic: Space Exploration and Space Stations
SciLinks code: HSM1430

Overview

This chapter discusses the development of rocket science. Students will learn about rockets, satellites, space probes, and space stations. In addition, the chapter discusses how the political climate after WWII led to the space race.

Assessing Prior Knowledge

Students should be familiar with the following topics:

- Newton's third law of motion
- the movement of bodies in the solar system

Identifying Misconceptions

Students may have some questions about the benefits of the space program. As students explore the concepts in this chapter, help them assess how developments in space technology benefit humanity. Many students also have misconceptions about rocket propulsion. Rockets do not move by pushing against air in the atmosphere. The reaction to the force and direction of the exhaust causes a rocket to move. Thus, rockets can accelerate in the vacuum of space, where there is nothing to push against. In fact, rockets accelerate more efficiently in space, where there is no friction. The Start-Up Activity will help address this misconception.

Exploring Space

About the PHOTO

Although the astronauts in the photo appear to be motionless, they are orbiting the Earth at almost 28,000 km/h! The astronauts reached orbit—about 300 km above the Earth's surface—in a space shuttle. Space shuttles are the first vehicles in a new generation of reusable spacecraft. They have opened an era of space exploration in which missions to space are more common than ever before.

PRE-READING ACTIVITY

Graphic Organizer

Chain-of-Events Chart Before you read the chapter, create the graphic organizer entitled "Chain-of-Events Chart" described in the **Study Skills** section of the Appendix. As you read the chapter, fill in the chart with a timeline that describes the exploration of space from the theories of Konstantin Tsiolkovsky to the future of space exploration.

Standards Correlations

National Science Education Standards

The following codes indicate the National Science Education Standards that correlate to this chapter. The full text of the standards is at the front of the book.

Chapter Opener
UCP 2; SAI 1

Section 1 Rocket Science
UCP 2, 3; HNS 1, 3

Section 2 Artificial Satellites
UCP 3, 4; SAI 1; ST 2; SPSP 5; HNS 3

Section 3 Space Probes
UCP 5; ST 2

Section 4 People in Space
UCP 4, 5; SAI 1; ST 2; SPSP 5; HNS 1, 3; *LabBook:* SAI 1; ST 1, 2

Chapter Lab
SAI 1; ST 1

Chapter Review
SAI 1; ST 2; HNS 1, 2, 3; SPSP 5

Science in Action
SPSP 5; HNS 1

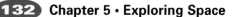

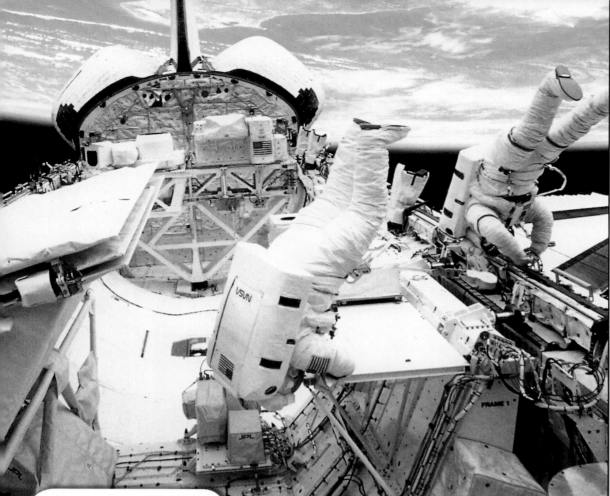

MATERIALS

FOR EACH GROUP
- balloon, large
- drinking straw
- meterstick
- poster board
- tape
- thread, 2 m

Teacher's Note: Discuss the limitations of using a balloon to model a launch vehicle. If the balloon were to "launch" something, what improvements should be made? (Students may suggest that the balloon would need to be more powerful and able to sustain thrust for a longer period of time. In addition, the balloon would need a steering or guidance device.)

Answers

1. Sample answer: The balloon traveled the same distance. The poster board did not affect the distance the balloon traveled because the balloon did not push off the poster board.

2. Sample answer: The balloon moved because of Newton's third law of motion. As air left the balloon, the balloon reacted by moving in the opposite direction. Hot gases escape from the bottom of a rocket, and the rocket reacts by moving in the opposite direction. Rockets do not push off a launch pad.

START-UP ACTIVITY

Balloon Rockets

In this activity you will launch a balloon "rocket" to learn about how rockets move.

Procedure

1. Insert a **2 m thread** through a **drinking straw,** and tie it between two objects that won't move, such as **chairs.** Make sure that the thread is tight.

2. Inflate a **large balloon.** Do not tie the neck of the balloon closed. Hold the neck of the balloon closed, and **tape** the balloon firmly to the straw, parallel to the thread.

3. Move the balloon to one end of the thread, and then release the neck of the balloon. Use a **meterstick** to record the distance the balloon traveled.

4. Repeat steps 2–3. This time, hold a piece of **poster board** behind the balloon.

Analysis

1. Did the poster board affect the distance that the balloon traveled? Explain your answer.

2. Newton's third law of motion states that for every action there is an equal and opposite reaction. Apply this idea and your observations of the balloon to explain how rockets accelerate. Do rockets move by "pushing off" a launch pad? Explain your answer.

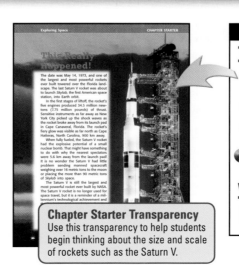

Chapter Starter Transparency
Use this transparency to help students begin thinking about the size and scale of rockets such as the Saturn V.

CHAPTER RESOURCES

Technology

- **Transparencies**
  - Chapter Starter Transparency

  READING SKILLS

- **Student Edition on CD-ROM**

- **Guided Reading Audio CD**
  - English or Spanish

- **Classroom Videos**
  - Brain Food Video Quiz

Workbooks

- **Science Puzzlers, Twisters & Teasers**
  - Exploring Space GENERAL

Overview

This section discusses the development of rocket technology and the establishment of NASA. The section also discusses the principles of rocket propulsion.

🔊 Bellringer

Ask students, "Why can't a commercial airplane be used for space exploration?" (Commercial airplanes cannot carry enough fuel for space exploration, their engines are not powerful enough to escape Earth's gravity, and they cannot withstand the extreme cold of space or the heat of reentry into Earth's atmosphere. Jet engines also rely on air for propulsion and for fuel combustion, and there is no air in space.)

Motivate

Demonstration —— GENERAL

Expanding Gas Attach a pre-stretched balloon over the mouth of a plastic bottle, and put the bottle in a bucket of hot water. Explain that as the gases in the balloon become hot, they expand and cause the balloon to inflate. The expansion of hot gases is powerful enough to launch rockets into space. As hot gases escape through the rocket nozzle, the rocket reacts by moving in the opposite direction—skyward. **LS Visual**

READING WARM-UP

Objectives

● Outline the development of rocket technology.

● Describe how a rocket accelerates.

● Explain the difference between orbital velocity and escape velocity.

Terms to Learn

rocket   thrust
NASA

READING STRATEGY

Discussion Read this section silently. Write down questions that you have about this section. Discuss your questions in a small group.

Figure 1 *Robert Goddard is known as the father of modern rocketry.*

CHAPTER RESOURCES

Chapter Resource File

- **Lesson Plan**
- **Directed Reading A** BASIC
- **Directed Reading B** SPECIAL NEEDS

Technology

🗄 **Transparencies**
- Bellringer

Rocket Science

If you could pack all of your friends in a car and drive to the moon, it would take about 165 days to get there. And that doesn't include stopping for gas or food!

The moon is incredibly far away, and years ago people could only dream of traveling into space. The problem was that no machine could generate enough force to overcome Earth's gravity and reach outer space. But about 100 years ago, a Russian high school teacher named Konstantin Tsiolkovsky (KAHN stuhn TEEN TSI uhl KAHV skee) proposed that machines called *rockets* could take people to outer space. A **rocket** is a machine that uses escaping gas to move. Tsiolkovsky stated, "The Earth is the cradle of mankind. But one does not have to live in the cradle forever." Rockets would become the key to leaving the cradle of Earth and starting the age of space exploration.

The Beginnings of Rocket Science

Tsiolkovsky's inspiration came from the imaginative stories of Jules Verne. In Verne's book *From the Earth to the Moon,* characters reached the moon in a capsule shot from an enormous cannon. Although this idea would not work, Tsiolkovsky proved—in theory—that rockets could generate enough force to reach outer space. He also suggested the use of liquid rocket fuel to increase a rocket's range. For his vision and careful work, Tsiolkovsky is known as the father of rocket theory.

A Boost for Modern Rocketry

Although Tsiolkovsky proved scientifically that rockets could reach outer space, he never built any rockets himself. That task was left to American physicist and inventor Robert Goddard, shown in **Figure 1.** Goddard launched the first successful liquid-fuel rocket in 1926. Goddard tested more than 150 rocket engines, and by the time of World War II, Goddard's work began to interest the United States military. His work drew much attention because of a terrifying new weapon that the German army had developed.

✓ **Reading Check** How did Tsiolkovsky and Goddard contribute to the development of rockets? (*See the Appendix for answers to Reading Checks.*)

Answer to Reading Check

Tsiolkovsky helped develop rocket theory. Goddard developed the first rockets.

From Rocket Bombs to Rocket Ships

Toward the end of World War II, Germany developed a new weapon known as the V-2 rocket. The V-2 rocket, shown in **Figure 2,** could deliver explosives from German military bases to London—a distance of about 350 km. The V-2 rocket was developed by a team led by Wernher von Braun, a young Ph.D. student whose research was supported by the German military. But in 1945, near the end of the war, von Braun and his entire research team surrendered to the advancing Americans. The United States thus gained 127 of the best German rocket scientists. With this gain, rocket research in the United States boomed in the 1950s.

The Birth of NASA

The end of World War II marked the beginning of the *Cold War*—a long period of political tension between the United States and the Soviet Union. The Cold War was marked by an arms race and by competition in space technology. In response to Soviet advances in space, the U.S. government formed the National Aeronautics and Space Administration, or **NASA,** in 1958. NASA combined all of the rocket-development teams in the United States. Their cooperation led to the development of many rockets, including those shown in **Figure 3.**

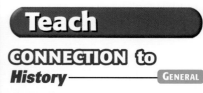

Figure 2 *The V-2 rocket is the ancestor of all modern rockets.*

rocket a machine that uses escaping gas from burning fuel to move

NASA the National Aeronautics and Space Administration

Figure 3	40 Years of NASA Rockets

A rocket's payload is the amount of material the rocket is able to carry into space.

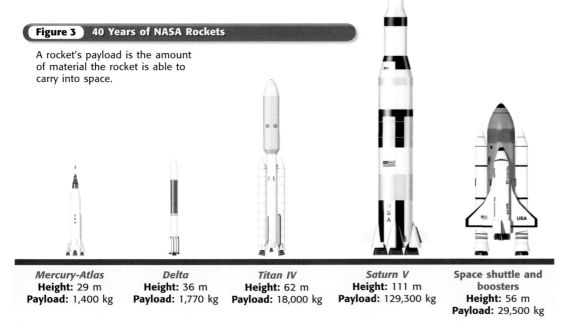

Mercury-Atlas	***Delta***	***Titan IV***	***Saturn V***	**Space shuttle and boosters**
Height: 29 m	**Height:** 36 m	**Height:** 62 m	**Height:** 111 m	
Payload: 1,400 kg	**Payload:** 1,770 kg	**Payload:** 18,000 kg	**Payload:** 129,300 kg	**Height:** 56 m
				Payload: 29,500 kg

Rocket Review Reproduce **Figure 4** on the board, and have student volunteers add descriptive labels that explain the action and reaction that accelerates a rocket. LS **Visual**

Quiz —————— GENERAL

1. How did NASA contribute to the United States' rocket program? (NASA coordinated the efforts of many rocket research teams, which led to the rapid development of a variety of rocket designs.)

2. How does Newton's third law of motion apply to rocket propulsion? (Newton's law states that for every action there is an equal and opposite reaction. When hot gases rush out of the bottom of a rocket, the rocket moves in the opposite direction.)

Alternative Assessment —— GENERAL

Poster Project Have students make a series of sketches that show the evolution of rocket design. Encourage students to use poster board so that they can make scale models to compare the size of the spacecraft. Have students write a brief description of each spacecraft. LS **Visual/Intrapersonal**

Answer to Reading Check

Rockets carry oxygen so that their fuel can be burned.

thrust the pushing or pulling force exerted by the engine of an aircraft or rocket

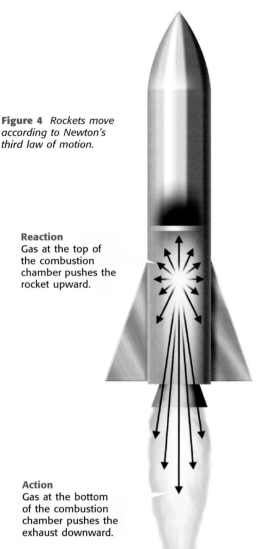

Figure 4 *Rockets move according to Newton's third law of motion.*

Reaction
Gas at the top of the combustion chamber pushes the rocket upward.

Action
Gas at the bottom of the combustion chamber pushes the exhaust downward.

How Rockets Work

If you are sitting in a chair that has wheels and you want to move, you would probably push away from a table or kick yourself along with your feet. Many people think that rockets move in a similar way—by pushing off of a launch pad. But if rockets moved in this way, how would they accelerate in the vacuum of space where there is nothing to push against?

For Every Action . . .

As you saw in the Start-Up Activity, the balloon moved according to Newton's third law of motion. This law states that for every action there is an equal and opposite reaction. For example, the air rushing backward from a balloon (the action) results in the forward motion of the balloon (the reaction). Rockets work in the same way. In fact, rockets were once called *reaction devices*.

However, in the case of rockets, the action and the reaction may not be obvious. The mass of a rocket—including all of the fuel it carries—is much greater than the mass of the hot gases that come out of the bottom of the rocket. But because the exhaust gases are under extreme pressure, they exert a huge amount of force. The force that accelerates a rocket is called **thrust.** Look at **Figure 4** to learn more about how rockets work.

You Need More Than Rocket Fuel

Rockets burn fuel to provide the thrust that propels them. In order for something to burn, oxygen must be present. Although oxygen is plentiful at the Earth's surface, there is little or no oxygen in the upper atmosphere and in outer space. For this reason, rockets that go into outer space must carry enough oxygen with them to be able to burn their fuel. The space shuttles, for example, carry hundreds of thousands of gallons of liquid oxygen. This oxygen is needed to burn the shuttle's rocket fuel.

✓ *Reading Check* Why do rockets carry oxygen in addition to fuel?

CONNECTION to
Physical Science — ADVANCED

Velocity Versus Speed It is important for students to understand the difference between velocity and speed. Speed is the rate at which an object moves. An object's velocity is the speed the object travels in a particular direction. Think of velocity as the rate of change in an object's position. This distinction is very important in rocket science. For example, rockets are always launched in the direction that Earth rotates. At the equator, Earth has a rotational velocity of nearly

0.5 km/s east. In other words, standing still, a rocket travels 0.5 km/s east, so, to reach orbital velocity, it has to increase its speed by only an additional 7.5 km/s. As an extension, show students the following locations on a world map: Cape Canaveral, Florida (28.5°N), and the ESA Spaceport in Korou, French Guiana (6°N). Ask students to explain why NASA chose a launch site in Florida, rather than a launch site in Maine or Alaska. (The Earth's rotational velocity is greatest at the equator, so rockets launched from areas close to the equator get the greatest boost possible.)

How to Leave the Earth

The gravitational pull of the Earth is the main factor that a rocket must overcome. As shown in **Figure 5**, a rocket must reach a certain *velocity*, or speed and direction, to orbit or escape the Earth.

Orbital Velocity and Escape Velocity

For a rocket to orbit the Earth, it must have enough thrust to reach orbital velocity. *Orbital velocity* is the speed and direction a rocket must travel in order to orbit a planet or moon. The lowest possible speed a rocket may go and still orbit the Earth is about 8 km/s (17,927 mi/h). If the rocket goes any slower, it will fall back to Earth. For a rocket to travel beyond Earth orbit, the rocket must achieve escape velocity. *Escape velocity* is the speed and direction a rocket must travel to completely break away from a planet's gravitational pull. The speed a rocket must reach to escape the Earth is about 11 km/s (24,606 mi/h).

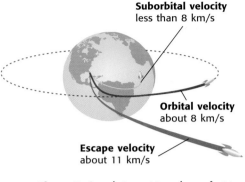

Suborbital velocity
less than 8 km/s

Orbital velocity
about 8 km/s

Escape velocity
about 11 km/s

Figure 5 *A rocket must travel very fast to escape the gravitational pull of the Earth.*

SECTION Review

Summary

- Tsiolkovsky and Goddard were pioneers of rocket science.

- The outcome of WWII and the political pressures of the Cold War helped advance rocket science.

- Rockets work according to Newton's third law of motion—for every action there is an equal and opposite reaction.

- Rockets need to reach different velocities to attain orbit and to escape a planet's gravitational attraction.

Using Key Terms

1. Use each of the following terms in a separate sentence: *rocket, thrust,* and *NASA.*

Understanding Key Ideas

2. What factor must a rocket overcome to reach escape velocity?
   a. Earth's axial tilt
   b. Earth's gravity
   c. the thrust of its engines
   d. Newton's third law of motion

3. Describe the contributions of Tsiolkovsky and Goddard to modern rocketry.

4. Use Newton's third law of motion to describe how rockets work.

5. What is the difference between orbital and escape velocity?

6. How did the Cold War accelerate the U.S. space program?

Math Skills

7. If you travel at 60 mi/h, it takes about 165 days to reach the moon. Approximately how far away is the moon?

Critical Thinking

8. **Applying Concepts** How do rockets accelerate in space?

9. **Making Inferences** Why does escape velocity vary depending on the planet from which a rocket is launched?

For a variety of links related to this chapter, go to www.scilinks.org

Topic: Rocket Technology
SciLinks code: HSM1323

CHAPTER RESOURCES

Chapter Resource File

- Section Quiz **GENERAL**
- Section Review **GENERAL**
- Vocabulary and Section Summary **GENERAL**
- Reinforcement Worksheet **BASIC**

Technology

Transparencies
- How a Rocket Works

SECTION

2

Focus

Overview

This section discusses the different kinds of satellites and satellite orbits. Students learn how remote sensing devices have helped us understand Earth as a global system.

🛰 Bellringer

Ask students to write two paragraphs describing the ways satellite technology affects their lives. (Students might mention the Global Positioning System, satellite television, satellite phones, or accurate weather forecasts.)

Motivate

CONNECTION ACTIVITY
Real World——————ADVANCED

TV Satellites Write the following information on the board, and ask students to use a globe to find the location of the geostationary communications satellites that relay the signals of their favorite television shows. Note that this information changes often because satellites are moved to different orbits.

- ABC uses *Telstar-5* at 97°E.

- CBS uses *Telstar-6* at 93°E.

- Discovery Channel and ESPN use *Galaxy-5* at 125°W.

- The Weather Channel uses *Satcom-C3* at 131°W.

READING WARM-UP

Objectives
- Identify the first satellites.
- Compare low Earth orbits with geostationary orbits.
- Explain the functions of military, communications, and weather satellites.
- Explain how remote sensing from satellites has helped us study Earth as a global system.

Terms to Learn
artificial satellite
low Earth orbit
geostationary orbit

READING STRATEGY

Reading Organizer As you read this section, make a table comparing the advantages and disadvantages of low Earth orbits and geostationary orbits.

Figure 1 *A model of Sputnik 1, the first satellite to orbit the Earth, is shown below. It started a revolution in modern life that led to technology such as the Global Positioning System.*

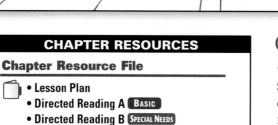

Artificial Satellites

You are watching TV, and suddenly a weather bulletin interrupts your favorite show. There is a HURRICANE WARNING! You grab a cell phone and call your friend—the hurricane is headed straight for where she lives!

In the story above, the TV show, the weather bulletin, and perhaps even the phone call were all made possible by artificial satellites orbiting thousands of miles above Earth! An **artificial satellite** is any human-made object placed in orbit around a body in space.

There are many kinds of artificial satellites. Weather satellites provide continuous updates on the movement of gases in the atmosphere so that we can predict weather on Earth's surface. Communications satellites relay TV programs, phone calls, and computer data. Remote-sensing satellites monitor changes in the environment. Perhaps more than the exploration of space, satellites have changed the way we live.

The First Satellites

The first artificial satellite, *Sputnik 1,* was launched by the Soviets in 1957. **Figure 1** shows a model of *Sputnik 1,* which orbited for 57 days before it fell back to Earth and burned up in the atmosphere. Two months later, *Sputnik 2* carried the first living being into space—a dog named Laika. The United States followed with the launch of its first satellite, *Explorer 1,* in 1958. The development of new satellites increased quickly. By 1964, communications satellite networks were able to send messages around the world. Today, thousands of satellites orbit the Earth, and more are launched every year.

CONNECTION to
Physical Science——GENERAL

Satellite Orbits Tell students that the orbit of satellites is determined by three major variables: gravity, velocity, and altitude. The force of gravity pulls satellites toward the Earth, so satellites must have a velocity great enough to remain in orbit. The closer a satellite is to Earth, the faster it must travel to remain in orbit. The result of these three variables is a curved path. **L5 Logical**

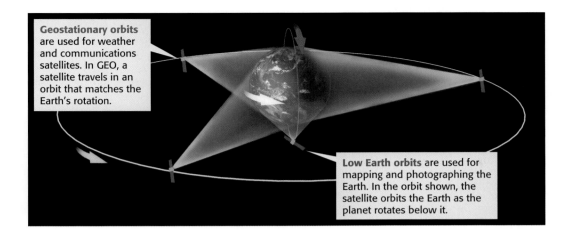

Geostationary orbits are used for weather and communications satellites. In GEO, a satellite travels in an orbit that matches the Earth's rotation.

Low Earth orbits are used for mapping and photographing the Earth. In the orbit shown, the satellite orbits the Earth as the planet rotates below it.

Figure 2 *Low Earth orbits are in the upper reaches of Earth's atmosphere, while geostationary orbits are about 36,000 km from Earth's surface.*

Choosing Your Orbit

Satellites are placed in different types of orbits, as shown in **Figure 2.** All of the early satellites were placed in **low Earth orbit** (LEO), which is a few hundred kilometers above the Earth's surface. LEO, while considered space, is still in the outermost part of Earth's atmosphere. A satellite in LEO moves around the Earth very quickly. This motion can place a satellite out of contact much of the time.

Communications satellites and most weather satellites orbit much farther from Earth. In this orbit, called a **geostationary orbit** (GEO), a satellite travels in an orbit that exactly matches the Earth's rotation. Thus, the satellite is always above the same spot on Earth. Ground stations are in continuous contact with these satellites so that TV programs and other communications will not be interrupted.

artificial satellite any human-made object placed in orbit around a body in space

low Earth orbit an orbit less than 1,500 km above the Earth's surface

geostationary orbit an orbit that is about 36,000 km above the Earth's surface and in which a satellite is above a fixed spot on the equator

✓ *Reading Check* **What is the difference between GEO and LEO?** *(See the Appendix for answers to Reading Checks.)*

Modeling LEO and GEO

1. Use a **length of thread** to measure 300 km on the scale of a **globe.**

2. Use another **length of thread** to measure 36,000 km on the globe's scale.

3. Use the short thread to measure the distance of LEO from the surface of the globe and the long thread to measure the distance of GEO from the surface of the globe.

4. Your teacher will turn off the lights. One student will spin the globe, while other students will hold **penlights** at LEO and GEO orbits.

5. Was more of the globe illuminated by the penlights in LEO or GEO?

6. Which orbit is better for communications satellites? Which orbit is better for spy satellites?

Answer to Math Practice

Atlanta, Georgia

Teacher's Notes: Students may need help measuring the distances on the map's scale. Show them how to use a piece of scrap paper with a straight-edge to measure the distances needed to adjust the compass.

Other Options: Roswell, New Mexico, is about 690 km from Oklahoma City, Oklahoma, 1,070 km from Salt Lake City, and 1,175 km from Monroe, Louisiana.

Discussion ———— ADVANCED

Polar Orbits Satellites such as those in the Landsat program are deployed in polar orbits; that is, they orbit Earth from pole to pole. Have students find out why polar orbits are best for mapping purposes. (One reason is that as the Landsat satellites orbit the Earth, the Earth rotates beneath them. In this way, the satellites can survey the entire planet without changing their orbit.)

Ask students to write a brief explanation of the advantages of polar orbits and draw a diagram showing the path a satellite in a polar orbit would take. **LS** Visual

Triangulation

GPS uses the principle of triangulation. To practice tri-angulation, use a drawing compass and a photocopy of a U.S. map that has a scale. Try to find a city that is 980 km from Detroit and Miami, and 950 km from Baltimore. For each city named, adjust the compass to the correct distance on the map's scale. Then, place the compass point on the city's location. Draw a circle with a radius equal to the given distance. Where do the circles overlap? Once you have solved this riddle, write one for a friend!

Figure 3 *This photo was taken in 1989 by a Soviet spy satellite in LEO about 220 km above San Francisco. Can you identify any objects on the ground?*

Military Satellites

Some satellites placed in LEO are equipped with cameras that can photograph the Earth's surface in amazing detail. It is possible to photograph objects as small as this book from LEO. While photographs taken by satellites are now used for everything from developing real estate to tracking the movements of dolphins, the technology was first developed by the military. Because satellites can take very detailed photos from hundreds of kilometers above the Earth's surface, they are ideal for defense purposes. The United States and the Soviet Union developed satellites to spy on each other right up to the end of the Cold War. **Figure 3,** for example, is a photo of San Francisco taken by a Soviet spy satellite in 1989. Even though the Cold War is over, spy satellites continue to play an important role in the military defense of many countries.

The Global Positioning System

In the past, people invented very complicated ways to keep from getting lost. Now, for less than $100, people can find out their exact location on Earth by using a Global Positioning System (GPS) receiver. GPS is another example of military satellite technology that has become a part of everyday life. The GPS consists of 27 solar-powered satellites that continuously send radio signals to Earth. From the amount of time it takes the signals to reach Earth, the hand-held receiver can calculate its distance from the satellites. Using the distance from four satellites, a GPS receiver can determine a person's location with great accuracy.

WEIRD SCIENCE

Geostationary satellites generally have a life span of 5 to 13 years. Because there are a limited number of locations in GEO, "dead" satellites must be disposed of in some way. Currently, satellites use their remaining propellant to navigate into a higher "graveyard orbit." Once in a grave-yard orbit, a dead satellite circles Earth every 2 to 5 days and does not interfere with operating satellites in geostationary orbits.

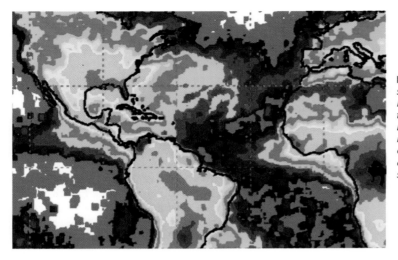

Figure 4 *This map shows average annual lightning strikes around the world. Red and black indicate a high number of strikes. Cooler colors, such as purple and blue, indicate fewer strikes.*

Weather Satellites

It is hard to imagine life without reliable weather forecasts. Every day, millions of people make decisions based on information provided by weather satellites. Weather satellites in GEO provide a big-picture view of the Earth's atmosphere. These satellites constantly monitor the atmosphere for the "triggers" that lead to severe weather conditions. Weather satellites in GEO created the map of world lightning strikes shown in **Figure 4.** Weather satellites in LEO are usually placed in polar orbits. Satellites in polar orbits revolve around the Earth in a north or south direction as the Earth rotates beneath them. These satellites, which orbit between 830 km and 870 km above the Earth, provide a much closer look at weather patterns.

Communications Satellites

Many types of modern communications use radio waves or microwaves to relay messages. Radio waves and microwaves are ideal for communications because they can travel through the air. The problem is that the Earth is round, but the waves travel in a straight line. So how do you send a message to someone on the other side of the Earth? Communications satellites in GEO solve this problem by relaying information from one point on Earth's surface to another. The signals are transmitted to a satellite and then sent to receivers around the world. Communications satellites relay computer data, and some television and radio broadcasts.

✓ Reading Check How do communications satellites relay information from one point on Earth's surface to another?

Tracking Satellites

A comfortable lawn chair and a clear night sky are all you need to track satellites. Just after sunset or before sunrise, satellites in LEO are easy to track. They look like slow-moving stars, and they generally move in a west to east direction. With a little practice, you should be able to find one or two satellites a minute. A pair of binoculars will help you get a closer look. Satellites in GEO are difficult to see because they do not appear to move. You and a parent can find out more about how to track specific satellites and space stations on the Internet.

CONNECTION to
Physical Science — GENERAL

The Heat of Reentry What causes objects to heat up as they enter the atmosphere? People generally assume that friction from high-speed collisions with air molecules generates this heat. Actually, friction plays a minor role relative to pressure. As an object, such as a spacecraft, enters our atmosphere, the object compresses a layer of air about a meter deep beneath it. Imagine a snowplow pushing a mound of snow before it. As the layer of air beneath the spacecraft reaches tremendous pressures, thermal energy is transferred to the surface of the spacecraft by conduction, which causes the surface to glow red-hot. If students have ever inflated a bike tire using a well-oiled pump, they have observed this effect: the repeated compression of air causes the pump to become hot. For this reason, LEO satellites orbit in the outer reaches of Earth's atmosphere at a point where air pressure is insignificant.
LS Logical

Answer to Reading Check

Information from one location is transmitted to a communications satellite. The satellite then sends the information to another location on Earth.

Notes for School-to-Home Activity:
Satellites do not produce their own light; they reflect sunlight and "Earthshine" off their surfaces. The best time to view satellites is about an hour after sunset or an hour before sunrise. Encourage students to lie on the ground or sit in a reclining chair and scan the skies for satellites. Students can use binoculars if binoculars are available. A satellite will look like a star moving across the sky in a straight line. The star will fade out of sight before it reaches the horizon because it will be in Earth's shadow. Satellites can be seen moving from west to east or from pole to pole. Geosynchronous satellites are difficult to detect because they do not appear to move and because they are so far away. There are many Web sites that will help you determine the location of satellites and other objects in orbit, including the space shuttles and the *International Space Station.*

Close

Reteaching —— BASIC

Orbital Review Have a student draw a diagram of GEO and LEO orbital paths. Ask students to help you add direction arrows that indicate the movement of satellites and the rotation of the Earth. Students should also help you fill in information about each orbit and the types of satellites that are placed in each orbit. Then, erase the information, and ask students to answer questions such as the following: "I want to place a spy satellite in an orbit in which the Earth rotates beneath it. What type of orbit should I use?" (polar LEO)
LS Visual

Quiz —— GENERAL

1. What was the name of the first satellite to be placed in orbit, and what nation launched it? (The Soviet Union launched *Sputnik 1*.)

2. What is a geostationary orbit? (In a geostationary orbit, a satellite travels at a speed that matches the rotational speed of Earth. A satellite completes an orbit in the same time that Earth completes one rotation.)

Alternative Assessment —— GENERAL

Satellite News Have each student bring in one article about a discovery made by a satellite using remote sensing. Have students share the articles with the class. **LS Verbal**

CONNECTION TO Environmental Science

WRITING SKILL **Space Junk** After more than 40 years of space launches, Earth orbits are getting cluttered with "space junk." The United States Space Command—a new branch of the military—tracks nearly 10,000 pieces of debris. Left uncontrolled, this debris may become a problem for space vehicles in the future. Write a creative illustrated proposal to clean up space junk.

Remote Sensing and Environmental Change

Using satellites, scientists have been able to study the Earth in ways that were never before possible. Satellites gather information by *remote sensing*. Remote sensing is the gathering of images and data from a distance. Remote-sensing satellites measure light and other forms of energy that are reflected from Earth. Some satellites use radar, which bounces high-frequency radio waves off the Earth and measures the returned signal.

Landsat: Monitoring the Earth from Orbit

One of the most successful remote-sensing projects is the Landsat program, which began in 1972 and continues today. It has given us the longest continuous record of Earth's surface as seen from space. Landsat satellites gather images in several wavelengths—from visible light to infrared. **Figure 5** shows Landsat images of part of the Mississippi Delta. One image was taken in 1973, and the other was taken in 2003. The two images reveal a pattern of environmental change over a 30-year period. The main change is a dramatic reduction in the amount of silt that is reaching the delta. A comparison of the images also reveals a large-scale loss of wetlands in the bottom left of the delta in 2003. The loss of wetlands affects plants and animals living on the delta and the fishing industry.

Figure 5 The Loss of Wetlands in the Mississippi Delta

Silt reaching the Mississippi Delta is shown in blue. In 1973 (left), the amount of silt reaching the delta was much greater than in 2003 (right). This reduction led to the rapid loss of wetlands, which are green in this image. Notice the lower left corner of the delta in both images. Areas of wetland loss are black.

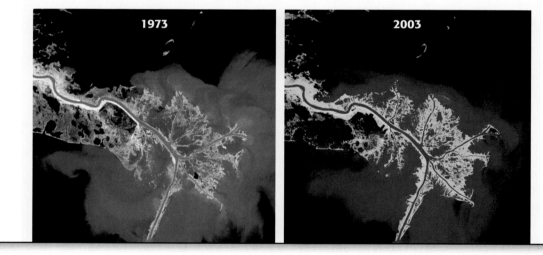

Cultural Awareness GENERAL

INTELSAT INTELSAT is an international not-for-profit communications satellite cooperative representing more than 140 nations. Decisions about the system's upkeep and future are reached by consensus among the member nations. INTELSAT satellites relay telephone calls, television broadcasts, and other telecommunications data. During the 1998 Winter Olympics, INTELSAT linked people from China, Germany, South Africa, the United States, Australia, and Japan to form an international 2,000-member chorus. Have groups of students work together to write a proposal to launch a satellite that would benefit the global community. **LS Interpersonal**

A New Generation of Remote-Sensing Satellites

The Landsat program has produced millions of images that are used to identify and track environmental change on Earth. Satellite remote sensing allows scientists to perform large-scale mapping, look at changes in patterns of vegetation growth, map the spread of urban development, and study the effect of humans on the global environment. In 1999, NASA launched *Terra 1*, the first satellite in NASA's Earth Observing System (EOS) program. Satellites in the EOS program are designed to work together so that they can gather integrated data on environmental change on the land, in the atmosphere, in the oceans, and on the icecaps.

Reading Check What is unique about the EOS program?

INTERNET ACTIVITY

For another activity related to this chapter, go to **go.hrw.com** and type in the keyword **HZ5EXPW**.

SECTION Review

Summary

- *Sputnik 1* was the first artificial satellite. *Explorer 1* was the first U.S. satellite.
- Low Earth orbits are used for making detailed images of the Earth.
- Geostationary orbits are used for communications, navigation, and weather satellites.
- Satellites with remote sensing technology have helped us understand the Earth as a global system.

Using Key Terms

1. Use each of the following terms in a separate sentence: *artificial satellite, low Earth orbit,* and *geostationary orbit.*

Understanding Key Ideas

2. In a low Earth orbit, the speed of a satellite is
   a. slower than the rotational speed of the Earth.
   b. equal to the rotational speed of the Earth.
   c. faster than the rotational speed of the Earth.
   d. None of the above

3. What was the name of the first satellite placed in orbit?

4. List three ways that satellites benefit human society.

5. What was the *Explorer 1*?

6. Explain the differences between LEO and GEO satellites.

7. How does the Global Positioning System work?

8. How do communications satellites relay signals around the curved surface of Earth?

Math Skills

9. The speed required to reach Earth orbit is 8 km/s. What does this equal in *meters per hour*?

Critical Thinking

10. **Applying Concepts** The *Hubble Space Telescope* is located in LEO. Does the telescope move faster or slower around the Earth compared with a geostationary weather satellite? Explain.

11. **Applying Concepts** To triangulate your location on a map, you need to know your distance from three points. If you knew your distance from two points, how many possible places could you occupy?

SCILINKS

Developed and maintained by the National Science Teachers Association

For a variety of links related to this chapter, go to www.scilinks.org

Topic: Artificial Satellites
SciLinks code: HSM0101

Answer to Reading Check

Satellites in the EOS program are designed to work together so that many different types of data can be integrated.

CHAPTER RESOURCES

Chapter Resource File
- Section Quiz GENERAL
- Section Review GENERAL
- Vocabulary and Section Summary GENERAL
- Datasheet for Quick Lab

Technology

Transparencies
- Landsat Data

Answers to Section Review

1. Sample answer: Artificial satellites are used for communications and for studying the Earth. Because the satellite was placed in low Earth orbit, it could make very detailed images of the Earth's surface. Communications satellites are placed in geostationary orbits.

2. c

3. *Sputnik 1*

4. Answers may vary. Students could note that satellites relay communications, help make accurate weather forecasts, and study environmental change using remote sensing.

5. *Explorer 1* was the first satellite launched by the United States.

6. Answers may vary. LEO is much closer to the Earth than GEO. Satellites in LEO are often used for studying the Earth in detail. Satellites in GEO are often used for communications and navigation.

7. The Global Positioning System uses a network of 27 satellites that sends signals to the Earth. A GPS receiver interprets the signals from four satellites and determines the location of the user based on the amount of time it takes the signals to reach the receiver. Using that information, the receiver can determine the position of the user using triangulation.

8. Answers may vary. Students should note that a signal is sent to a satellite in GEO. Then, the satellite transmits the signal back to a ground station on Earth.

9. 8,000 m/s × 3,600 s/h = 28,800,000 m/h

10. The *Hubble Space Telescope* will orbit faster than a satellite in GEO because objects in LEO orbit at a much faster speed than objects in GEO.

11. If you knew your distance from two points on a map, you could be in two possible locations.

SECTION

3

Focus

Overview

This section describes some of the discoveries made by the earliest space probes. Students will also learn about the data gathered by recent probes that have visited the inner and outer planets. Finally, students will learn about a new approach to space exploration—"faster, cheaper, and better."

Bellringer

Ask students to consider the following questions: "Does exploring other planets benefit us here on Earth? Why or why not?"

Motivate

ACTIVITY ——————— GENERAL

Design Your Own Space Mission Have students imagine that they could send a space probe anywhere in the solar system. In their **science journal**, have students write a paragraph about where they would send their probe, what kind of instruments it would carry, what kind of data it would collect, and what its primary mission would be. Invite volunteers to read their paragraphs to the class and elaborate on their choices. **LS Verbal/Visual**

READING WARM-UP

Objectives

- Describe five discoveries made by space probes.
- Explain how space-probe missions help us better understand the Earth.
- Describe how NASA's new strategy of "faster, cheaper, and better" relates to space probes.

Terms to Learn

space probe

READING STRATEGY

Reading Organizer As you read this section, make a concept map showing the space probes, the planetary bodies they visited, and their discoveries.

space probe an uncrewed vehicle that carries scientific instruments into space to collect scientific data

Space Probes

What does the surface of Mars look like? Does life exist anywhere else in the solar system?

To answer questions like these, scientists send space probes to explore the solar system. A **space probe** is an uncrewed vehicle that carries scientific instruments to planets or other bodies in space. Unlike satellites, which stay in Earth orbit, space probes travel away from the Earth. Space probes are valuable because they can complete missions that would be very dangerous and expensive for humans to undertake.

Visits to the Inner Solar System

Because Earth's moon and the inner planets are much closer than the other planets and moons in the solar system, they were the first to be explored by space probes. Let's take a closer look at some missions to the moon, Venus, and Mars.

Luna and Clementine: Missions to the Moon

Luna 1, the first space probe, was launched by the Soviets in 1959 to fly past the moon. In 1966, *Luna 9* made the first soft landing on the moon's surface. During the next 10 years, the United States and the Soviet Union completed more than 30 lunar missions. Thousands of images of the moon's surface were taken. In 1994, the United States probe *Clementine* discovered that craters of the moon may contain water left by comet impacts. In 1998, the *Lunar Prospector* confirmed that frozen water exists on the moon. This ice would be very valuable to a human colony on the moon.

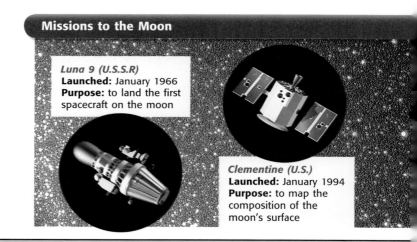

Missions to the Moon

Luna 9 (U.S.S.R)
Launched: January 1966
Purpose: to land the first spacecraft on the moon

Clementine (U.S.)
Launched: January 1994
Purpose: to map the composition of the moon's surface

CHAPTER RESOURCES

Chapter Resource File

- **Lesson Plan**
- **Directed Reading A** BASIC
- **Directed Reading B** SPECIAL NEEDS

Technology

Transparencies
- Bellringer

Is That a Fact!

As a rocket moves away from Earth, gravitational pull and air resistance decrease. In addition, the rocket's mass decreases as the rocket burns fuel. Thus, rockets accelerate as they travel from Earth's surface. The space shuttle accelerates from 0 km/h to 27,000 km/h in a little more than 8 min!

Venera 9: The First Probe to Land on Venus

The Soviet probe *Venera 9* was the first probe to land on Venus. The probe parachuted into Venus's atmosphere and transmitted images of the surface to Earth. *Venera 9* found that the surface temperature and atmospheric pressure on Venus are much higher than on Earth. The surface temperature of Venus is an average of 464°C—hot enough to melt lead! *Venera 9* also found that the chemistry of the surface rocks on Venus is similar to that of Earth rocks. Perhaps most important, *Venera 9* and earlier missions revealed that Venus has a severe greenhouse effect. Scientists study Venus's atmosphere to learn about the effects of increased greenhouse gases in Earth's atmosphere.

The Magellan Mission: Mapping Venus

In 1989, the United States launched the *Magellan* probe, which used radar to map 98% of the surface of Venus. The radar data were transmitted back to Earth where computers used the data to generate three-dimensional images like the one shown in **Figure 1.** The Magellan mission showed that, in many ways, the geology of Venus is similar to that of Earth. Venus has features that suggest plate tectonics occurs there, as it does on Earth. Venus also has volcanoes, and some of them may be active.

✓ Reading Check What discoveries were made by *Magellan?* (*See the Appendix for answers to Reading Checks.*)

Missions to Venus

Venera 9 (U.S.S.R.)
Launched: June 1975
Purpose: to record the surface conditions of Venus

Magellan (U.S.)
Launched: May 1989
Purpose: to make a global map of the surface of Venus

Figure 1 *This false-color image of volcanoes on the surface of Venus was made with radar data transmitted to Earth by* Magellan.

Science Bloopers

Life on Venus? For the first half of the 20th century, a popular theory suggested that every planet had once supported life or would support it in the future. The theory was based on the idea that the sun has gradually cooled since its formation. Thus, Mars had once harbored life; it was currently Earth's turn, and Venus would be next. Some scientists thought that Venus might already be home to primitive life-forms equivalent to those of Earth's Cambrian period. However, the 1962 *Mariner 2* flyby showed that the surface temperature of Venus was a constant 464°C. We also know now that the sun is hotter now than when it formed 4.5 billion years ago.

CONNECTION to
Physical Science — GENERAL

Gravity Assists *Voyagers 1* and *2, Galileo,* and *Ulysses* have used a maneuver called a *gravity assist* to explore the solar system. In a gravity assist, spacecraft make use of a planet's gravitational pull to accelerate, slow down, or change direction. Accomplishing such maneuvers using gravity is a triumph of Newtonian physics and saves a tremendous amount of fuel. A slingshot analogy is sometimes used to describe gravity assists because of the way the spacecraft swings around the planet and is released, slingshotting out into space in a new flight path. *Voyagers 1* and *2* gained momentum with gravity assists from Jupiter, Saturn, Uranus, and Neptune. The *Ulysses* space probe obtained a gravity assist from Jupiter that sent it into a highly inclined trajectory, which eventually placed it in a polar orbit around the sun.

Missions to Mars

Viking 2 (U.S.)
Launched: September 1975
Purpose: to search for life on the surface of Mars

Mars Pathfinder (U.S.)
Launched: December 1996
Purpose: to use inexpensive technology to study the surface of Mars

Figure 2 *The Mars Pathfinder took detailed photographs of the Martian surface. Photographs, such as this one, revealed evidence of massive flooding.*

The Viking Missions: Exploring Mars

In 1975, the United States sent a pair of probes—*Viking 1* and *Viking 2*—to Mars. The surface of Mars is more like the Earth's surface than that of any other planet. For this reason, one of the main goals of the Viking missions was to look for signs of life. The probes contained instruments designed to gather soil and test it for evidence of life. However, no hard evidence was found. The Viking missions did find evidence that Mars was once much warmer and wetter than it is now. This discovery led scientists to ask even more questions about Mars. Did the Martian climate once support life? Why and when did the Martian climate change?

The Mars Pathfinder Mission: Revisiting Mars

More than 20 years later, in 1997, the surface of Mars was visited again by a NASA space probe. The goal of the Mars Pathfinder mission was to show that Martian exploration is possible at a much lower cost than the Viking missions. The probe sent back detailed images of dry water channels on the planet's surface. These images, such as the one shown in **Figure 2,** suggest that massive floods flowed across the surface of Mars relatively recently in the planet's past. The *Mars Pathfinder* successfully landed on Mars and deployed the *Sojourner* rover. *Sojourner* traveled across the surface of Mars for almost three months, collecting data and recording images. The European Space Agency and NASA have many more Mars missions planned for the near future. These missions will pave the way for a crewed mission to Mars that may occur in your lifetime!

✓ **Reading Check** What discoveries were made by the Mars Pathfinder mission?

CONNECTION ACTIVITY
History — GENERAL

Writing **The United Nations Outer Space Treaty** On January 23, 1967, the United Nations Outer Space Treaty was signed. The treaty guarantees all nations the freedom to explore and use space. The treaty emphasizes a humanistic and pacifist philosophy, which governs the actions of countries as they explore outer space, the moon, and other objects in space. Have students research this treaty and outline its major points. **LS** Intrapersonal

Answer to Reading Check
The Mars Pathfinder mission found evidence that water once flowed across the surface of Mars.

Visits to the Outer Solar System

The planets in the outer solar system—Jupiter, Saturn, Uranus, Neptune, and Pluto—are very far away. Probes such as those described below can take 10 years or more to complete their missions.

Pioneer and Voyager: To Jupiter and Beyond

The *Pioneer 10* and *Pioneer 11* space probes were the first to visit the outer planets. Among other things, these probes sampled the *solar wind*—the flow of particles coming from the sun. The Pioneer probes also found that the dark belts on Jupiter provide deep views into Jupiter's atmosphere. In 1983, *Pioneer 10* became the first probe to travel past the orbit of Pluto, the outermost planet.

The Voyager space probes were the first to detect Jupiter's faint rings, and *Voyager 2* was the first probe to fly by the four gas giants—Jupiter, Saturn, Uranus, and Neptune. The paths of the Pioneer and Voyager space probes are shown in **Figure 3.** Today, they are near the solar system's edge and some are still sending back data.

The Galileo Mission: A Return to Jupiter

The *Galileo* probe arrived at Jupiter in 1995. While *Galileo* itself began a long tour of Jupiter's moons, it sent a smaller probe into Jupiter's atmosphere to measure its composition, density, temperature, and cloud structure. *Galileo* gathered data about the geology of Jupiter's major moons and Jupiter's magnetic properties. The moons of Jupiter proved to be far more exciting than the earlier Pioneer and Voyager images had suggested. *Galileo* discovered that two of Jupiter's moons have magnetic fields and that one of its moons, Europa, may have an ocean of liquid water under its icy surface.

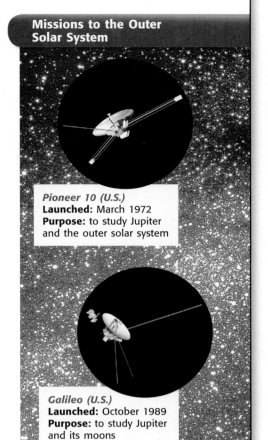

Missions to the Outer Solar System

Pioneer 10 (U.S.)
Launched: March 1972
Purpose: to study Jupiter and the outer solar system

Galileo (U.S.)
Launched: October 1989
Purpose: to study Jupiter and its moons

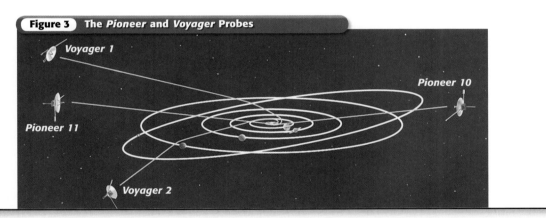

Figure 3 The *Pioneer* and *Voyager* Probes

Voyager 1
Pioneer 10
Pioneer 11
Voyager 2

CHAPTER RESOURCES

Technology

Transparencies
• Space Probes in the Outer Solar System

The Voyager Probes When the Voyager space probes were launched in 1977, each carried two gold-plated phonograph records with a variety of sounds and music intended as a message for intelligent life in the universe. The probes also carried a variety of images and written messages. Share the following list with students, and discuss the kinds of historical, scientific, and cultural information that the Voyagers carry:

• greetings from Earth spoken in 55 languages, ranging from ancient Sumerian to English

• printed messages from U.S. President Jimmy Carter and U.N. Secretary General Kurt Waldheim

• the sound of surf, wind, thunder, birdcalls, cricket chirps, and a whale song

• musical selections ranging from bagpipe music from Azerbaijan to "Johnny B. Goode" by Chuck Berry

• a diagram of the solar system

• anatomical drawings of a man and a woman

• a variety of images, ranging from the Great Wall of China to rush-hour traffic in India to a house in New England

A complete description of the contents of the Voyager probes can be found on NASA's Web site using the keyword "Golden Record."

Note: Some pictures, such as a photograph of a nursing mother, may be inappropriate for children. You may want to print selected photographs for students.
LS Visual/Auditory

Reteaching — BASIC

Space Probe Profiles Have students create a poster of each space probe that includes pictures of the probe and information about its mission and discoveries. **LS Visual**

Quiz — GENERAL

1. How is a space probe different from an artificial satellite? (A space probe carries scientific instruments to other bodies in space. Probes do not orbit Earth. Students may also note that some probes become satellites of other planets.)

2. What was the main goal of the Viking missions? (to look for signs of life on Mars)

Alternative Assessment — GENERAL

Debate: Faster, Cheaper, and Better? Some scientists think that traditional space probe projects are obsolete. These projects are expensive and require considerable support staff to receive and interpret the data. Others feel that large, complex probes are the best way to explore the solar system. Larger probes can carry more equipment and can gather more information than smaller, cheaper probes. Have students form two groups to research this debate. Student groups should present their findings as if they were NASA scientists going before Congress to secure funding for their projects. **LS Verbal**

Figure 4 *This artist's view shows the* Huygens *probe parachuting to the surface of Saturn's moon Titan. Saturn and* Cassini *are in the background.*

CONNECTION TO Social Studies

Cosmic Message in a Bottle When the Voyager space probes were launched in 1977, they carried a variety of messages intended for alien civilizations that might find them. In addition to greetings spoken in 55 different languages, a variety of songs, nature sounds, a diagram of the solar system, and photographs of life on Earth were included. Find out more about the message carried by the Voyager missions, and then create your own cosmic message in a bottle. **ACTIVITY**

The Cassini Mission: Exploring Saturn's Moons

In 1997, the *Cassini* space probe was launched on a seven-year journey to Saturn where it will make a grand tour of Saturn's moons. As shown in **Figure 4,** a smaller probe, called the *Huygens* probe, will detach itself from *Cassini* and descend into the atmosphere of Saturn's moon Titan. Scientists are interested in Titan's atmosphere because it may be similar to the Earth's early atmosphere. Titan's atmosphere may reveal clues about how life developed on Earth.

Faster, Cheaper, and Better

The early space probe missions were very large and costly. Probes such as *Voyager 2* and *Galileo* took years to develop. Now, NASA has a vision for missions that are "faster, cheaper, and better." One new program, called Discovery, seeks proposals for smaller science programs. The first six approved Discovery missions include sending small space probes to asteroids, landing on Mars again, studying the moon, and returning comet dust to Earth.

Stardust: Comet Detective

Launched in 1999, the *Stardust* space probe is the first probe to focus only on a comet. The probe will arrive at the comet in 2004 and gather samples of the comet's dust tail. It will return the samples to Earth in 2006. For the first time, material from beyond the orbit of the moon will be brought back to Earth. The comet dust should help scientists better understand how the solar system formed.

✓ **Reading Check** What is the mission of the *Stardust* probe?

Notes for Connection to Social Studies: Have students consider these questions: What important events or discoveries have occurred since 1977? What information would you consider most important? What kinds of music or images would you want aliens to know about? How would you design an intelligible message for alien civilizations?

Answer to Reading Check

The mission of the Stardust probe is to gather samples from a comet's tail and return them to Earth.

CONNECTION to Physical Science — GENERAL

Ion Propulsion In *Deep Space 1*, xenon gas is bombarded with electrons to create ions. Positive ions are drawn toward a high voltage grid and are expelled at 30 km/s. This may sound fast, but the thrust generated by *Deep Space 1* is 10,000 times weaker than that generated by a typical space probe. Although the thrust is equivalent to the weight of one sheet of paper on Earth, the probe gradually accelerates to incredible speeds over many months.

Deep Space 1: Testing Ion Propulsion

Another NASA project is the New Millennium program. Its purpose is to test new technologies that can be used in the future. *Deep Space 1,* shown in **Figure 5,** is the first mission of this program. It is a space probe with an ion-propulsion system. Instead of burning chemical fuel, an ion rocket uses charged particles that exit the vehicle at high speed. An ion rocket follows Newton's third law of motion, but it does so using a unique source of propulsion. Ion propulsion is like sitting on the back of a truck and shooting peas out of a straw. If there were no friction, the truck would gradually accelerate to tremendous speeds.

Figure 5 Deep Space 1 *uses a revolutionary type of propulsion— an ion rocket.*

SECTION Review

Summary

- Exploration with space probes began with missions to the moon. Space probes then explored other bodies in the inner solar system.
- Space-probe missions to Mars have focused on the search for signs of water and life.
- The Pioneer and Voyager programs explored the outer solar system.
- Space probe missions have helped us understand Earth's formation and environment.
- NASA's new strategy of "faster, cheaper, and better" seeks to create space-probe missions that are smaller than those of the past.

Using Key Terms

The statements below are false. For each statement, replace the underlined term to make a true statement.

1. <u>Luna 1</u> discovered evidence of water on the moon.
2. <u>Venera 9</u> helped map 98% of Venus's surface.
3. <u>Stardust</u> uses ion propulsion to accelerate.

Understanding Key Ideas

4. What is the significance of the discovery of evidence of water on the moon?
   a. Water is responsible for the formation of craters.
   b. Water was left by early space probes.
   c. Water could be used by future moon colonies.
   d. The existence of water proves that there is life on the moon.
5. Describe three discoveries that have been made by space probes.
6. How do missions to Venus, Mars, and Titan help us understand Earth's environment?

Math Skills

7. Traveling at the speed of light, signals from *Voyager 1* take about 12 h to reach Earth. The speed of light is about 299,793 km/s, how far away is the probe?

Critical Thinking

8. **Making Inferences** Why did we need space probes to discover water channels on Mars and evidence of ice on Europa?
9. **Expressing Opinions** What are the advantages of the new Discovery program over the older space-probe missions, and what are the disadvantages?
10. **Applying Concepts** How does *Deep Space 1* use Newton's third law of motion to accelerate?

SCILINKS
Developed and maintained by the National Science Teachers Association

For a variety of links related to this chapter, go to www.scilinks.org

Topic: Space Probes
SciLinks code: HSM1342

CONNECTION to Physical Science—GENERAL

Forming Ions *Deep Space 1* carried 81 kg of xenon gas propellant, which is enough to operate the thruster at one-half throttle for more than 20 months. Use the teaching transparency "Forming Positive and Negative Ions" to show students how the xenon ions that propel the probe are formed.

Focus

Overview

This section explores the political rivalry that led to the Apollo program. The section discusses how reusable space shuttles have revolutionized space travel and research. Students also learn about *Skylab*, *Mir*, and the *International Space Station*.

🔔 Bellringer

Ask students to write a letter from a space station orbiting Earth. They should describe the station, their mission, and their day-to-day lives.

Motivate

Discussion ——— GENERAL

A "Walk" on the Moon Read students *Apollo 12* astronaut Alan Bean's description of his 1969 moon walk:

> "After pushing off on one foot, there will be a long wait until you land on the other, exactly like running in slow motion. . . . As you run, you'll feel as if you're leaping long, impossible distances. And in fact you are."

This and other quotes from Apollo astronauts may help students experience the excitement of space exploration by humans.
LS Verbal

READING WARM-UP

Objectives
- Summarize the history and future of human spaceflight.
- Explain the benefits of crewed space programs.
- Identify five "space-age spinoffs" that are used in everyday life.

Terms to Learn
space shuttle
space station

READING STRATEGY

Reading Organizer As you read this section, make a flowchart that shows the events of the space race.

People in Space

One April morning in 1961, a rocket stood on a launch pad in a remote part of the Soviet Union. Inside, a 27-year-old cosmonaut named Yuri Gagarin sat and waited. He was about to do what no human had done before—travel to outer space. No one knew if the human brain would function in space or if he would be instantly killed by radiation.

On April 12, 1961, Yuri Gagarin, shown in **Figure 1,** became the first human to orbit Earth. The flight lasted 108 minutes. An old woman, her granddaughter, and a cow were the first to see Gagarin as he safely parachuted back to Earth, but the news of his success was quickly broadcast around the world.

The Race Is On

The Soviets were first once again, and the Americans were concerned that their rivals were winning the space race. Therefore, on May 25, 1961, President Kennedy announced, "I believe that the nation should commit itself to achieving the goal, before this decade is out, of landing a man on the moon and returning him safely to the Earth. No single project in this period will be more impressive to mankind, or more important for the long range exploration of space."

Kennedy's speech took everyone by surprise—even NASA's leaders. Go to the moon? We had not even reached orbit yet! In response to Kennedy's challenge, a new spaceport called Kennedy Space Center was built in Florida and Mission Control was established in Houston, Texas. In February 1962, John Glenn became the first American to orbit the Earth.

Figure 1 *In 1961, Yuri Gagarin (left) became the first person in space. In 1962, John Glenn (right) became the first American to orbit the Earth.*

CHAPTER RESOURCES

Chapter Resource File

- **Lesson Plan**
- **Directed Reading A** BASIC
- **Directed Reading B** SPECIAL NEEDS

Technology

🗄 **Transparencies**
- Bellringer

CONNECTION to
Life Science ——— GENERAL

The Smallest Astronauts During the second crewed mission to the moon (*Apollo 12,* November 14–24, 1969), astronauts retrieved some pieces of the space probe *Surveyor.* Amazingly, the pieces of *Surveyor* harbored bacteria from Earth that survived for 2 1/2 years in the moon's dry, near-vacuum environment!

"The Eagle Has Landed"

Seven years later, on July 20, 1969, Kennedy's challenge was met. The world watched on television as the *Apollo 11* landing module—the *Eagle,* shown in **Figure 2**—landed on the moon. Neil Armstrong became the first human to set foot on a world other than Earth. This moment forever changed the way we view ourselves and our planet. The Apollo missions also contributed to the advancement of science. *Apollo 11* returned moon rocks to Earth for study. Its crew also put devices on the moon to study moonquakes and the solar wind.

The Space Shuttle

The Saturn V rockets, which carried the Apollo astronauts to the moon, were huge and very expensive. They were longer than a football field, and each could be used only once. To save money, NASA began to develop the space shuttle program in 1972. A **space shuttle** is a reusable space vehicle that takes off like a rocket and lands like an airplane.

The Space Shuttle Gets off the Ground

Columbia, the first space shuttle, was launched on April 12, 1981. Since then, NASA has completed more than 100 successful shuttle missions. If you look at the shuttle *Endeavour* in **Figure 3,** you can see its main parts. The orbiter is about the size of an airplane. It carries the astronauts and payload into space. The liquid-fuel tank is the large red column. Two white solid-fuel booster rockets help the shuttle reach orbit. Then they fall back to Earth along with the fuel tank. The booster rockets are reused, the fuel tank is not. After completing a mission, the orbiter returns to Earth and lands like an airplane.

Reading Check What are the main parts of the shuttle? (*See the Appendix for answers to Reading Checks.*)

Shuttle Tragedies

On January 28, 1986, the booster rocket on the space shuttle *Challenger* exploded just after takeoff, killing all seven of its astronauts. On board was Christa McAuliffe, who would have been the first teacher in space. Investigations found that cold weather on the morning of the launch had caused rubber gaskets in the solid fuel booster rockets to stiffen and fail. The failure of the gaskets led to the explosion. The shuttle program resumed in 1988. In 2003, however, the space shuttle *Columbia* exploded as it reentered the atmosphere. All seven astronauts onboard were killed. These disasters emphasize the dangers of space exploration that continue to challenge scientists and engineers.

Figure 2 *Neil Armstrong took this photo of Edwin "Buzz" Aldrin as Aldrin was about to become the second human to set foot on the moon.*

space shuttle a reusable space vehicle that takes off like a rocket and lands like an airplane

Figure 3 *The space shuttles are the first reusable space vehicles.*

As *Apollo 11* approached the moon, a special mechanism kept it rotating slowly. Had the spacecraft not been rotating, the side exposed to the sun would have quickly overheated. The Apollo astronauts comically referred to this rotisserie-like movement as the "barbecue mode." When Neil Armstrong hopped off the ladder onto the moon's surface and made his historic speech, he bungled his line. The astronaut meant to say, "One small leap for *a* man, one giant leap for mankind." This flub inspired numerous jokes, including one by Apollo astronaut Pete Conrad, who was known for his sense of humor. As the relatively short astronaut stepped onto the moon, he joked, "Whoopie! Man, that may have been a small one for Neil, but that's a long one for me." When the poet Joseph Brodsky won the Nobel Prize for literature in 1987, he quipped, "A big step for me, a small one for mankind."

Teach

READING STRATEGY ── GENERAL

Prediction Guide Have students predict whether the following statements are true or false:

- There is only one space shuttle. (false)
- Both the United States and the Soviet Union have launched space stations. (true)
- Many nations are collaborating on an international space station. (true)

LS Logical

Demonstration ── BASIC

O-Ring Failure The day of the Challenger disaster was unseasonably cold for Florida, with temperatures below freezing. The investigation that followed the tragedy traced the explosion to the failure of O-ring seals that were designed to prevent hot exhaust gases from leaking out of the spacecraft's rocket boosters. Due to the cold temperatures, the rings had stiffened and failed to seal effectively. As a result, one of the rocket boosters leaked, which led to a catastrophic explosion 73 s after liftoff. To replicate physicist Richard Feynman's dramatic demonstration of how the cold weather contributed to the O-ring failure, pass a rubber gasket from a hardware store around class. Allow students to note the gasket's pliability. Then, place the gasket in a glass of ice water for a minute. Pass the gasket around again, and allow students to examine how inflexible the gasket became.

LS Kinesthetic

Answer to Reading Check
the orbiter, the liquid-fuel tank, and the solid-fuel booster rockets

Debate — GENERAL

Space Exploration: Does the Expense Outweigh the Benefits?
Encourage students to debate the costs of space exploration versus the potential benefits to humankind. Students should recognize that although space exploration is very expensive, measuring or predicting how much could be gained by exploration is difficult. The space industry also employs many people, including scientists, engineers, and support personnel. On the other hand, there are many immediate problems on Earth, such as hunger, disease, homelessness, and illiteracy.
LS Verbal

Group ACTIVITY — GENERAL

Skylab Results Have students find out about the experiments conducted on *Skylab* or *Mir*. Groups should focus on biological, medical, space manufacturing, or astronomical experiments. Have groups share their findings in an oral presentation. **LS** Verbal

Figure 4 *As this illustration shows, space planes may provide transportation to outer space and around the world.*

space station a long-term orbiting platform from which other vehicles can be launched or scientific research can be carried out

CONNECTION TO Biology

Effects of Weightlessness
When a human body stays in space for long periods of time without having to work against gravity, the bones lose mass and muscles become weaker. Find out about the exercises to reduce the loss of bone mass used by astronauts aboard the *International Space Station*. Create an "Astronaut Exercise Book" to share with your friends.

Space Planes: The Shuttles of the Future?

NASA is working to develop a space plane called the *X-33*. This craft will fly like a normal airplane, but it will have rocket engines for use in space. Once in operation, space planes, such as the one shown in **Figure 4,** may lower the cost of getting material to LEO by 90%. Private companies are also becoming interested in developing space vehicles for commercial use and to make space travel cheaper, easier, and safer.

Space Stations—People Working in Space

A long-term orbiting platform in space is called a **space station.** On April 19, 1971, the Soviets became the first to successfully place a space station in orbit. A crew of three Soviet cosmonauts conducted a 23-day mission aboard the station, which was called *Salyut 1*. By 1982, the Soviets had put up seven space stations. Because of this experience, the Soviet Union became a leader in space-station development and in the study of the effects of weightlessness on humans. Their discoveries will be important for future flights to other planets—journeys that will take years to complete.

Skylab and *Mir*

Skylab, the United States' first space station, was a science and engineering lab used to conduct a wide variety of scientific studies. These studies included experiments in biology and space manufacturing and astronomical observations. Three different crews spent a total of 171 days on *Skylab* before it was abandoned. In 1986, the Soviets began to launch the pieces for a much more ambitious space station called *Mir* (meaning "peace"). Astronauts on *Mir* conducted a wide range of experiments, made many astronomical observations, and studied manufacturing in space. After 15 years, *Mir* was abandoned and it burned up in the Earth's atmosphere in 2001.

Homework — GENERAL

Privatized Space Exploration Many space technology innovations are made by private companies and individuals. Several awards are currently offered for the first private group to launch an inexpensive rocket into space. Ask students to find out more about the role of private companies in the future of space exploration, and suggest that they learn about experimental projects such as the *Roton* rocket. **LS** Intrapersonal

WEIRD SCIENCE

As astronaut William Pogue exercised on *Skylab,* his sweat "just sort of slithered around" instead of pooling on the floor beneath him, he reported. After exercising, he corralled the hovering sweat with a towel so that it wouldn't interfere with the spacecraft's equipment!

The *International Space Station*

The *International Space Station (ISS)*, the newest space station, is being constructed in LEO. Russia, the United States, and 14 other countries are designing and building different parts of the station. **Figure 5** shows what the *ISS* will look like when it is completed. The *ISS* is being built with materials brought up on the space shuttles and by Russian rockets. The United States is providing lab modules, the supporting frame, solar panels, living quarters, and a biomedical laboratory. The Russians are contributing a service module, docking modules, life-support and research modules, and transportation to and from the station. Other components will come from Japan, Canada, and several European countries.

✓ Reading Check What contributions are the Americans and Russians making to the *ISS*?

Research on the *International Space Station*

The *ISS* will provide many benefits—some of which we cannot predict. What scientists do know is that it will be an ideal place to perform space-science experiments and to test new technologies. Much of the space race involved political and military rivalry between the Soviet Union and the United States. Hopefully, the *ISS* will promote cooperation between countries while continuing the pioneering spirit of the first astronauts and cosmonauts.

Oral Histories The exciting times of the Apollo moon missions thrilled the nation. Interview adults in your community about their memories of those times. Prepare a list of questions first, and have your questions and contacts approved by your teacher. If possible, use a tape recorder or video camera to record the interviews. As a class, create a library of your oral histories for future students.

ACTIVITY

Figure 5 *When the* International Space Station *is completed, it will be about the size of a soccer field and will weigh about 500 tons.*

Answer to Reading Check

The Russians are supplying a service module, docking modules, life-support and research modules, and transportation to and from the station. The Americans are providing lab modules, the supporting frame, solar panels, living quarters, and a biomedical laboratory.

SCIENCE HUMOR

The news that *Skylab* would re-enter the atmosphere and plummet to Earth generated mild panic among some people. Despite NASA's assurances that the debris would fall in oceans or unpopulated areas, a few quick profits were made from the sale of hard hats billed as "*Skylab* Survival Kits." Much of the debris fell into the Indian Ocean, but some charred fragments were found strewn across the Australian Outback, which prompted Australian officials to present the United States with a $400 fine for littering!

Space Timeline As a class, make a timeline on the board that shows major events in the history of space exploration. Have students help you add notes that are related to concepts in the section.
LS Interpersonal

Quiz —— GENERAL

1. What kind of vehicle has been proposed that would make space travel cheaper?
(a space plane)

2. What is a space station?
(It is a long-term orbiting platform from which science and technology research can be conducted by resident astronauts.)

Alternative Assessment —— GENERAL

People in Space Science

Have a round-table discussion in which students impersonate the following people after having researched their lives and careers:

Robert Goddard, Konstantin Tsiolkovsky, Yuri Gagarin, John Glenn, Rita Mae Jemison, Valentina Tereshkova, Arnaldo Tamayo-Mendez, Alexi Leonov, Edward White, Chuck Yeager, Helen Sharman

Have students discuss the history of the space program and then speculate on future developments.
LS Interpersonal Co-op Learning

Figure 6 As shown in this illustration, humans may eventually establish a colony on the moon or on Mars.

Figure 7 Aerogel is the lightest solid on Earth. Aerogel is only 3 times heavier than air, and has 39 times the insulating properties of the best fiberglass insulation.

To the Moon, Mars, and Beyond

We may eventually need resources beyond what Earth can offer. Space offers many such resources. For example, a rare form of helium is found on the moon. If this helium could be used in nuclear fusion reactors, it would produce no radioactive waste! A base on the moon similar to the one shown in **Figure 6** could be used to manufacture materials in low gravity or in a vacuum. A colony on the moon or on Mars could be an important link to bringing space resources to Earth. It would also be a good base for exploring the rest of the solar system. The key will be to make these missions economically worthwhile.

The Benefits of the Space Program

Space exploration is expensive, and it has cost several human lives since the time that Yuri Gagarin and John Glenn first left the Earth more than 40 years ago. We have visited the moon, and we have sent probes outside the solar system. So why should we continue to explore space? There are many answers to this question. Space exploration has expanded our scientific knowledge of everything from the most massive stars to the smallest particles. Life-saving technologies have also resulted from the space missions. For example, artificial heart pumps use a turbine developed to pump fuel in the space shuttles. NASA's aerogel, shown in **Figure 7,** may become an energy-saving replacement for windows in the future. All of the scientific benefits of the space programs cannot be predicted. However, the exploration of space is also a challenge to human courage and a quest for new knowledge of ourselves and the universe.

PORTFOLIO **Space-Age Spinoff Comic Books** Encourage students to research the development of a few of the spinoffs listed in **Table 1.** Students may also find out about other spinoffs of their choice. Then, have students write and illustrate a space-age spinoff comic book that describes how a few spinoffs were developed and how the technology made its way into our daily lives. **LS** Visual

Space-Age Spinoffs

Technologies that were developed for the space programs but are now used in everyday life are called space-age spinoffs. There are dozens of examples of common items that were first developed for the space programs. Cordless power tools, for example, were first developed for use on the moon by the Apollo astronauts. Hand-held cameras that were developed to study the heat emitted from the space shuttle are used by firefighters to detect dangerous hot spots in fires. A few other examples of space-age spinoffs are shown in **Table 1.**

Reading Check What are space-age spinoffs?

Table 1   Space-Age Spinoffs
smoke detectors
bar coding on merchandise
pacemakers
artificial heart pumps
land mine removal devices
medical lasers
fire fighting equipment
invisible braces
video game joysticks
ear thermometers

SECTION Review

Summary

- In 1961, the Soviet cosmonaut Yuri Gagarin became the first person in space. In 1969, Neil Armstrong became the first person on the moon.
- During the 1970s, the United States focused on developing the space shuttle. The Soviets focused on developing space stations.
- The United States, Russia, and 14 other countries are currently developing the *International Space Station.*
- There have been many scientific, economic, and social benefits of the space programs.

Using Key Terms

1. Use each of the following terms in a separate sentence: *space shuttle* and *space station.*

Using Key Ideas

2. What is the main difference between the space shuttles and other space vehicles?
   a. The space shuttles are powered by liquid rocket fuel.
   b. The space shuttles take off like a plane and land like a rocket.
   c. The space shuttles are reusable.
   d. The space shuttles are not reusable.

3. Describe the history and future of human spaceflight. How was the race to explore space influenced by the Cold War?

4. Describe five "space-age spinoffs."

5. How will space stations help in the exploration of space?

6. In the 1970s, what was the main difference in the focus of the space programs in the United States and in the Soviet Union?

Math Skills

7. When it is fueled, a space shuttle has a mass of about 2,000,000 kg. About 80% of that mass is fuel and oxygen. Calculate the mass of a space shuttle's fuel and oxygen.

Critical Thinking

8. **Making Inferences** Why did the United States stop sending people to the moon after the Apollo program ended?

9. **Expressing Opinions** Imagine that you are a U.S. senator reviewing NASA's proposed budget. Write a two-paragraph position statement expressing your opinion about increasing or decreasing funding for NASA.

SCILINKS

NSTA

Developed and maintained by the National Science Teachers Association

For a variety of links related to this chapter, go to www.scilinks.org

Topic: Space Exploration and Space Stations

SciLinks code: HSM1340

Answers to Section Review

1. Sample answer: NASA has completed more than 100 successful space shuttle missions. The International Space Station is being built with the help of many nations.

2. c

3. Answers may vary. Students should note that Cold War tensions greatly accelerated the space programs of the United States and the Soviet Union.

4. Answers may vary. Students can describe any of the space-age spinoffs that are described in the text or in Table 1.

5. Answers may vary. Space stations will serve as refueling, construction, and research stations.

6. Answers may vary. The United States was mainly focused on space shuttle development, while the Soviet Union was focused on space station development.

7. 2 million kg × 0.80 = 1.6 million kg

8. Answers may vary. After the Apollo program, the United States focused on developing a reusable space shuttle.

9. Answers may vary. Accept any well-supported answer. Students may describe the technology that has come from the space program to support increasing NASA funding. Students may also support a position to decrease funding based on the expense of the space program and the tragedies that have occurred.

Answer to Reading Check

Space-age spinoffs are technologies that were developed for the space program but are now used in everyday life.

CHAPTER RESOURCES
Chapter Resource File
• Section Quiz **GENERAL**
• Section Review **GENERAL**
• Vocabulary and Section Summary **GENERAL**
• SciLinks Activity **GENERAL**

Inquiry Lab

Water Rockets Save the Day!

Teacher's Notes

Time Required

Two 45-minute class periods

Lab Ratings

EASY ———————————→ HARD

Teacher Prep 🧪🧪🧪
Student Set-Up 🧪🧪🧪
Concept Level 🧪🧪
Clean Up 🧪🧪

MATERIALS

The materials listed in this lab are best for groups of 3 or 4.

Safety Caution

Remind students to review all safety cautions and icons before beginning this lab activity. Make sure that students are several meters away from the launch site when the rockets are being launched.

OBJECTIVES

Predict which design features would improve a rocket's flight.

Design and build a rocket that includes your design features.

Test your rocket design, and evaluate your results.

MATERIALS

- bottle, soda, 2 L
- clay, modeling
- foam board
- rocket launcher
- scissors
- tape, duct
- watch or clock that indicates seconds
- water

SAFETY

Water Rockets Save the Day!

Imagine that for the big Fourth of July celebration, you and your friends had planned a full day of swimming, volleyball, and fireworks at the lake. You've just learned, however, that the city passed a law that bans all fireworks within city limits. But you do not give up so easily on having fun. Last year at summer camp, you learned how to build water rockets. And you have kept the launcher in your garage since then. With a little bit of creativity, you and your friends are going to celebrate with a splash!

Ask a Question

1 What is the most efficient design for a water rocket?

Form a Hypothesis

2 Write a hypothesis that provides a possible answer to the question above.

Test the Hypothesis

3 Decide how your rocket will look, and then draw a sketch.

4 Using only the materials listed, decide how to build your rocket. Write a description of your plan, and have your teacher approve your plan. Keep in mind that you will need to leave the opening of your bottle clear. The bottle opening will be placed over a rubber stopper on the rocket launcher.

5 Fins are often used to stabilize rockets. Do you want fins on your water rocket? Decide on the best shape for the fins, and then decide how many fins your rocket needs. Use the foam board to construct the fins.

6 Your rocket must be heavy enough to fly in a controlled manner. Consider using clay in the body of your rocket to provide some additional weight and stability.

7 Pour water into your rocket until the rocket is one-third to one-half full.

8 Your teacher will provide the launcher and will assist you during blastoff. Attach your rocket to the launcher by placing the opening of the bottle on the rubber stopper.

Preparation Notes

Each student or group of students will need a 2 L soda bottle to construct their rocket. You can ask students a week or two ahead of time to save any 2 L soda bottles they have at home. Water rockets require a launcher that you will need to make or purchase in advance. Launchers are the key to a successful and safe liftoff. They can be purchased at most hobby shops or from Science Kit®. You can expect to spend at least $20 for a launcher or launcher kit at a hobby shop. You need only one launcher for your class. When buying a launcher, make sure the launcher is compatible with the size of your water-rocket bottle. This lab is a popular activity, and water-bottle rocket hobbyists have posted detailed information on the Internet.

⑨ When the rocket is in place, clear the immediate area and begin pumping air into your rocket. Watch the pump gauge, and take note of how much pressure is needed for liftoff. **Caution:** Be sure to step back from the launch site. You should be several meters away from the bottle when you launch it.

⑩ Use the watch to time your rocket's flight. How long was your rocket in the air?

⑪ Make small changes in your rocket design that you think will improve the rocket's performance. Consider using different amounts of water and clay or experimenting with different fins. You may also want to compare your design with those of your classmates.

Analyze the Results

❶ Describing Events How did your rocket perform? If you used fins, do you think they helped your flight? Explain your answer.

❷ Explaining Results What do you think propelled your rocket? Use Newton's third law of motion to explain your answer.

❸ Analyzing Results How did the amount of water in your rocket affect the launch?

Draw Conclusions

❹ Drawing Conclusions What modifications made your rocket fly for the longest time? How did the design help the rockets fly so far?

❺ Evaluating Results Which group's rocket was the most stable? How did the design help the rocket fly straight?

❻ Making Predictions How can you improve your design to make your rocket perform even better?

Analyze the Results

1. Answers may vary. Fins might help stabilize the rocket when it is in flight.
2. Answers may vary. Students should note that the water in the bottle was under pressure. When the rocket was released, the water escaped out of the opening. The bottle reacted by moving in the opposite direction—upward.
3. Answers may vary. Water is the propellant for the rocket. Pressurized air provides the force to launch the rocket. The amount of water in a rocket determines how much space air can occupy. The ideal rocket should expel all of the water at the maximum pressure.

Draw Conclusions

4. Answers may vary. Modifying the fins, adjusting the water-to-air ratio, increasing the air pressure inside the rocket, and changing the amount of modeling clay used should affect the duration of the rocket's flight.
5. Answers may vary.
6. Answers may vary.

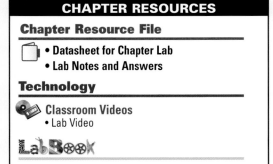

CHAPTER RESOURCES

Chapter Resource File
- • Datasheet for Chapter Lab
- • Lab Notes and Answers

Technology
- **Classroom Videos**
  - • Lab Video

LabBook
- • Reach for the Stars

CHAPTER RESOURCES

Workbooks
- **Whiz-Bang Demonstrations**
  - • Rocket Science **GENERAL**
- **Inquiry Labs**
  - • Crash Landing **ADVANCED**
  - • Space Fitness **ADVANCED**
- **EcoLabs & Field Activities**
  - • There's a Space for Us **GENERAL**
- **Long-Term Projects & Research Ideas**
  - • Space Voyage **ADVANCED**

Alyson Mike
East Valley Middle School
East Helena, Montana

Assignment Guide

Section	Questions
1	6–7, 16, 21–22, 24–27
2	1, 3, 8–10, 17, 19
3	4–7, 11, 13–15
4	12, 18, 23
3 and 4	2
1, 2, 3, and 4	20

ANSWERS

Using Key Terms

1. Sample answer: A geostationary orbit is about 36,000 km above the Earth's surface. In order to maintain GEO, satellites must orbit at a speed that exactly matches the speed of Earth's rotation. LEO is less than 1,500 km above the Earth's surface. To remain in LEO, satellites must travel much faster than the Earth rotates.

2. Sample answer: A space probe is a vehicle that carries scientific instruments to planets or other bodies in space. Space probes do not carry people. A space station is a long-term orbiting platform with scientific research labs and living quarters for astronauts.

3. Sample answer: An artificial satellite is any human-made object placed in orbit around a body in space. A moon is a natural satellite of a planet.

4. thrust

5. oxygen

Chapter Review

USING KEY TERMS

For each pair of terms, explain how the meanings of the terms differ.

1 *geostationary orbit* and *low Earth orbit*

2 *space probe* and *space station*

3 *artificial satellite* and *moon*

Complete each of the following sentences by choosing the correct term from the word bank.

> escape velocity   oxygen
> nitrogen   thrust

4 The force that accelerates a rocket is called ___.

5 Rockets need to have ___ in order to burn fuel.

UNDERSTANDING KEY IDEAS

Multiple Choice

6 Whose rocket research team surrendered to the Americans at the end of World War II?
   a. Konstantin Tsiolkovsky's
   b. Robert Goddard's
   c. Wernher von Braun's
   d. Yuri Gargarin's

7 Rockets work according to Newton's
   a. first law of motion.
   b. second law of motion.
   c. third law of motion.
   d. law of universal gravitation.

8 The first artificial satellite to orbit the Earth was
   a. *Pioneer 4.*
   b. *Explorer 1.*
   c. *Voyager 2.*
   d. *Sputnik 1.*

9 Communications satellites are able to transfer TV signals between continents because communications satellites
   a. are located in LEO.
   b. relay signals past the horizon.
   c. travel quickly around Earth.
   d. can be used during the day and night.

10 GEO is a better orbit for communications satellites because satellites that are in GEO
   a. remain in position over one spot.
   b. have polar orbits.
   c. do not revolve around the Earth.
   d. orbit a few hundred kilometers above the Earth.

11 Which space probe discovered evidence of water at the moon's south pole?
   a. *Luna 9*
   b. *Viking 1*
   c. *Clementine*
   d. *Magellan*

12 When did humans first set foot on the moon?
   a. 1959
   b. 1964
   c. 1969
   d. 1973

13 Which of the following planets has not yet been visited by space probes?
   a. Venus
   b. Neptune
   c. Mars
   d. Pluto

Understanding Key Ideas

6. c
7. c
8. d
9. b
10. a
11. c
12. c
13. d
14. d
15. c

14 Which of the following space probes has left our solar system?

   a. *Galileo*   **c.** *Viking 10*

   b. *Magellan*   **d.** *Pioneer 10*

15 Based on space-probe data, which of the following is the most likely place in our solar system to find liquid water?

   a. the moon   **c.** Europa

   b. Mercury   **d.** Venus

Short Answer

16 Describe how Newton's third law of motion relates to the movement of rockets.

17 What is one disadvantage that objects in LEO have?

18 Why did the United States develop the space shuttle?

19 How does data from satellites help us understand the Earth's environment?

CRITICAL THINKING

20 Concept Mapping Use the following terms to create a concept map: *orbital velocity, thrust, LEO, artificial satellites, escape velocity, space probes, GEO,* and *rockets.*

21 Making Inferences What is the difference between speed and velocity?

22 Applying Concepts Why must rockets that travel in outer space carry oxygen with them?

23 Expressing Opinions What impact has space research had on scientific thought, on society, and on the environment?

INTERPRETING GRAPHICS

The diagram below illustrates suborbital velocity, orbital velocity, and escape velocity. Use the diagram below to answer the questions that follow.

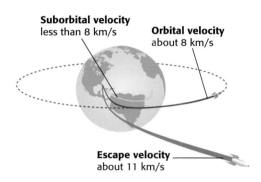

Suborbital velocity less than 8 km/s

Orbital velocity about 8 km/s

Escape velocity about 11 km/s

24 Could a rocket traveling at 6 km/s reach orbital velocity?

25 If a rocket traveled for 3 days at the minimum escape velocity, how far would the rocket travel?

26 How much faster would a rocket traveling in orbital velocity need to travel to reach escape velocity?

27 If the escape velocity for a planet was 9 km/s, would you assume that the mass of the planet was more or less than the mass of Earth?

Critical Thinking

20. An answer to this exercise can be found at the end of this book.

21. Speed is a measure of how fast an object travels. Velocity is the speed and direction an object travels.

22. Rockets must carry oxygen with them because there is little or no oxygen in the Earth's upper atmosphere or in outer space. Without oxygen, rocket fuel cannot burn.

23. Answers may vary. Students should note that space research has helped develop many areas of human knowledge, including medicine, physics, engineering, biology, geology, and astronomy. The space program is also a challenge to human courage and can promote international cooperation and peace. Students may also mention space-age spinoffs. Finally, space probes enable us to study the atmospheres and surfaces of other bodies in the solar system. Scientists use these data to understand changes in Earth's environment.

Interpreting Graphics

24. no

25. 60 s/min × 60 min/h × 24 h/day × 3 days = 259,200 s

259,200 s × 11 km/s = 2,851,200 km

26. 11 km/s − 8 km/s = 3 km/s

A rocket traveling in orbital velocity would need to travel 3 km/s faster to reach escape velocity.

27. If a planet's escape velocity was 9 km/s, the planet's mass would be less than the mass of Earth.

16. Answers may vary. Sample answer: Newton's third law of motion states that for every action, there is an equal reaction in the opposite direction. As the gases escape through the rocket nozzle, the rocket reacts by moving in the opposite direction.

17. Answers may vary. One disadvantage of LEO is that a satellite cannot maintain constant communication with a ground station.

18. Answers may vary. One reason the United States developed the space shuttle was to have a reusable space vehicle that would be less expensive than the Apollo rockets.

19. Answers may vary.

CHAPTER RESOURCES

Chapter Resource File

- Chapter Review **GENERAL**
- Chapter Test A **GENERAL**
- Chapter Test B **ADVANCED**
- Chapter Test C **SPECIAL NEEDS**
- Vocabulary Activity **GENERAL**

Workbooks

Study Guide
- Assessment resources are also available in Spanish.

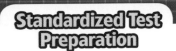

Standardized Test Preparation

Teacher's Note

To provide practice under more realistic testing conditions, give students 20 minutes to answer all of the questions in this Standardized Test Preparation.

MISCONCEPTION ALERT

Answers to the standardized test preparation can help you identify student misconceptions and misunderstandings.

READING

Passage 1

1. C
2. H
3. B

TEST DOCTOR

Question 1: Students may be confused by the fact that none of the answer options come directly from the passage. They may think none of the answers are correct. Tell students that the main idea of a paragraph is not always stated in the paragraph. In this case, students would need to choose the best answer that most fully encompasses the main idea of the paragraph.

READING

Read each of the passages below. Then, answer the questions that follow each passage.

Passage 1 One of the strange things about living in space is free fall, the reduced effect of gravity. Everything inside the *International Space Station* that is not fastened down will float! The engineers who designed the space station have come up with some <u>intriguing</u> solutions to this problem. For example, each astronaut sleeps in a sack similar to a sleeping bag that is fastened to the module. The sack keeps the astronauts from floating around while they sleep. Astronauts shower with a hand-held nozzle. Afterward, the water droplets are vacuumed up. Other problems that are being studied include how to prepare and serve food, how to design an effective toilet, and how to dispose of waste.

1. What is the main idea of the passage?
   A There is no gravity in space.
   B Astronauts will stay aboard the space station for long periods of time.
   C Living in free fall presents interesting problems.
   D Sleeping bags are needed to keep astronauts warm in space.

2. Which of the following is a problem mentioned in the passage?
   F how to dissipate the heat of reentry
   G how to maintain air pressure
   H how to serve food
   I how to listen to music

3. Which of the following words is the best antonym for *intriguing*?
   A authentic
   B boring
   C interesting
   D unsolvable

Passage 2 In 1999, the crew of the space station *Mir* tried to place a large, umbrella-like mirror in orbit. The mirror was designed to reflect sunlight to Siberia. The experiment failed because the crew was unable to unfold the mirror. If things had gone as planned, the beam of reflected sunlight would have been 5 to 10 times brighter than the light from the moon! If the first space mirror had worked, Russia was planning to place many more mirrors in orbit to lengthen winter days in Siberia, extend the growing season, and even reduce the amount of electricity needed for lighting. Luckily, the experiment failed. If it had succeeded, the environmental effects of extra daylight in Siberia would have been catastrophic. Astronomers were concerned that the mirrors would cause light pollution and obstruct their view of the universe. Outer space should belong to all of humanity, and any project of this kind, including placing advertisements on the moon, should be banned.

1. Which of the following is a statement of opinion?
   A Astronomers were concerned about the effects of the space mirror.
   B Outer space should belong to all of humanity.
   C The experiment failed because the mirror could not unfold.
   D Russia was planning to place many more mirrors in orbit.

2. What can you infer about the location of Siberia?
   F It is near the equator.
   G It is closer to the equator than it is to the North Pole.
   H It is closer to the North Pole than it is to the equator.
   I It is the same distance from the equator as it is from the North Pole.

Passage 2

1. B
2. H

TEST DOCTOR

Question 2: This question may be difficult for students because they must infer information that is not stated in the passage. Emphasize that some standardized test questions may ask students to make inferences. In this case, students must infer that Siberia is closer to the North Pole than it is to the equator, because the passage discusses a strategy to provide more solar energy for the region. Areas that are close to the poles receive less solar energy than areas close to the equator do.

The diagram below shows the location of satellites in LEO and GEO. Use the diagram below to answer the questions that follow.

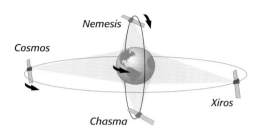

1. Which satellites are always located over the same spot on Earth?

A *Xiros* and *Chasma*

B *Xiros* and *Cosmos*

C *Nemesis* and *Cosmos*

D *Chasma* and *Nemesis*

2. Which satellites are likely to be spy satellites?

F *Xiros* and *Cosmos*

G *Xiros* and *Chasma*

H *Chasma* and *Nemesis*

I *Nemesis* and *Cosmos*

3. Which satellites are likely to be communications satellites?

A *Chasma* and *Nemesis*

B *Nemesis* and *Cosmos*

C *Xiros* and *Cosmos*

D *Xiros* and *Chasma*

4. Which satellites are traveling in an orbit that is 90° with respect to the direction of Earth's rotation?

F *Nemesis* and *Cosmos*

G *Xiros* and *Chasma*

H *Nemesis* and *Chasma*

I *Xiros* and *Cosmos*

MATH

Read each question below, and choose the best answer.

1. To escape Earth's gravity, a rocket must travel at least 11 km/s. About how many hours would it take to get to the moon at this speed? (On average, the moon is about 384,500 km away from Earth.)

A 1 h

B 7 h

C 8 h

D 10 h

2. The Saturn V launch vehicle, which carried the Apollo astronauts into space, had a mass of about 2.7 million kilograms and carried about 2.5 million kilograms of propellant. What percentage of Saturn V's mass was propellant?

F 9.25%

G 9%

H 92.5%

I 90%

3. Scientists discovered that when a person is in orbit, bone mass in the lower hip and spine is lost at a rate of 1.2% per month. At that rate, how long would it take for 7.2% of bone mass to be lost?

A 4 months

B 6 months

C 7.2 months

D 8 months

4. The space shuttle can carry 25,400 kg of cargo into orbit. Assume that the average astronaut has a mass of 75 kg and that each satellite has a mass of 4,300 kg. If a shuttle mission is already carrying 9,000 kg of equipment and 10 astronauts, how many satellites can the shuttle carry?

F 2

G 3

H 4

I 5

✚ TEST DOCTOR

Question 4: Students may not know what a 90° angle is. Tell students that a 90° angle makes up the corner of a square, or, in this case, a cross.

MATH

1. D

2. H

3. B

4. G

✚ TEST DOCTOR

Question 2: To derive the percentage, students should divide 2.5 million kilograms by 2.7 million kilograms and then multiply by 100. Students might get an answer of 108% if they divide 2.7 by 2.5 and then multiply by 100. If students accidentally multiply by 10, they will get answer F, 9.25%.

Question 4: This question requires several steps. First, students must find the total weight of the cargo plus astronauts. Multiplying 75 times 10 to get 750 gives the weight of the astronauts. Adding 750 to 9,000 will give the weight of the cargo plus astronauts, which is 9,750. Now, students must subtract this weight from 25,400 to find the amount that the shuttle can still hold. Finally, dividing this amount by 4,300 will give students the number of satellites that will fit on the shuttle. Remind students to ignore the remainder in this situation, because the question is asking about whole satellites.

Standardized Test Preparation

CHAPTER RESOURCES

Chapter Resource File

📁 • Standardized Test Preparation GENERAL

State Resources

 For specific resources for your state, visit **go.hrw.com** and type in the keyword **HSMSTR**.

Science, Technology, and Society

Background

NASA is also working with the European Space Agency and the Italian Space Agency on a mission called Mars Express. In June 2003, the *Mars Express* probe was launched to explore the polar regions of Mars. The spacecraft's lander is called the *Beagle 2* after Charles Darwin's ship. The *Beagle 2* carries equipment to explore Mars's atmosphere, surface, and subsurface. In fact, the *Beagle 2* has a radar instrument capable of finding water and ice that might be buried up to 5 km below the surface of the planet. The *Beagle 2* is expected to land on the surface of Mars on Christmas day, 2003.

Weird Science

Discussion ——— GENERAL

The volunteers at the Flashline Mars Arctic Research Station (FMARS) have already learned many things that will help NASA prepare for a trip to Mars. For example, they learned that they can use much less water if they skip a few baths. Also, it is important to have backup equipment and to travel in groups. Have students discuss other ideas that might help NASA plan for a trip to Mars.

Science in Action

Science, Technology, and Society

Mission to Mars

In spring 2003, two cutting-edge NASA rovers were sent on a mission to Mars. When they reach their destination, they will parachute through the thin Martian atmosphere and land on the surface. First, the rovers will use video and infrared cameras to look around. Then, for at least 92 Earth days (90 Martian days), the rovers will explore the surface of Mars. They will gather geologic evidence of liquid water because liquid water may have enabled Mars to support life in the past. Each rover will carry five scientific tools and a Rock Abrasion Tool, or "RAT," which will grind away rock surfaces to expose the rock interiors for scientific tests. Stay tuned for more news from Mars!

Language Arts ACTIVITY

Watch for stories about this mission in newspapers and magazines. If you read about a discovery on Mars, bring a copy of the article to share with your class. As a class, compile a scrapbook entitled "Mars in the News."

Weird Science

Flashline Mars Arctic Research Station

If you wanted to visit a place on Earth that is like the surface of Mars, where would you go? You might head to an impact crater on Devon Island, close to the Arctic circle. The rugged terrain and harsh weather there resemble what explorers will find on Mars, although Mars has no breathable air and is a lot colder. In the summer, volunteers from the Mars Society live in an experimental base in the crater and test technology that might be used on Mars. The volunteers try to simulate the experience of explorers on Mars. For example, the volunteers wear spacesuits when they go outside, and they explore the landscape by using rovers. They even communicate with the outside world using types of technology likely to be used on Mars. These dedicated volunteers have already made discoveries that will help NASA plan a crewed mission to Mars!

Social Studies ACTIVITY

A Mars mission could require astronauts to endure nearly two years of extreme isolation. Research how NASA would prepare astronauts for the psychological pressures of a mission to Mars.

Answer to Language Arts Activity

Answers may vary. Scrapbooks may include pictures taken by the Mars rovers and NASA press releases. Students can edit and publish their findings in a *Mars Tribune* newspaper to be distributed to other classes or posted on the Internet.

Answer to Social Studies Activity

Answers may vary. Popular science magazines, such as *Discover*, have published articles concerning the psychological pressures of a Mars mission.

Franklin Chang-Diaz

Astronaut You have to wear a suit, but the commute is not too long. In fact, it is only about eight and a half minutes, and what a view on your way to work! Astronauts, such as Franklin Chang-Diaz, have one of the most exciting jobs on Earth—or in space. Chang-Diaz has flown on seven space shuttle missions and has completed three space walks. Since the time he became an astronaut in 1981, Chang-Diaz has spent more than 1,601 hours (66 days) in space.

Chang-Diaz was born in San Jose, Costa Rica. He earned a degree in mechanical engineering in 1973 and received a doctorate in applied plasma physics from the Massachusetts Institute of Technology (MIT) in 1977. His work in physics attracted the attention of NASA, and he began training at the Johnson Space Center in Houston, Texas. In addition to doing research on the space shuttle, Chang-Diaz has worked on developing plasma propulsion systems for long space flights. He has also helped create closer ties between astronauts and scientists by starting organizations such as the Astronaut Science Colloquium Program and the Astronaut Science Support Group. If you want to find out more about what it takes to be an astronaut, look on NASA's Web site.

Math ACTIVITY

If 1 out of 120 people interviewed by NASA is selected for astronaut training, how many people will be selected for training if 10,680 people are interviewed?

As this mission patch shows, Chang-Diaz flew on the 111th space shuttle mission.

To learn more about these Science in Action topics, visit **go.hrw.com** and type in the keyword **HZ5EXPF.**

Current Science

Check out Current Science® articles related to this chapter by visiting go.hrw.com. Just type in the keyword **HZ5CS22.**

Careers

Teaching Strategy— GENERAL

There are a number of resources available on the Internet for students interested in becoming astronauts. Have teams of students research different aspects of this exciting career and present their findings to the class.

SCIENCE HUMOR

Astronaut Jerry Linenger listed the following skills on his astronaut application: woodworking, drafting, carpentry, small-engine repair, electrical wiring, sprinkling-system installation, heavy cement work, plumbing, and bricklaying. After spending time on *Mir*, he joked, "With the exception of the cement work and bricklaying, all of these skills proved indispensable on *Mir*."

Answer to Math Activity
$10,680 \div 120 = 89$

The Sun's Yearly Trip Through the Zodiac

Teacher's Notes

Time Required
Two 45-minute class periods

Lab Ratings

EASY ————————→ HARD

Teacher Prep 🧪🧪🧪
Student Set-Up 🧪🧪
Concept Level 🧪🧪🧪
Clean Up 🧪

MATERIALS
The materials listed on the student page are enough for a group of 12 students. However, you may choose to use this activity as a demonstration for the entire class.

Preparation Notes

One week before the activity, collect large cardboard boxes. Designate a large, clear area for the activity, such as a gym, cafeteria, playground, or large classroom. You may wish to get a basketball from the physical education instructor, or ask students to bring basketballs from home. Folding chairs work best for this activity because of their portability.

Review the terms *clockwise* and *counterclockwise* to ensure consistency of student results.

The Sun's Yearly Trip Through the Zodiac

During the course of a year, the sun appears to move through a circle of 12 constellations in the sky. The 12 constellations make up a "belt" in the sky called the *zodiac.* Each month, the sun appears to be in a different constellation. The ancient Babylonians developed a 12-month calendar based on the idea that the sun moved through this circle of constellations as it revolved around the Earth. They believed that the constellations of stars were fixed in position and that the sun and planets moved past the stars. Later, Copernicus developed a model of the solar system in which the Earth and the planets revolve around the sun. But how can Copernicus's model of the solar system be correct when the sun appears to move through the zodiac?

MATERIALS
- ball, inflated
- box, cardboard, large
- cards, index (12)
- chairs (12)
- tape, masking (1 roll)

Ask a Question

1 If the sun is at the center of the solar system, why does it appear to move with respect to the stars in the sky?

Form a Hypothesis

2 Write a possible answer to the question above. Explain your reasoning.

Test the Hypothesis

3 Set the chairs in a large circle so that the backs of the chairs all face the center of the circle. Make sure that the chairs are equally spaced, like the numbers on the face of a clock.

4 Write the name of each constellation in the zodiac on the index cards. You should have one card for each constellation.

5 Stand inside the circle with the masking tape and the index cards. Moving counterclockwise, attach the cards to the backs of the chairs in the following order: Aries, Taurus, Gemini, Cancer, Leo, Virgo, Libra, Scorpio, Sagittarius, Capricorn, Aquarius, and Pisces.

6 Use masking tape to label the ball "Sun."

7 Place the large, closed box in the center of the circle. Set the roll of masking tape flat on top of the box.

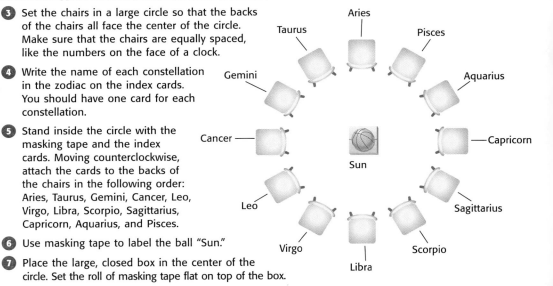

8. Place the ball on top of the roll of masking tape so that the ball stays in place.

9. Stand inside the circle of chairs. You will represent the Earth. As you move around the ball, you will model the Earth's orbit around the sun. Notice that even though only the "Earth" is moving, as seen from the Earth, the sun appears to move through the entire zodiac!

10. Stand in front of the chair labeled "Aries." Look at the ball representing the sun. Then, look past the ball to the chair at the opposite side of the circle. Where in the zodiac does the sun appear to be?

11. Move to the next chair on your right (counterclockwise). Where does the sun appear to be? Is it in the same constellation? Explain your answer.

12. Repeat step 10 until you have observed the position of the sun from each chair in the circle.

Analyze the Results

1. Did the sun appear to move through the 12 constellations, even though the Earth was orbiting around the sun? How can you explain this apparent movement?

Draw Conclusions

2. How does Copernicus's model of the solar system explain the apparent movement of the sun through the constellations of the zodiac?

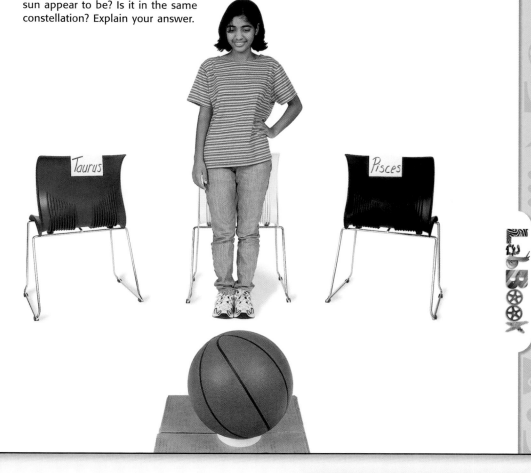

Background

Begin the activity by asking students if they are familiar with the constellations of the zodiac. Have they seen them depicted in a list or in a circle? How did the mythology of the zodiac begin? (The 12 familiar signs of the zodiac were adopted by the Babylonians about 3,000 years ago. The word *zodiac* means "circle," and the Babylonians thought that the sun and the planets moved in a circle through 12 fixed constellations in the night sky.)

Ask each student group to list as many zodiac constellations as they can remember. (The constellations and their corresponding signs are as follows: Aries, the ram; Taurus, the bull; Gemini, the twins; Cancer, the crab; Leo, the lion; Virgo, the virgin; Libra, the scales; Scorpio, the scorpion; Sagittarius, the hunter; Capricorn, the mountain goat; Aquarius, the water bearer; and Pisces, the fish.)

Analyze the Results

1. When students stand in front of a chair, the sun appears to be in the constellation opposite the chair. As they move outside the circle counterclockwise, the sun appears to shift through the constellations counterclockwise. As the Earth (the student) orbits the sun (the ball), the sun never appears in the same constellation because of the Earth's perspective relative to the fixed constellations.

Draw Conclusions

2. In Copernicus's heliocentric model of the solar system, the Earth orbits the sun. As the Earth moves around the sun, the position of the sun in relation to the constellations changes.

Skills Practice Lab

I See the Light!

Teacher's Notes

Time Required

One to two 45-minute class periods

Lab Ratings

EASY —————→ HARD

Teacher Prep 🧪🧪
Student Set-Up 🧪🧪🧪
Concept Level 🧪🧪🧪🧪
Clean Up 🧪

MATERIALS

The materials listed on the student page are enough for a group of 1 to 2 students.

Safety Caution

Remind students to review all safety cautions and icons before beginning this lab activity. Remind students to be careful about traffic hazards around your school's flagpole.

Skills Practice Lab

I See the Light!

How do you find the distance to an object you can't reach? You can do it by measuring something you can reach, finding a few angles, and using mathematics. In this activity, you'll practice measuring the distances of objects here on Earth. When you get used to it, you can take your skills to the stars!

Ask a Question

1 How can you measure the distance to a star?

Form a Hypothesis

2 Write a hypothesis that might answer this question. Explain your reasoning.

Test the Hypothesis

3 Draw a line 4 cm away from the edge of one side of the piece of poster board. Fold the poster board along this line.

4 Tape the protractor to the poster board with its flat edge against the fold, as shown in the photo below.

5 Use a pencil to carefully punch a hole through the poster board along its folded edge at the center of the protractor.

6 Thread the string through the hole, and tape one end to the underside of the poster board. The other end should be long enough to hang off the far end of the poster board.

7 Carefully punch a second hole in the smaller area of the poster board halfway between its short sides. The hole should be directly above the first hole and should be large enough for the pencil to fit through. This hole is the viewing hole of your new parallax device. This device will allow you to measure the distance of faraway objects.

8 Find a location that is at least 50 steps away from a tall, narrow object, such as the school's flagpole or a tall tree. (This object will represent background stars.) Set the meterstick on the ground with one of its long edges facing the flagpole.

9 Ask your partner, who represents a nearby star, to take 10 steps toward the flagpole, starting at the left end of the meterstick. You will be the observer. When you stand at the left end of the meterstick, which represents the location of the sun, your partner's nose should be lined up with the flagpole.

MATERIALS

- calculator, scientific
- meterstick
- pencil, sharp
- poster board, 16 × 16 cm
- protractor
- ruler, metric
- scissors
- string, 30 cm
- tape measure, metric
- tape, transparent

SAFETY

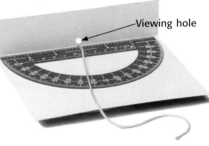

Viewing hole

Preparation Notes

Students may need an introduction to angles and the use of protractors. Be sure they understand how to use protractors *before* you perform this activity. Explain that astronomers use trigonometry, which is the measurement of triangles, to calculate distances to nearby stars. By using the TAN function on a calculator, students can find the length of the unknown leg of the triangle formed by the sun, Earth, and a star. If students are unfamiliar with the TAN function, they may prefer to use the table provided.

Angle	Tangent	Angle	Tangent
1°	0.0175	6°	0.1051
2°	0.0349	7°	0.1228
3°	0.0524	8°	0.1405
4°	0.0699	9°	0.1584
5°	0.0875	10°	0.1763

10 Move to the other end of the meterstick, which represents the location of Earth. Does your partner appear to the left or right of the flagpole? Record your observations.

11 Hold the string so that it runs straight from the viewing hole to the 90° mark on the protractor. Using one eye, look through the viewing hole along the string, and point the device at your partner's nose.

12 Holding the device still, slowly move your head until you can see the flagpole through the viewing hole. Move the string so that it lines up between your eye and the flagpole. Make sure the string is taut, and hold it tightly against the protractor.

13 Read and record the angle made by the string and the string's original position at 90° (count the number of degrees between 90° and the string's new position).

14 Use the measuring tape to find and record the distance from the left end of the meterstick to your partner's nose.

15 Now, find a place outside that is at least 100 steps away from the flagpole. Set the meterstick on the ground as before, and repeat steps 9–14.

Analyze the Results

1 The angle you recorded in step 13 is called the *parallax angle.* The distance from one end of the meterstick to the other is called the *baseline.* With this angle and the length of your baseline, you can calculate the distance to your partner.

2 To calculate the distance (*d*) to your partner, use the following equation:

$$d = b/\tan A$$

In this equation, *A* is the parallax angle, and *b* is the length of the baseline (1 m). (Tan *A* means the tangent of angle *A*, which you will learn more about in math classes.)

3 To find *d*, enter 1 (the length of your baseline in meters) into the calculator, press the division key, enter the value of *A* (the parallax angle you recorded), then press the tan key. Finally, press the equals key.

4 Record this result. It is the distance in meters between the left end of the meterstick and your partner. You may want to use a table like the one below.

5 How close is this calculated distance to the distance you measured?

6 Repeat steps 1–3 under Analyze the Results using the angle you found when the flagpole was 100 steps away.

Draw Conclusions

7 At which position, 50 steps or 100 steps from the flagpole, did your calculated distance better match the actual distance as measured?

8 What do you think would happen if you were even farther from the flagpole?

9 When astronomers use parallax, their "flagpoles" are distant stars. Might this affect the accuracy of their parallax readings?

Distance by Parallax Versus Measuring Tape		
	At 50 steps	At 100 steps
Parallax angle		
Distance (calculated)		
Distance (measured)		

CHAPTER RESOURCES

Chapter Resource File

- Datasheet for LabBook
- Lab Notes and Answers

Susan Gorman
North Ridge Middle School
North Richmond Hills, Texas

Lab Notes

If the school's flagpole is not in a convenient spot, a tall tree or lamppost will work. If students have difficulty keeping the device still while moving their head, you might have them try steadying the device on a tripod or on the end of a meterstick.

Some students may have difficulty understanding the relationship between measuring stars with parallax and the parallax effects on measurements (for example, reading a dial off to the side can give a different value). The viewing hole in the parallax device used in this activity reduces such errors. Thus, the parallax effect in this activity measures only faraway distances.

Students should realize that as the distance to a reference point (such as a flagpole or background stars) increases, the angle measured by the parallax device becomes closer to the actual parallax angle. Therefore, students should find that their calculation of the distance to their partner should be more accurate when they move 100 steps from the tree. They should also realize that astronomers use the "fixed stars," which are essentially at optical infinity, as a reference.

Analyze the Results

5. Answers may vary but should approximate the actual distance given in meters.

6. Answers may vary due to differences in technique.

Draw Conclusions

7. At 100 steps, the distance calculated should be closer to the distance measured.

8. Accuracy should increase.

9. Their calculations should be very close to the actual distance (if it could be measured).

Model-Making Lab

Why Do They Wander?

Teacher's Notes

Time Required

This activity will take approximately 30 minutes. But it may take as much as one 45-minute class period to instruct students on how to use a compass.

Lab Ratings

EASY ——————→ HARD

Teacher Prep 🧪🧪
Student Set-Up 🧪
Concept Level 🧪🧪🧪🧪
Clean Up 🧪

MATERIALS

The materials listed on the student page are enough for 1 to 2 students. The compasses, rulers, and colored pencils may be shared among several groups.

Safety Caution

Remind students to review all safety cautions and icons before beginning this lab activity.

Preparation Notes

Students may need instruction on how to use a drawing compass. This activity works best when students work individually or in pairs. Each group will need a compass, a piece of white paper, and a metric ruler.

Model-Making Lab

Why Do They Wander?

Before the discoveries of Nicholas Copernicus in the early 1500s, most people thought that the planets and the sun revolved around the Earth and that the Earth was the center of the solar system. But Copernicus observed that the sun is the center of the solar system and that all the planets, including Earth, revolve around the sun. He also explained a puzzling aspect of the movement of planets across the night sky.

If you watch a planet every night for several months, you'll notice that it appears to "wander" among the stars. While the stars remain in fixed positions relative to each other, the planets appear to move independently of the stars. Mars first travels to the left, then back to the right, and then again to the left.

In this lab, you will make your own model of part of the solar system to find out how Copernicus's model of the solar system explained this zigzag motion of the planets.

MATERIALS

- compass, drawing
- paper, white
- pencils, colored
- ruler, metric

SAFETY

Ask a Question

1 Why do the planets appear to move back and forth in the Earth's night sky?

Form a Hypothesis

2 Write a possible answer to the question above.

Test the Hypothesis

3 Use the compass to draw a circle with a diameter of 9 cm on the paper. This circle will represent the orbit of the Earth around the sun. (Note: The orbits of the planets are actually slightly elliptical, but circles will work for this activity.)

4 Using the same center point, draw a circle with a diameter of 12 cm. This circle will represent the orbit of Mars.

5 Using a blue pencil, draw three parallel lines diagonally across one end of your paper, as shown at right. These lines will help you plot the path Mars appears to travel in Earth's night sky. Turn your paper so the diagonal lines are at the top of the page.

Lab Notes

Plan View Versus Sky View: It is important to note that the circles represent a plan view of part of the solar system, and the diagonal lines represent a view of the apparent motion of Mars in Earth's night sky. Students are asked to jump from line to line as they draw their dots in order to show them the apparent path of Mars in the sky.

Note on Scale: Notice that, according to the drawing, it appears that for less than half of its orbit, Mars travels more than a year of Earth's time. In fact, an Earth year is actually more than half of a Martian year. Mars's period of revolution is 1.88 Earth years. The drawing on this page is not to scale in this respect; if it were to scale, the wandering motion of the planets could not be depicted on one page.

6 Place 11 dots on your Earth orbit, as shown below, and number them 1 through 11. These dots will represent Earth's position from month to month.

7 Now, place 11 dots along the top of your Mars orbit, as shown below. Number the dots as shown. These dots will represent the position of Mars at the same time intervals. Notice that Mars travels slower than Earth.

8 Draw a green line to connect the first dot on Earth's orbit to the first dot on Mars's orbit. Extend this line to the first diagonal line at the top of your paper. Place a green dot where the green line meets the first blue diagonal line. Label the green dot "1."

9 Now, connect the second dot on Earth's orbit to the second dot on Mars's orbit, and extend the line all the way to the first diagonal at the top of your paper. Place a green dot where this line meets the first blue diagonal line, and label this dot "2."

10 Continue drawing green lines from Earth's orbit through Mars's orbit and finally to the blue diagonal lines. Pay attention to the pattern of dots you are adding to the diagonal lines. When the direction of the dots changes, extend the green line to the next diagonal, and add the dots to that line instead.

11 When you are finished adding green lines, draw a red line to connect all the green dots on the blue diagonal lines in the order you drew them.

Analyze the Results

1 What do the green lines connecting points along Earth's orbit and Mars's orbit represent?

2 What does the red line connecting the dots along the diagonal lines look like? How can you explain this?

Draw Conclusions

3 What does this demonstration show about the motion of Mars?

4 Why do planets appear to move back and forth across the sky?

5 Were the Greeks justified in calling the planets *wanderers*? Explain.

Analyze the Results

1. The green lines connecting Earth's orbit and Mars's orbit represent the students' line of sight as they stand on Earth and look at Mars.

2. Sample answer: The red line along the diagonals changed direction at the fifth and seventh points; This happened because Mars was behind Earth in its orbit. To a person on Earth, Mars would seem to have changed direction at those moments.

Draw Conclusions

3. Sample answer: When Earth catches up to Mars, Mars appears to reverse its direction. As Earth passes Mars, Mars appears to revert to its original direction.

4. Sample answer: Planets appear to move back and forth because they travel around the sun at different speeds and at different distances. When the Earth overtakes a slower planet, such as Mars, that planet appears to move backward in Earth's sky.

5. Sample answer: Although the planets do not actually wander, it does appear as if they do to people on Earth. Students may consider this enough justification for calling the planets wanderers. Accept all well-supported responses.

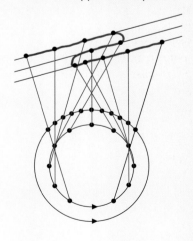

CHAPTER RESOURCES

Chapter Resource File

- **Datasheet for LabBook**
- **Lab Notes and Answers**

CLASSROOM TESTED & APPROVED

Joseph W. Price
H. M. Browne Junior High
Washington, D.C.

MISCONCEPTION ALERT

The apparent motion of Mars illustrated in this lab is called *retrograde motion*. It should not be confused with retrograde orbit or retrograde rotation. In addition, this lab does not accurately portray the actual *positions* of Earth and Mars during their orbits; it merely shows how their relative positions change.

Eclipses

Teacher's Notes

Time Required

One 45-minute class period

Lab Ratings

EASY ———————————→ HARD

Teacher Prep 🧪
Student Set-Up 🧪🧪
Concept Level 🧪🧪🧪
Clean Up 🧪

MATERIALS

The materials listed on the student page are enough for each student or for students working in groups of 2 to 3.

Analyze the Results

1. The flashlight represents the sun.

2. Step 4 modeled a lunar eclipse, as viewed from Earth.

3. Step 4 modeled a solar eclipse, as viewed from the moon.

4. Step 5 modeled a solar eclipse, as viewed from Earth.

5. Step 5 modeled an eclipse of Earth, as viewed from the moon.

Model-Making Lab

Eclipses

As the Earth and the moon revolve around the sun, they both cast shadows into space. An eclipse occurs when one planetary body passes through the shadow of another. You can demonstrate how an eclipse occurs by using clay models of planetary bodies.

MATERIALS

- clay, modeling
- flashlight, small
- paper, notebook (1 sheet)
- ruler, metric

Procedure

1. Make two balls out of the modeling clay. One ball should have a diameter of about 4 cm and will represent the Earth. The other should have a diameter of about 1 cm and will represent the moon.

2. Place the two balls about 15 cm apart on the sheet of paper. (You may want to prop the smaller ball up on folded paper or on clay so that the centers of the two balls are at the same level.)

3. Hold the flashlight approximately 15 cm away from the large ball. The flashlight and the two balls should be in a straight line. Keep the flashlight at about the same level as the clay. When the whole class is ready, your teacher will turn off the lights.

4. Turn on your flashlight. Shine the light on the larger ball, and sketch your model. Include the beam of light in your drawing.

5. Move the flashlight to the opposite side of the paper. The flashlight should now be approximately 15 cm away from the smaller clay ball. Repeat step 4.

Analyze the Results

1. What does the flashlight in your model represent?

2. As viewed from Earth, what event did your model represent in step 4?

3. As viewed from the moon, what event did your model represent in step 4?

4. As viewed from Earth, what event did your model represent in step 5?

5. As viewed from the moon, what event did your model represent in step 5?

6. According to your model, how often would solar and lunar eclipses occur? Is this accurate? Explain.

6. There would be a lunar and a solar eclipse each month. This reason is that the model shows Earth and the moon orbiting in exactly the same plane around the sun. However, the planes are usually above or below the shadow of the other, so an eclipse does not always occur.

CHAPTER RESOURCES

Chapter Resource File

- Datasheet for LabBook
- Lab Notes and Answers

CLASSROOM TESTED & APPROVED

Joseph W. Price
H. M. Browne Junior High
Washington, D.C.

Skills Practice Lab

Phases of the Moon

It's easy to see when the moon is full. But you may have wondered exactly what happens when the moon appears as a crescent or when you cannot see the moon at all. Does the Earth cast its shadow on the moon? In this activity, you will discover how and why the moon appears as it does in each phase.

MATERIALS

- ball, plastic-foam
- globe, world
- light source

SAFETY

Procedure

1 Place your globe near the light source. Be sure that the north pole is tilted toward the light. Rotate the globe so that your state faces the light.

2 Using the ball as your model of the moon, move the moon between the Earth (the globe) and the sun (the light). The side of the moon that faces the Earth will be in darkness. Write your observations of this new-moon phase.

3 Continue to move the moon in its orbit around the Earth. When part of the moon is illuminated by the light, as viewed from Earth, the moon is in the crescent phase. Record your observations.

4 If you have time, you may draw your own moon-phase diagram.

Analyze the Results

1 About 2 weeks after the new moon appears, the entire moon is visible in the sky. Move the ball to show this event.

2 What other phases can you add to your diagram? For example, when does the quarter moon appear?

3 Explain why the moon sometimes appears as a crescent to viewers on Earth.

CHAPTER RESOURCES

Chapter Resource File

- **Datasheet for LabBook**
- **Lab Notes and Answers**

CLASSROOM TESTED & APPROVED

Joseph W. Price
H. M. Browne Junior High
Washington, D.C.

Phases of the Moon

Teacher's Notes

Time Required
One 45-minute class period

Lab Ratings

EASY ─────────────→ HARD

Teacher Prep 🧪🧪
Student Set-Up 🧪🧪
Concept Level 🧪
Clean Up 🧪

MATERIALS

The materials listed on the student page are enough for a group of 3 to 4 students. You can use a lamp or a flashlight as the light source.

Analyze the Results

1. At full moon, Earth is between the sun and the moon. To represent this phase, students should move the plastic-foam ball to the opposite side of the globe from the light source.

2. In the model, students should move the plastic-foam ball one-quarter of the way around the globe and three-quarters of the way around the globe. These positions represent the first-quarter phase and last-quarter phase.

3. Sample answer: As the moon continues to move in its orbit around Earth, part of its illuminated half becomes visible. When a sliver of the moon is visible from Earth, the moon enters a crescent phase.

Reach for the Stars

Teacher's Notes

Time Required
Two 45-minute class periods

Lab Ratings

EASY ——————————→ HARD

Teacher Prep &
Student Set-Up &&&
Concept Level &&
Clean Up &&

MATERIALS
The materials listed on the student page are enough for 3 to 4 students.

Safety Caution
Remind students to review all safety cautions and icons before beginning this lab activity.

Model-Making Lab

Reach for the Stars

Have you ever thought about living and working in space? Well, in order for you to do so, you would have to learn to cope with the new environment and surroundings. At the same time that astronauts are adjusting to the topsy-turvy conditions of space travel, they are also dealing with special tools used to repair and build space stations. In this activity, you will get the chance to model one tool that might help astronauts work in space.

MATERIALS

- ball, plastic-foam
- box, cardboard
- hole punch
- paper brads (2)
- paper clips, jumbo (2)
- ruler, metric
- scissors
- wire, metal

SAFETY

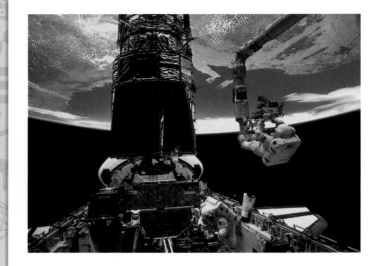

Ask a Question

1. How can I build a piece of equipment that models how astronauts work in space?

Form a Hypothesis

2. Write a possible answer for the question above. Describe a possible tool that would help astronauts work in space.

Test the Hypothesis

3. Cut three strips from the cardboard box. Each strip should be about 5 cm wide. The strips should be at least 20 cm long but not longer than 40 cm.

Alyson Mike
East Valley Middle School
East Helena, Montana

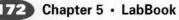

④ Punch holes near the center of each end of the three cardboard strips. The holes should be about 3 cm from the end of each strip.

⑤ Lay the strips end to end along your table. Slide the second strip toward the first strip so that a hole in the first strip lines up with a hole in the second strip. Slip a paper brad through the holes, and bend its ends out to attach the cardboard strips.

⑥ Use another brad to attach the third cardboard strip to the free end of the second strip. Now, you have your mechanical arm. The paper brads create joints where the cardboard strips meet.

⑦ Straighten the wire, and slide it through the hole in one end of your mechanical arm. Bend about 3 cm of the wire in a 90° angle so that it will not slide back out of the hole.

⑧ Now, try to move the arm by holding the free ends of the cardboard and wire. The arm should bend and straighten at the joints. If it is difficult to move your mechanical arm, adjust the design. Consider loosening the brads, for example.

⑨ Your mechanical arm now needs a hand. Otherwise, it won't be able to pick things up! Straighten one paper clip, and slide it through the hole where you attached the wire in step 7. Bend one end of the paper clip to form a loop around the cardboard and the other end to form a hook. You will use this hook to pick things up.

⑩ Bend a second paper clip into a U shape. Stick the straight end of this paper clip into the foam ball. Leave the ball on your desk.

⑪ Move your mechanical arm so that you can lift the foam ball. The paper-clip hook on the mechanical arm will have to catch the paper clip on the ball.

Analyze the Results

❶ Did you have any trouble moving the mechanical arm in step 8? What adjustments did you make?

❷ Did you have trouble picking up the foam ball? What might have made picking up the ball easier?

Draw Conclusions

❸ What improvements could you make to your mechanical arm that might make it easier to use?

❹ How would a tool like this one help astronauts work in space?

Applying Your Data

Adjust the design for your mechanical arm. Can you find a way to lift objects other than the foam ball? For example, can you lift heavier objects or objects that do not have a loop attached? How?

Research the tools that astronauts use on space stations and on the space shuttle. How do their tools help them work in the special conditions of space?

Analyze the Results

1. Answers may vary. Students may have loosened the paper brads.

2. Answers may vary. Altering the paper-clip loop on the ball or changing the shape of the hook on the arm could make the task easier.

Draw Conclusions

3. Answers may vary. Students may suggest using different materials, changing the length of different arm segments, or mounting the arm on a secure base.

4. Answers may vary. This device could help astronauts manipulate objects outside a spacecraft without having to go on a spacewalk. Also, if the arm were mechanized, it could allow astronauts to move massive objects with precision.

Applying Your Data

Answers may vary.

✓ *Reading Check* Answers

Chapter 1   Studying Space
Section 1

Page 5: Copernicus believed in a sun-centered universe.

Page 6: Newton's law of gravity helped explain why the planets orbit the sun and moons orbit planets.

Section 2

Page 8: The objective lens collects light and forms an image at the back of the telescope. The eyepiece magnifies the image produced by the objective lens.

Page 10: Air pollution, water vapor, and light pollution distort the images produced by optical telescopes.

Page 13: because the atmosphere blocks most X-ray radiation from space

Section 3

Page 15: Different constellations are visible in the Northern and Southern Hemispheres because different portions of the sky are visible from the Northern and Southern hemispheres.

Page 17: The apparent movement of the sun and stars is caused by the Earth's rotation on its axis.

Page 18: 9.46 trillion kilometers

Page 20: One might conclude that all of the galaxies are traveling toward the Earth and that the universe is contracting.

Chapter 2   Stars, Galaxies, and the Universe
Section 1

Page 32: Rigel is hotter than Betelgeuse because blue stars are hotter than red stars.

Page 34: A star's absorption spectrum indicates some of the elements that are in the star's atmosphere.

Page 36: Apparent magnitude is the brightness of a light or star.

Page 37: A light-year is the distance that light travels in 1 year.

Page 38: The actual motion of stars is hard to see because the stars are so distant.

Section 2

Page 41: A red giant star is a star that expands and cools once it uses all of its hydrogen. As the center of a star continues to shrink a red giant star can become a red supergiant star.

Page 45: A black hole is an object that is so massive that even light cannot escape its gravity. A black hole can be detected when it gives off X rays.

Section 3

Page 46: Spiral galaxies have a bulge at the center and spiral arms. The arms of spiral galaxies are made up of gas, dust, and new stars.

Page 48: A globular cluster is a tight group of up to 1 million stars that looks like a ball. An open cluster is a group of closely grouped stars that are usually located along the spiral disk of a galaxy.

Page 49: Quasars are starlike sources of light that are extremely far away. Some scientists think that quasars may be the core of young galaxies that are in the process of forming.

Section 4

Page 51: Cosmic background radiation is radiation that is left over from the big bang. After the big bang, cosmic background radiation was distributed everywhere and filled all of space.

Page 52: One way to calculate the age of the universe is to measure the distance from Earth to various galaxies.

Page 53: If gravity stops the expansion of the universe, the universe might collapse. If the expansion of the universe continues forever, stars will age and die and the universe will eventually become cold and dark.

Chapter 3   Formation of the Solar System
Section 1

Page 65: The solar nebula is the cloud of gas and dust that formed our solar system.

Page 67: Jupiter, Saturn, Uranus, and Neptune

Section 2

Page 69: Energy from gravity is not enough to power the sun, because if all of the sun's gravitational energy were released, the sun would last for only 45 million years.

Page 71: The nuclei of hydrogen atoms repel each other because they are positively charged and like charges repel each other.

Page 72: Sunspots are cooler, dark spots on the sun. Sunspots occur because when activity slows down in the convective zone, areas of the photosphere become cooler.

Section 3

Page 74: During Earth's early formation, planetesimals collided with the Earth. The energy of their motion heated the planet.

Page 76: Scientists think that the Earth's first atmosphere was a steamy mixture of carbon dioxide and water vapor.

Page 78: When photosynthetic organisms appeared on Earth, they released oxygen into the Earth's atmosphere. Over several million years, more and more oxygen was added to the atmosphere, which helped form Earth's current atmosphere.

Section 4

Page 81: Kepler's third law of motion states that planets that are farther away from the sun take longer to orbit the sun.

Page 82: Newton's law of universal gravitation states that the force of gravity depends on the product of the masses of the objects divided by the square of the distance between the objects.

Chapter 4   A Family of Planets
Section 1

Page 95: Light travels about 300,000 km/s.

Page 97: Jupiter, Saturn, Uranus, Neptune, and Pluto are in the outer solar system.

Section 2

Page 99: Radar technology was used to map the surface of Venus.

Page 100: Earth's global system includes the atmosphere, the oceans, and the biosphere.

Page 102: Mars' crust is chemically different from Earth's crust, so the Martian crust does not move. As a result, volcanoes build up in the same spots on Mars.

Section 3

Page 105: Saturn's rings are made of icy particles ranging in size from a few centimeters to several meters wide.

Page 107: Neptune's interior releases energy to its outer layers, which creates belts of clouds in Neptune's atmosphere.

Section 4

Page 111: The moon formed from a piece of Earth's mantle, which broke off during a collision between Earth and a large object.

Page 113: During a solar eclipse, the moon blocks out the sun and casts a shadow on Earth.

Page 114: We don't see solar and lunar eclipses every month because the moon's orbit around Earth is tilted.

Page 115: Because Titan's atmosphere is similar to the atmosphere on Earth before life evolved, scientists can study Titan's atmosphere to learn how life began.

Page 116: Pluto is eclipsed by Charon every 120 years.

Section 5

Page 119: Comets come from the Oort cloud and the Kuiper belt.

Page 121: The major types of meteorites are stony, metallic, and stony-iron meteorites.

Page 122: Large objects strike Earth every few thousand years.

Chapter 5   Exploring Space
Section 1

Page 134: Tsiolkovsky helped develop rocket theory. Goddard developed the first rockets.

Page 136: Rockets carry oxygen so that their fuel can be burned.

Section 2

Page 139: Answers may vary. LEO is much closer to the Earth than GEO.

Page 141: Information from one location is transmitted to a communications satellite. The satellite then sends the information to another location on Earth.

Page 143: Satellites in the EOS program are designed to work together so that many different types of data can be integrated.

Section 3

Page 145: The Magellan mission showed that, in many ways, the surface of Venus is similar to the surface of Earth.

Page 146: The Mars Pathfinder mission found evidence that water once flowed across the surface of Mars.

Page 148: The mission of the Stardust probe is to gather samples from a comet's tail and return them to Earth.

Section 4

Page 151: the orbiter, the liquid-fuel tank, and the solid-fuel booster rockets

Page 153: The Russians are supplying a service module, docking modules, life-support and research modules, and transportation to and from the station. The Americans are providing lab modules, the supporting frame, solar panels, living quarters, and a biomedical laboratory.

Page 155: Space-age spinoffs are technologies that were developed for the space program but are now used in everyday life.

Appendix

Study Skills

FoldNote Instructions

Have you ever tried to study for a test or quiz but didn't know where to start? Or have you read a chapter and found that you can remember only a few ideas? Well, FoldNotes are a fun and exciting way to help you learn and remember the ideas you encounter as you learn science!

FoldNotes are tools that you can use to organize concepts. By focusing on a few main concepts, FoldNotes help you learn and remember how the concepts fit together. They can help you see the "big picture." Below you will find instructions for building 10 different FoldNotes.

Pyramid

1. Place a sheet of paper in front of you. Fold the lower left-hand corner of the paper diagonally to the opposite edge of the paper.

2. Cut off the tab of paper created by the fold (at the top).

3. Open the paper so that it is a square. Fold the lower right-hand corner of the paper diagonally to the opposite corner to form a triangle.

4. Open the paper. The creases of the two folds will have created an X.

5. Using scissors, cut along one of the creases. Start from any corner, and stop at the center point to create two flaps. Use tape or glue to attach one of the flaps on top of the other flap.

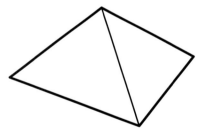

Double Door

1. Fold a sheet of paper in half from the top to the bottom. Then, unfold the paper.

2. Fold the top and bottom edges of the paper to the crease.

Booklet

1. Fold a sheet of paper in half from left to right. Then, unfold the paper.

2. Fold the sheet of paper in half again from the top to the bottom. Then, unfold the paper.

3. Refold the sheet of paper in half from left to right.

4. Fold the top and bottom edges to the center crease.

5. Completely unfold the paper.

6. Refold the paper from top to bottom.

7. Using scissors, cut a slit along the center crease of the sheet from the folded edge to the creases made in step 4. Do not cut the entire sheet in half.

8. Fold the sheet of paper in half from left to right. While holding the bottom and top edges of the paper, push the bottom and top edges together so that the center collapses at the center slit. Fold the four flaps to form a four-page book.

Layered Book

1. Lay one sheet of paper on top of another sheet. Slide the top sheet up so that 2 cm of the bottom sheet is showing.

2. Hold the two sheets together, fold down the top of the two sheets so that you see four 2 cm tabs along the bottom.

3. Using a stapler, staple the top of the FoldNote.

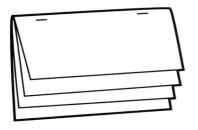

Appendix

Key-Term Fold

1. Fold a sheet of lined notebook paper in half from left to right.

2. Using scissors, cut along every third line from the right edge of the paper to the center fold to make tabs.

Four-Corner Fold

1. Fold a sheet of paper in half from left to right. Then, unfold the paper.

2. Fold each side of the paper to the crease in the center of the paper.

3. Fold the paper in half from the top to the bottom. Then, unfold the paper.

4. Using scissors, cut the top flap creases made in step 3 to form four flaps.

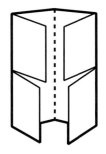

Three-Panel Flip Chart

1. Fold a piece of paper in half from the top to the bottom.

2. Fold the paper in thirds from side to side. Then, unfold the paper so that you can see the three sections.

3. From the top of the paper, cut along each of the vertical fold lines to the fold in the middle of the paper. You will now have three flaps.

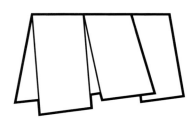

Table Fold

1. Fold a piece of paper in half from the top to the bottom. Then, fold the paper in half again.

2. Fold the paper in thirds from side to side.

3. Unfold the paper completely. Carefully trace the fold lines by using a pen or pencil.

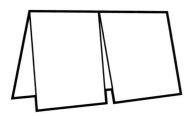

Two-Panel Flip Chart

1. Fold a piece of paper in half from the top to the bottom.

2. Fold the paper in half from side to side. Then, unfold the paper so that you can see the two sections.

3. From the top of the paper, cut along the vertical fold line to the fold in the middle of the paper. You will now have two flaps.

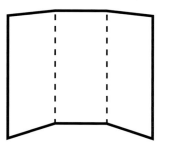

Tri-Fold

1. Fold a piece a paper in thirds from the top to the bottom.

2. Unfold the paper so that you can see the three sections. Then, turn the paper sideways so that the three sections form vertical columns.

3. Trace the fold lines by using a pen or pencil. Label the columns "Know," "Want," and "Learn."

Graphic Organizer Instructions

Graphic Organizer Have you ever wished that you could "draw out" the many concepts you learn in your science class? Sometimes, being able to *see* how concepts are related really helps you remember what you've learned. Graphic Organizers do just that! They give you a way to draw or map out concepts.

All you need to make a Graphic Organizer is a piece of paper and a pencil. Below you will find instructions for four different Graphic Organizers designed to help you organize the concepts you'll learn in this book.

Spider Map

1. Draw a diagram like the one shown. In the circle, write the main topic.

2. From the circle, draw legs to represent different categories of the main topic. You can have as many categories as you want.

3. From the category legs, draw horizontal lines. As you read the chapter, write details about each category on the horizontal lines.

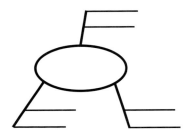

Comparison Table

1. Draw a chart like the one shown. Your chart can have as many columns and rows as you want.

2. In the top row, write the topics that you want to compare.

3. In the left column, write characteristics of the topics that you want to compare. As you read the chapter, fill in the characteristics for each topic in the appropriate boxes.

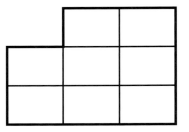

Chain-of-Events-Chart

1. Draw a box. In the box, write the first step of a process or the first event of a timeline.

2. Under the box, draw another box, and use an arrow to connect the two boxes. In the second box, write the next step of the process or the next event in the timeline.

3. Continue adding boxes until the process or timeline is finished.

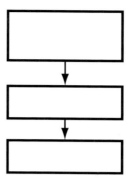

Concept Map

1. Draw a circle in the center of a piece of paper. Write the main idea of the chapter in the center of the circle.

2. From the circle, draw other circles. In those circles, write characteristics of the main idea. Draw arrows from the center circle to the circles that contain the characteristics.

3. From each circle that contains a characteristic, draw other circles. In those circles, write specific details about the characteristic. Draw arrows from each circle that contains a characteristic to the circles that contain specific details. You may draw as many circles as you want.

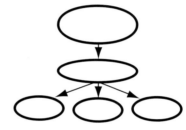

SI Measurement

The International System of Units, or SI, is the standard system of measurement used by many scientists. Using the same standards of measurement makes it easier for scientists to communicate with one another.

SI works by combining prefixes and base units. Each base unit can be used with different prefixes to define smaller and larger quantities. The table below lists common SI prefixes.

SI Prefixes

Prefix	Symbol	Factor	Example
kilo-	k	1,000	kilogram, 1 kg = 1,000 g
hecto-	h	100	hectoliter, 1 hL = 100 L
deka-	da	10	dekameter, 1 dam = 10 m
		1	meter, liter, gram
deci-	d	0.1	decigram, 1 dg = 0.1 g
centi-	c	0.01	centimeter, 1 cm = 0.01 m
milli-	m	0.001	milliliter, 1 mL = 0.001 L
micro-	μ	0.000 001	micrometer, 1 μm = 0.000 001 m

SI Conversion Table

SI units	From SI to English	From English to SI
Length		
kilometer (km) = 1,000 m	1 km = 0.621 mi	1 mi = 1.609 km
meter (m) = 100 cm	1 m = 3.281 ft	1 ft = 0.305 m
centimeter (cm) = 0.01 m	1 cm = 0.394 in.	1 in. = 2.540 cm
millimeter (mm) = 0.001 m	1 mm = 0.039 in.	
micrometer (μm) = 0.000 001 m		
nanometer (nm) = 0.000 000 001 m		
Area		
square kilometer (km^2) = 100 hectares	1 km^2 = 0.386 mi^2	1 mi^2 = 2.590 km^2
hectare (ha) = 10,000 m^2	1 ha = 2.471 acres	1 acre = 0.405 ha
square meter (m^2) = 10,000 cm^2	1 m^2 = 10.764 ft^2	1 ft^2 = 0.093 m^2
square centimeter (cm^2) = 100 mm^2	1 cm^2 = 0.155 in.^{2}	1 in.^{2} = 6.452 cm^2
Volume		
liter (L) = 1,000 mL = 1 dm^3	1 L = 1.057 fl qt	1 fl qt = 0.946 L
milliliter (mL) = 0.001 L = 1 cm^3	1 mL = 0.034 fl oz	1 fl oz = 29.574 mL
microliter (μL) = 0.000 001 L		
Mass		
kilogram (kg) = 1,000 g	1 kg = 2.205 lb	1 lb = 0.454 kg
gram (g) = 1,000 mg	1 g = 0.035 oz	1 oz = 28.350 g
milligram (mg) = 0.001 g		
microgram (μg) = 0.000 001 g		

Temperature Scales

Temperature can be expressed by using three different scales: Fahrenheit, Celsius, and Kelvin. The SI unit for temperature is the kelvin (K).

Although 0 K is much colder than 0°C, a change of 1 K is equal to a change of 1°C.

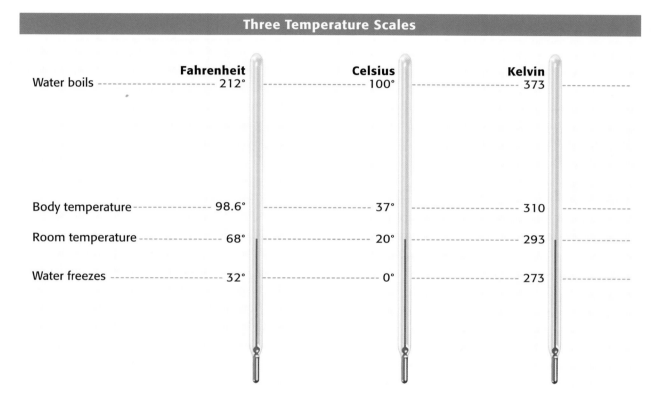

Three Temperature Scales

	Fahrenheit	Celsius	Kelvin
Water boils	212°	100°	373
Body temperature	98.6°	37°	310
Room temperature	68°	20°	293
Water freezes	32°	0°	273

Temperature Conversions Table

To convert	Use this equation:	Example
Celsius to Fahrenheit $°C \rightarrow °F$	$°F = \left(\dfrac{9}{5} \times °C\right) + 32$	Convert 45°C to °F. $°F = \left(\dfrac{9}{5} \times 45°C\right) + 32 = 113°F$
Fahrenheit to Celsius $°F \rightarrow °C$	$°C = \dfrac{5}{9} \times (°F - 32)$	Convert 68°F to °C. $°C = \dfrac{5}{9} \times (68°F - 32) = 20°C$
Celsius to Kelvin $°C \rightarrow K$	$K = °C + 273$	Convert 45°C to K. $K = 45°C + 273 = 318 \ K$
Kelvin to Celsius $K \rightarrow °C$	$°C = K - 273$	Convert 32 K to °C. $°C = 32K - 273 = -241°C$

Appendix

Measuring Skills

Using a Graduated Cylinder

When using a graduated cylinder to measure volume, keep the following procedures in mind:

1 Place the cylinder on a flat, level surface before measuring liquid.

2 Move your head so that your eye is level with the surface of the liquid.

3 Read the mark closest to the liquid level. On glass graduated cylinders, read the mark closest to the center of the curve in the liquid's surface.

Using a Meterstick or Metric Ruler

When using a meterstick or metric ruler to measure length, keep the following procedures in mind:

1 Place the ruler firmly against the object that you are measuring.

2 Align one edge of the object exactly with the 0 end of the ruler.

3 Look at the other edge of the object to see which of the marks on the ruler is closest to that edge. (Note: Each small slash between the centimeters represents a millimeter, which is one-tenth of a centimeter.)

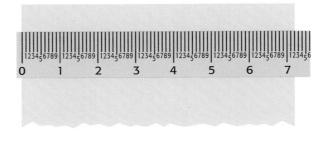

Using a Triple-Beam Balance

When using a triple-beam balance to measure mass, keep the following procedures in mind:

1 Make sure the balance is on a level surface.

2 Place all of the countermasses at 0. Adjust the balancing knob until the pointer rests at 0.

3 Place the object you wish to measure on the pan. **Caution:** Do not place hot objects or chemicals directly on the balance pan.

4 Move the largest countermass along the beam to the right until it is at the last notch that does not tip the balance. Follow the same procedure with the next-largest countermass. Then, move the smallest countermass until the pointer rests at 0.

5 Add the readings from the three beams together to determine the mass of the object.

6 When determining the mass of crystals or powders, first find the mass of a piece of filter paper. Then, add the crystals or powder to the paper, and remeasure. The actual mass of the crystals or powder is the total mass minus the mass of the paper. When finding the mass of liquids, first find the mass of the empty container. Then, find the combined mass of the liquid and container. The mass of the liquid is the total mass minus the mass of the container.

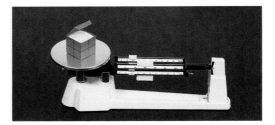

Scientific Methods

The ways in which scientists answer questions and solve problems are called **scientific methods.** The same steps are often used by scientists as they look for answers. However, there is more than one way to use these steps. Scientists may use all of the steps or just some of the steps during an investigation. They may even repeat some of the steps. The goal of using scientific methods is to come up with reliable answers and solutions.

Six Steps of Scientific Methods

1 Ask a Question Good questions come from careful **observations.** You make observations by using your senses to gather information. Sometimes, you may use instruments, such as microscopes and telescopes, to extend the range of your senses. As you observe the natural world, you will discover that you have many more questions than answers. These questions drive investigations.

Questions beginning with *what, why, how,* and *when* are important in focusing an investigation. Here is an example of a question that could lead to an investigation.

> **Question:** How does acid rain affect plant growth?

2 Form a Hypothesis After you ask a question, you need to form a **hypothesis.** A hypothesis is a clear statement of what you expect the answer to your question to be. Your hypothesis will represent your best "educated guess" based on what you have observed and what you already know. A good hypothesis is testable. Otherwise, the investigation can go no further. Here is a hypothesis based on the question, "How does acid rain affect plant growth?"

> **Hypothesis:** Acid rain slows plant growth.

The hypothesis can lead to predictions. A prediction is what you think the outcome of your experiment or data collection will be. Predictions are usually stated in an if-then format. Here is a sample prediction for the hypothesis that acid rain slows plant growth.

> **Prediction:** If a plant is watered with only acid rain (which has a pH of 4), then the plant will grow at half its normal rate.

3 Test the Hypothesis After you have formed a hypothesis and made a prediction, your hypothesis should be tested. One way to test a hypothesis is with a controlled experiment. A **controlled experiment** tests only one factor at a time. In an experiment to test the effect of acid rain on plant growth, the **control group** would be watered with normal rain water. The **experimental group** would be watered with acid rain. All of the plants should receive the same amount of sunlight and water each day. The air temperature should be the same for all groups. However, the acidity of the water will be a variable. In fact, any factor that is different from one group to another is a **variable.** If your hypothesis is correct, then the acidity of the water and plant growth are *dependant variables.* The amount a plant grows is dependent on the acidity of the water. However, the amount of water each plant receives and the amount of sunlight each plant receives are *independent variables.* Either of these factors could change without affecting the other factor.

Sometimes, the nature of an investigation makes a controlled experiment impossible. For example, the Earth's core is surrounded by thousands of meters of rock. Under such circumstances, a hypothesis may be tested by making detailed observations.

4 Analyze the Results After you have completed your experiments, made your observations, and collected your data, you must analyze all the information you have gathered. Tables and graphs are often used in this step to organize the data.

 **5 Draw Conclusions** After analyzing your data, you can determine if your results support your hypothesis. If your hypothesis is supported, you (or others) might want to repeat the observations or experiments to verify your results. If your hypothesis is not supported by the data, you may have to check your procedure for errors. You may even have to reject your hypothesis and make a new one. If you cannot draw a conclusion from your results, you may have to try the investigation again or carry out further observations or experiments.

 **6 Communicate Results** After any scientific investigation, you should report your results. By preparing a written or oral report, you let others know what you have learned. They may repeat your investigation to see if they get the same results. Your report may even lead to another question and then to another investigation.

Scientific Methods in Action

Scientific methods contain loops in which several steps may be repeated over and over again. In some cases, certain steps are unnecessary. Thus, there is not a "straight line" of steps. For example, sometimes scientists find that testing one hypothesis raises new questions and new hypotheses to be tested. And sometimes, testing the hypothesis leads directly to a conclusion. Furthermore, the steps in scientific methods are not always used in the same order. Follow the steps in the diagram, and see how many different directions scientific methods can take you.

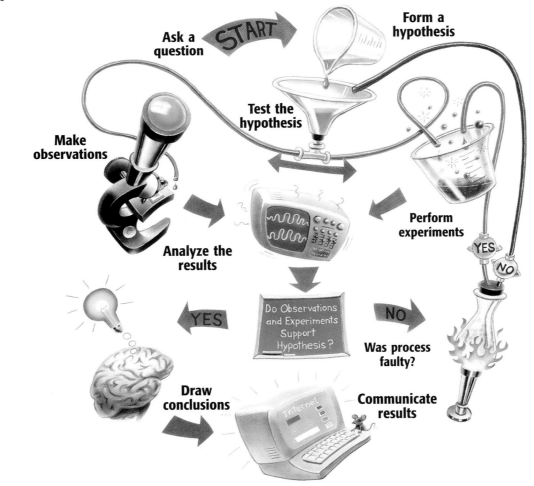

Making Charts and Graphs

Pie Charts

A pie chart shows how each group of data relates to all of the data. Each part of the circle forming the chart represents a category of the data. The entire circle represents all of the data. For example, a biologist studying a hardwood forest in Wisconsin found that there were five different types of trees. The data table at right summarizes the biologist's findings.

Wisconsin Hardwood Trees	
Type of tree	Number found
Oak	600
Maple	750
Beech	300
Birch	1,200
Hickory	150
Total	3,000

How to Make a Pie Chart

1 To make a pie chart of these data, first find the percentage of each type of tree. Divide the number of trees of each type by the total number of trees, and multiply by 100.

$$\frac{600 \text{ oak}}{3,000 \text{ trees}} \times 100 = 20\%$$

$$\frac{750 \text{ maple}}{3,000 \text{ trees}} \times 100 = 25\%$$

$$\frac{300 \text{ beech}}{3,000 \text{ trees}} \times 100 = 10\%$$

$$\frac{1,200 \text{ birch}}{3,000 \text{ trees}} \times 100 = 40\%$$

$$\frac{150 \text{ hickory}}{3,000 \text{ trees}} \times 100 = 5\%$$

2 Now, determine the size of the wedges that make up the pie chart. Multiply each percentage by 360°. Remember that a circle contains 360°.

$20\% \times 360° = 72°$ $25\% \times 360° = 90°$

$10\% \times 360° = 36°$ $40\% \times 360° = 144°$

$5\% \times 360° = 18°$

3 Check that the sum of the percentages is 100 and the sum of the degrees is 360.

$20\% + 25\% + 10\% + 40\% + 5\% = 100\%$

$72° + 90° + 36° + 144° + 18° = 360°$

4 Use a compass to draw a circle and mark the center of the circle.

5 Then, use a protractor to draw angles of 72°, 90°, 36°, 144°, and 18° in the circle.

6 Finally, label each part of the chart, and choose an appropriate title.

A Community of Wisconsin Hardwood Trees

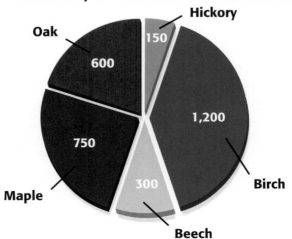

Line Graphs

Line graphs are most often used to demonstrate continuous change. For example, Mr. Smith's students analyzed the population records for their hometown, Appleton, between 1900 and 2000. Examine the data at right.

Because the year and the population change, they are the *variables*. The population is determined by, or dependent on, the year. Therefore, the population is called the **dependent variable,** and the year is called the **independent variable.** Each set of data is called a **data pair.** To prepare a line graph, you must first organize data pairs into a table like the one at right.

| Population of Appleton, 1900–2000 ||
Year	Population
1900	1,800
1920	2,500
1940	3,200
1960	3,900
1980	4,600
2000	5,300

How to Make a Line Graph

1 Place the independent variable along the horizontal (*x*) axis. Place the dependent variable along the vertical (*y*) axis.

2 Label the *x*-axis "Year" and the *y*-axis "Population." Look at your largest and smallest values for the population. For the *y*-axis, determine a scale that will provide enough space to show these values. You must use the same scale for the entire length of the axis. Next, find an appropriate scale for the *x*-axis.

3 Choose reasonable starting points for each axis.

4 Plot the data pairs as accurately as possible.

5 Choose a title that accurately represents the data.

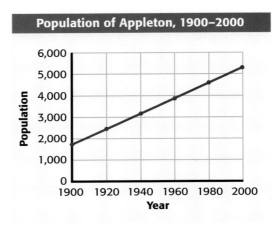

How to Determine Slope

Slope is the ratio of the change in the *y*-value to the change in the *x*-value, or "rise over run."

1 Choose two points on the line graph. For example, the population of Appleton in 2000 was 5,300 people. Therefore, you can define point *a* as (2000, 5,300). In 1900, the population was 1,800 people. You can define point *b* as (1900, 1,800).

2 Find the change in the *y*-value.
(*y* at point *a*) − (*y* at point *b*) =
5,300 people − 1,800 people =
3,500 people

3 Find the change in the *x*-value.
(*x* at point *a*) − (*x* at point *b*) =
2000 − 1900 = 100 years

4 Calculate the slope of the graph by dividing the change in *y* by the change in *x*.

$$slope = \frac{change\ in\ y}{change\ in\ x}$$

$$slope = \frac{3,500\ people}{100\ years}$$

$$slope = 35\ people\ per\ year$$

In this example, the population in Appleton increased by a fixed amount each year. The graph of these data is a straight line. Therefore, the relationship is **linear.** When the graph of a set of data is not a straight line, the relationship is **nonlinear.**

Using Algebra to Determine Slope

The equation in step 4 may also be arranged to be

$$y = kx$$

where y represents the change in the y-value, k represents the slope, and x represents the change in the x-value.

$$slope = \frac{change\ in\ y}{change\ in\ x}$$

$$k = \frac{y}{x}$$

$$k \times x = \frac{y \times x}{x}$$

$$kx = y$$

Bar Graphs

Bar graphs are used to demonstrate change that is not continuous. These graphs can be used to indicate trends when the data cover a long period of time. A meteorologist gathered the precipitation data shown here for Hartford, Connecticut, for April 1–15, 1996, and used a bar graph to represent the data.

Precipitation in Hartford, Connecticut April 1–15, 1996			
Date	Precipitation (cm)	Date	Precipitation (cm)
April 1	0.5	April 9	0.25
April 2	1.25	April 10	0.0
April 3	0.0	April 11	1.0
April 4	0.0	April 12	0.0
April 5	0.0	April 13	0.25
April 6	0.0	April 14	0.0
April 7	0.0	April 15	6.50
April 8	1.75		

How to Make a Bar Graph

1 Use an appropriate scale and a reasonable starting point for each axis.

2 Label the axes, and plot the data.

3 Choose a title that accurately represents the data.

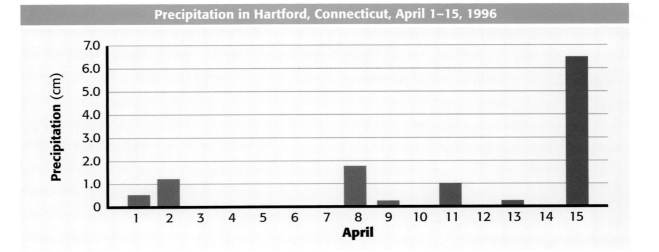

Precipitation in Hartford, Connecticut, April 1–15, 1996

Math Refresher

Science requires an understanding of many math concepts. The following pages will help you review some important math skills.

Averages

An **average,** or **mean,** simplifies a set of numbers into a single number that *approximates* the value of the set.

Example: Find the average of the following set of numbers: 5, 4, 7, and 8.

Step 1: Find the sum.
$$5 + 4 + 7 + 8 = 24$$

Step 2: Divide the sum by the number of numbers in your set. Because there are four numbers in this example, divide the sum by 4.
$$\frac{24}{4} = 6$$

The average, or mean, is **6.**

Ratios

A **ratio** is a comparison between numbers, and it is usually written as a fraction.

Example: Find the ratio of thermometers to students if you have 36 thermometers and 48 students in your class.

Step 1: Make the ratio.
$$\frac{36 \text{ thermometers}}{48 \text{ students}}$$

Step 2: Reduce the fraction to its simplest form.
$$\frac{36}{48} = \frac{36 \div 12}{48 \div 12} = \frac{3}{4}$$

The ratio of thermometers to students is **3 to 4,** or $\frac{3}{4}$. The ratio may also be written in the form 3:4.

Proportions

A **proportion** is an equation that states that two ratios are equal.
$$\frac{3}{1} = \frac{12}{4}$$

To solve a proportion, first multiply across the equal sign. This is called *cross-multiplication*. If you know three of the quantities in a proportion, you can use cross-multiplication to find the fourth.

Example: Imagine that you are making a scale model of the solar system for your science project. The diameter of Jupiter is 11.2 times the diameter of the Earth. If you are using a plastic-foam ball that has a diameter of 2 cm to represent the Earth, what must the diameter of the ball representing Jupiter be?
$$\frac{11.2}{1} = \frac{x}{2 \text{ cm}}$$

Step 1: Cross-multiply.
$$\frac{11.2}{1} \times \frac{x}{2}$$
$$11.2 \times 2 = x \times 1$$

Step 2: Multiply.
$$22.4 = x \times 1$$

Step 3: Isolate the variable by dividing both sides by 1.
$$x = \frac{22.4}{1}$$
$$x = 22.4 \text{ cm}$$

You will need to use a ball that has a diameter of **22.4** cm to represent Jupiter.

Percentages

A **percentage** is a ratio of a given number to 100.

> **Example:** What is 85% of 40?

Step 1: Rewrite the percentage by moving the decimal point two places to the left.

$$0.\underset{\smile}{85}$$

Step 2: Multiply the decimal by the number that you are calculating the percentage of.

$$0.85 \times 40 = 34$$

85% of 40 is **34.**

Decimals

To **add** or **subtract decimals,** line up the digits vertically so that the decimal points line up. Then, add or subtract the columns from right to left. Carry or borrow numbers as necessary.

> **Example:** Add the following numbers: 3.1415 and 2.96.

Step 1: Line up the digits vertically so that the decimal points line up.

$$\begin{array}{r} 3.1415 \\ + 2.96 \\ \hline \end{array}$$

Step 2: Add the columns from right to left, and carry when necessary.

$$\begin{array}{r} {}^{1}{}^{1} \\ 3.1415 \\ + 2.96 \\ \hline 6.1015 \end{array}$$

The sum is **6.1015.**

Fractions

Numbers tell you how many; **fractions** tell you *how much of a whole*.

> **Example:** Your class has 24 plants. Your teacher instructs you to put 5 plants in a shady spot. What fraction of the plants in your class will you put in a shady spot?

Step 1: In the denominator, write the total number of parts in the whole.

$$\frac{?}{24}$$

Step 2: In the numerator, write the number of parts of the whole that are being considered.

$$\frac{5}{24}$$

So, $\frac{5}{24}$ of the plants will be in the shade.

Reducing Fractions

It is usually best to express a fraction in its simplest form. Expressing a fraction in its simplest form is called *reducing* a fraction.

> **Example:** Reduce the fraction $\frac{30}{45}$ to its simplest form.

Step 1: Find the largest whole number that will divide evenly into both the numerator and denominator. This number is called the *greatest common factor* (GCF).

Factors of the numerator 30:
1, 2, 3, 5, 6, 10, **15,** 30

Factors of the denominator 45:
1, 3, 5, 9, **15,** 45

Step 2: Divide both the numerator and the denominator by the GCF, which in this case is 15.

$$\frac{30}{45} = \frac{30 \div 15}{45 \div 15} = \frac{2}{3}$$

Thus, $\frac{30}{45}$ reduced to its simplest form is $\frac{2}{3}$.

Adding and Subtracting Fractions

To **add** or **subtract fractions** that have the **same denominator,** simply add or subtract the numerators.

Examples:

$$\frac{3}{5} + \frac{1}{5} = ? \quad \text{and} \quad \frac{3}{4} - \frac{1}{4} = ?$$

Step 1: Add or subtract the numerators.

$$\frac{3}{5} + \frac{1}{5} = \frac{4}{} \quad \text{and} \quad \frac{3}{4} - \frac{1}{4} = \frac{2}{}$$

Step 2: Write the sum or difference over the denominator.

$$\frac{3}{5} + \frac{1}{5} = \frac{4}{5} \quad \text{and} \quad \frac{3}{4} - \frac{1}{4} = \frac{2}{4}$$

Step 3: If necessary, reduce the fraction to its simplest form.

$$\frac{4}{5} \text{ cannot be reduced, and } \frac{2}{4} = \frac{1}{2}.$$

To **add** or **subtract fractions** that have **different denominators,** first find the least common denominator (LCD).

Examples:

$$\frac{1}{2} + \frac{1}{6} = ? \quad \text{and} \quad \frac{3}{4} - \frac{2}{3} = ?$$

Step 1: Write the equivalent fractions that have a common denominator.

$$\frac{3}{6} + \frac{1}{6} = ? \quad \text{and} \quad \frac{9}{12} - \frac{8}{12} = ?$$

Step 2: Add or subtract the fractions.

$$\frac{3}{6} + \frac{1}{6} = \frac{4}{6} \quad \text{and} \quad \frac{9}{12} - \frac{8}{12} = \frac{1}{12}$$

Step 3: If necessary, reduce the fraction to its simplest form.

The fraction $\frac{4}{6} = \frac{2}{3}$, and $\frac{1}{12}$ cannot be reduced.

Multiplying Fractions

To **multiply fractions,** multiply the numerators and the denominators together, and then reduce the fraction to its simplest form.

Example:

$$\frac{5}{9} \times \frac{7}{10} = ?$$

Step 1: Multiply the numerators and denominators.

$$\frac{5}{9} \times \frac{7}{10} = \frac{5 \times 7}{9 \times 10} = \frac{35}{90}$$

Step 2: Reduce the fraction.

$$\frac{35}{90} = \frac{35 \div 5}{90 \div 5} = \frac{7}{18}$$

Dividing Fractions

To **divide fractions,** first rewrite the divisor (the number you divide by) upside down. This number is called the *reciprocal* of the divisor. Then multiply and reduce if necessary.

Example:

$$\frac{5}{8} \div \frac{3}{2} = ?$$

Step 1: Rewrite the divisor as its reciprocal.

$$\frac{3}{2} \rightarrow \frac{2}{3}$$

Step 2: Multiply the fractions.

$$\frac{5}{8} \times \frac{2}{3} = \frac{5 \times 2}{8 \times 3} = \frac{10}{24}$$

Step 3: Reduce the fraction.

$$\frac{10}{24} = \frac{10 \div 2}{24 \div 2} = \frac{5}{12}$$

Appendix

Scientific Notation

Scientific notation is a short way of representing very large and very small numbers without writing all of the place-holding zeros.

Example: Write 653,000,000 in scientific notation.

Step 1: Write the number without the place-holding zeros.

653

Step 2: Place the decimal point after the first digit.

6.53

Step 3: Find the exponent by counting the number of places that you moved the decimal point.

6.53000000

The decimal point was moved eight places to the left. Therefore, the exponent of 10 is positive 8. If you had moved the decimal point to the right, the exponent would be negative.

Step 4: Write the number in scientific notation.

6.53×10^8

Area

Area is the number of square units needed to cover the surface of an object.

Formulas:

area of a square = side × side
area of a rectangle = length × width
area of a triangle = $\frac{1}{2}$ × base × height

Examples: Find the areas.

Triangle

$area = \frac{1}{2} \times base \times height$

$area = \frac{1}{2} \times 3 \text{ cm} \times 4 \text{ cm}$

$area = $ **6 cm²**

4 cm

3 cm

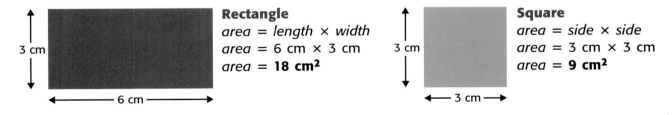

Rectangle
area = length × width
area = 6 cm × 3 cm
area = **18 cm²**

3 cm

6 cm

Square
area = side × side
area = 3 cm × 3 cm
area = **9 cm²**

3 cm

3 cm

Volume

Volume is the amount of space that something occupies.

Formulas:

volume of a cube =
side × side × side

volume of a prism =
area of base × height

Examples:

Find the volume of the solids.

Cube

volume = side × side × side
volume = 4 cm × 4 cm × 4 cm
volume = **64 cm³**

4 cm

4 cm   4 cm

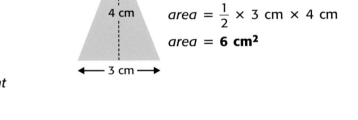

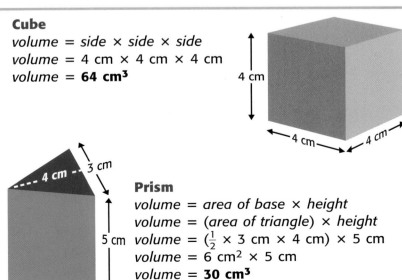

3 cm

4 cm

5 cm

Prism

volume = area of base × height
volume = (area of triangle) × height
volume = ($\frac{1}{2}$ × 3 cm × 4 cm) × 5 cm
volume = 6 cm² × 5 cm
volume = **30 cm³**

Physical Science Refresher

Atoms and Elements

Every object in the universe is made up of particles of some kind of matter. **Matter** is anything that takes up space and has mass. All matter is made up of elements. An **element** is a substance that cannot be separated into simpler components by ordinary chemical means. This is because each element consists of only one kind of atom. An **atom** is the smallest unit of an element that has all of the properties of that element.

Atomic Structure

Atoms are made up of small particles called subatomic particles. The three major types of subatomic particles are **electrons, protons, and neutrons.** Electrons have a negative electric charge, protons have a positive charge, and neutrons have no electric charge. The protons and neutrons are packed close to one another to form the **nucleus.** The protons give the nucleus a positive charge. Electrons are most likely to be found in regions around the nucleus called **electron clouds.** The negatively charged electrons are attracted to the positively charged nucleus. An atom may have several energy levels in which electrons are located.

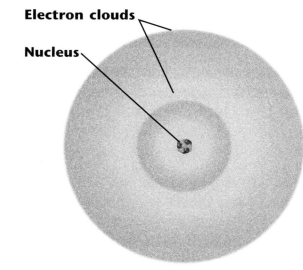

Electron clouds

Nucleus

Atomic Number

To help in the identification of elements, scientists have assigned an **atomic number** to each kind of atom. The atomic number is the number of protons in the atom. Atoms with the same number of protons are all the same kind of element. In an uncharged, or electrically neutral, atom there are an equal number of protons and electrons. Therefore, the atomic number equals the number of electrons in an uncharged atom. The number of neutrons, however, can vary for a given element. Atoms of the same element that have different numbers of neutrons are called **isotopes.**

Periodic Table of the Elements

In the periodic table, the elements are arranged from left to right in order of increasing atomic number. Each element in the table is in a separate box. An uncharged atom of each element has one more electron and one more proton than an uncharged atom of the element to its left. Each horizontal row of the table is called a **period.** Changes in chemical properties of elements across a period correspond to changes in the electron arrangements of their atoms. Each vertical column of the table, known as a **group,** lists elements with similar properties. The elements in a group have similar chemical properties because their atoms have the same number of electrons in their outer energy level. For example, the elements helium, neon, argon, krypton, xenon, and radon all have similar properties and are known as the noble gases.

Molecules and Compounds

When two or more elements are joined chemically, the resulting substance is called a **compound.** A compound is a new substance with properties different from those of the elements that compose it. For example, water, H_2O, is a compound formed when hydrogen (H) and oxygen (O) combine. The smallest complete unit of a compound that has the properties of that compound is called a **molecule.** A chemical formula indicates the elements in a compound. It also indicates the relative number of atoms of each element present. The chemical formula for water is H_2O, which indicates that each water molecule consists of two atoms of hydrogen and one atom of oxygen. The subscript number after the symbol for an element indicates how many atoms of that element are in a single molecule of the compound.

Acids, Bases, and pH

An ion is an atom or group of atoms that has an electric charge because it has lost or gained one or more electrons. When an acid, such as hydrochloric acid, HCl, is mixed with water, it separates into ions. An **acid** is a compound that produces hydrogen ions, H+, in water. The hydrogen ions then combine with a water molecule to form a hydronium ion, H_3O^+. A **base,** on the other hand, is a substance that produces hydroxide ions, OH^-, in water.

To determine whether a solution is acidic or basic, scientists use pH. The **pH** is a measure of the hydronium ion concentration in a solution. The pH scale ranges from 0 to 14. The middle point, pH = 7, is neutral, neither acidic nor basic. Acids have a pH less than 7; bases have a pH greater than 7. The lower the number is, the more acidic the solution. The higher the number is, the more basic the solution.

Chemical Equations

A chemical reaction occurs when a chemical change takes place. (In a chemical change, new substances with new properties are formed.) A chemical equation is a useful way of describing a chemical reaction by means of chemical formulas. The equation indicates what substances react and what the products are. For example, when carbon and oxygen combine, they can form carbon dioxide. The equation for the reaction is as follows: $C + O_2 \rightarrow CO_2$.

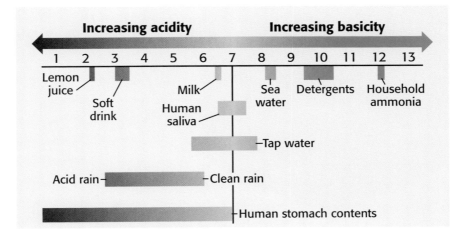

Sky Maps

Spring

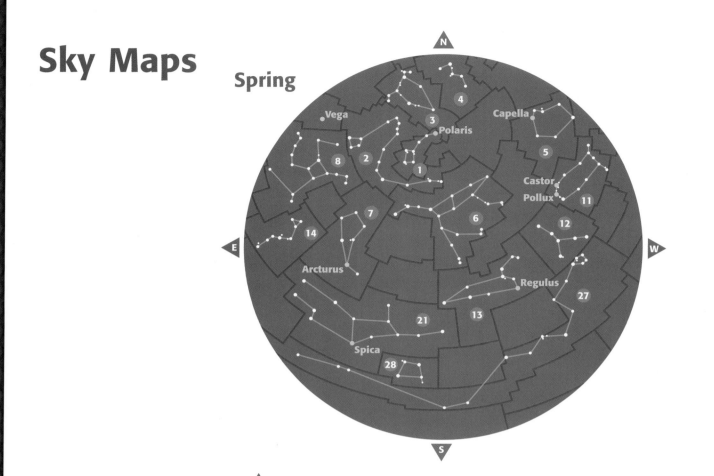

Summer

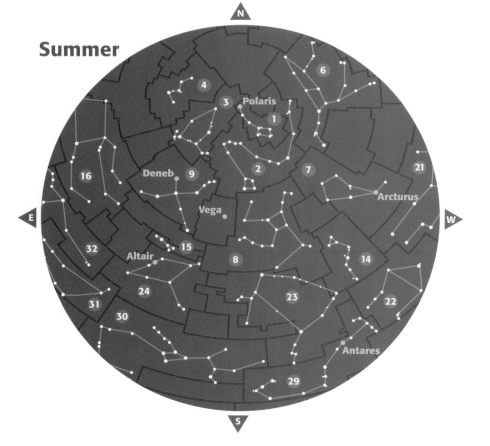

Constellations

1 **Ursa Minor**
2 **Draco**
3 **Cepheus**
4 **Cassiopeia**
5 **Auriga**
6 **Ursa Major**
7 **Bootes**
8 **Hercules**
9 **Cygnus**
10 **Perseus**
11 **Gemini**
12 **Cancer**
13 **Leo**
14 **Serpens**
15 **Sagitta**
16 **Pegasus**
17 **Pisces**

Autumn

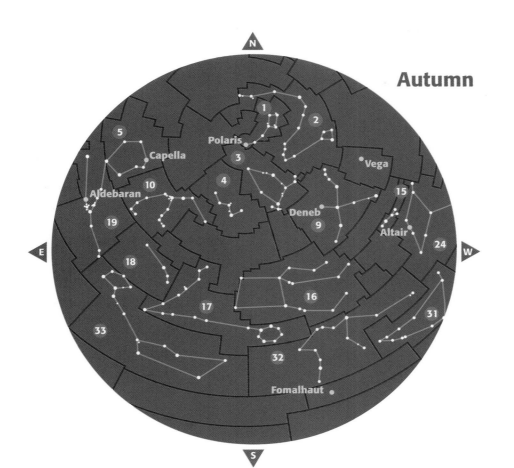

Constellations

18 Aries
19 Taurus
20 Orion
21 Virgo
22 Libra
23 Ophiuchus
24 Aquila
25 Lepus
26 Canis Major
27 Hydra
28 Corvus
29 Scorpius
30 Sagittarius
31 Capricornus
32 Aquarius
33 Cetus
34 Columba

Winter

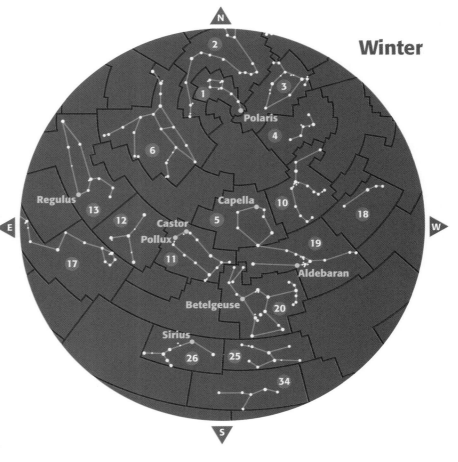

Glossary

A

absolute magnitude the brightness that a star would have at a distance of 32.6 light-years from Earth (36)

altitude the angle between an object in the sky and the horizon (16)

apparent magnitude the brightness of a star as seen from the Earth (36)

artificial satellite any human-made object placed in orbit around a body in space (138)

asteroid a small, rocky object that orbits the sun, usually in a band between the orbits of Mars and Jupiter (120)

asteroid belt the region of the solar system that is between the orbits of Mars and Jupiter and in which most asteroids orbit (120)

astronomical unit the average distance between the Earth and the sun; approximately 150 million kilometers (symbol, AU) (95)

astronomy the study of the universe (4)

B

big bang theory the theory that states that the universe began with a tremendous explosion about 13.7 billion years ago (51)

black hole an object so massive and dense that even light cannot escape its gravity (45)

C

comet a small body of ice, rock, and cosmic dust that follows an elliptical orbit around the sun and that gives off gas and dust in the form of a tail as it passes close to the sun (118)

constellation a region of the sky that contains a recognizable star pattern and that is used to describe the location of objects in space (14)

core the central part of the Earth below the mantle (75)

cosmology the study of the origin, properties, processes, and evolution of the universe (50)

crust the thin and solid outermost layer of the Earth above the mantle (75)

D

day the time required for Earth to rotate once on its axis (4)

E

eclipse an event in which the shadow of one celestial body falls on another (113)

electromagnetic spectrum all of the frequencies or wavelengths of electromagnetic radiation (11)

G

galaxy a collection of stars, dust, and gas bound together by gravity (46)

gas giant a planet that has a deep, massive atmosphere, such as Jupiter, Saturn, Uranus, or Neptune (104)

geostationary orbit an orbit that is about 36,000 km above the Earth's surface and in which a satellite is above a fixed spot on the equator (139)

globular cluster a tight group of stars that looks like a ball and contains up to 1 million stars (48)

H

horizon the line where the sky and the Earth appear to meet (16)

H-R diagram Hertzsprung-Russell diagram, a graph that shows the relationship between a star's surface temperature and absolute magnitude (42)

L

light-year the distance that light travels in one year; about 9.46 trillion kilometers (18, 37)

low earth orbit an orbit that is less than 1,500 km above the Earth's surface (139)

M

main sequence the location on the H-R diagram where most stars lie; it has a diagonal pattern from the lower right (low temperature and luminosity) to the upper left (high temperature and luminosity) (43)

mantle the layer of rock between the Earth's crust and core (75)

S

satellite/satélite un cuerpo natural o artificial que gira alrededor de un planeta (110)

solar nebula/nebulosa solar la nube de gas y polvo que formó nuestro Sistema Solar (65)

space probe/sonda espacial un vehículo no tripulado que lleva instrumentos científicos al espacio con el fin de recopilar información científica (144)

space shuttle/transbordador espacial un vehículo espacial reutilizable que despega como un cohete y aterriza como un avión (151)

space station/estación espacial una plataforma orbital de largo plazo desde la cual pueden lanzarse otros vehículos o en la que pueden realizarse investigaciones científicas (152)

spectrum/espectro la banda de colores que se produce cuando la luz blanca pasa a través de un prisma (33)

sunspot/mancha solar un área oscura en la fotosfera del Sol que es más fría que las áreas que la rodean y que tiene un campo magnético fuerte (72)

supernova/supernova una explosión gigantesca en la que una estrella masiva se colapsa y lanza sus capas externas hacia el espacio (44)

T

telescope/telescopio un instrumento que capta la radiación electromagnética del cielo y la concentra para mejorar la observación (8)

terrestrial planet/planeta terrestre uno de los planetas muy densos que se encuentran más cerca del Sol; Mercurio, Venus, Marte y la Tierra (98)

thrust/empuje la fuerza de empuje o arrastre ejercida por el motor de un avión o cohete (136)

W

white dwarf/enana blanca una estrella pequeña, caliente y tenue que es el centro sobrante de una estrella vieja (41)

Y

year/año el tiempo que se requiere para que la Tierra le dé la vuelta al Sol una vez (4)

Z

zenith/cenit el punto del cielo situado directamente sobre un observador en la Tierra (16)

Index

Boldface page numbers refer to illustrative material, such as figures, tables, margin elements, photographs, and illustrations.

Index

Index

Index

Index

Index

Credits

Abbreviations used: (t) top, (c) center, (b) bottom, (l) left, (r) right, (bkgd) background

PHOTOGRAPHY

Front Cover Larry Landolfi/Photo Researchers, Inc.

Skills Practice Lab Teens Sam Dudgeon/HRW

Connection to Astrology Corbis Images; **Connection to Biology** David M. Phillips/Visuals Unlimited; **Connection to Chemistry** Digital Image copyright © 2005 PhotoDisc; **Connection to Environment** Digital Image copyright © 2005 PhotoDisc; **Connection to Geology** Letraset Phototone; **Connection to Language Arts** Digital Image copyright © 2005 PhotoDisc; **Connection to Meteorology** Digital Image copyright © 2005 PhotoDisc; **Connection to Oceanography** © ICONOTEC; **Connection to Physics** Digital Image copyright © 2005 PhotoDisc

Table of Contents iv (bl), Peter Van Steen/HRW; iv (b), Bill & Sally Fletcher/Tom Stack & Associates; v (t), NASA/Peter Arnold, Inc.; vi–vii, Victoria Smith/HRW; x (bl), Sam Dudgeon/HRW; xi (tl), John Langford/HRW; xi (b), Sam Dudgeon/HRW; xii (tl), Victoria Smith/HRW; xii (bl), Stephanie Morris/HRW; xii (br), Sam Dudgeon/HRW; xiii (tl), Patti Murray/Animals, Animals; xiii (tr), Jana Birchum/HRW; xiii (b), Peter Van Steen/HRW

Chapter One 2–3, Roger Ressmeyer/CORBIS; 4, David L. Brown/Tom Stack & Associates; 6, The Bridgeman Art Library; 7, Roger Ressmeyer/Corbis; 8 (bl), Peter Van Steen/HRW; 8 (r), Fred Espenek; 10 (tl), Simon Fraser/Science Photo Library/Photo Researchers, Inc.; 10 (b), NASA; 10 (inset), Roger Ressmeyer/Corbis; 11 (radio), Sam Dudgeon/HRW; 11 (microwave) Sam Dudgeon/HRW; 11 (keyboard), Chuck O'Rear/Woodfin Camp & Associates, Inc.; 11 (sunburn), HRW; 11 (x–ray), David M. Dennis/Tom Stack & Associates; 11 (head), Michael Scott/Getty Images/Stone; 11 (tea), Tony McConnell/SPL/Photo Researchers, Inc.; 12 (gamma), NASA; 12 (radio), NASA; 12 (x–ray), NASA; 12 (infrared), NASA; 13, MSFC/NASA; 16, Peter Van Steen/HRW; 16 (bkgd), Frank Zullo/Photo Researchers, Inc.; 19 (tl), Jim Cummings/Getty Images/Taxi; 19 (tc), Mike Yamashita/Woodfin Camp/Picture Quest; 19 (tr), NASA; 19 (cr), Nozomi MSI Team/ISAS; 19 (bc), Jerry Lodriguss/Photo Researchers, Inc.; 19 (br), Tony & Daphne Hallas/Science Photo Library/Photo Researchers, Inc.; 20 (b), Jane C. Charlton, Penn State/HST/ESA/NASA; 20 (tc), NCAR/Tom Stack & Associates; 22, Peter Van Steen/HRW; 23, Peter Van Steen/HRW; 24 (t), MSFC/NASA; 24 (tea),Tony McConnell/SPL/Photo Researchers, Inc.; 28 (t), Craig Matthew and Robert Simmon/NASA/GSFC/DMSP; 29 (t), American Museum of Natural History; 29 (bl), Richard Berenholtz/CORBIS

Chapter Two 30–31, NASA; 582 (bl), Phil Degginger/Color–Pic, Inc.; 32 (br), John Sanford/Astrostock; 33, Sam Dudgeon/HRW; 35, Roger Ressmeyer/CORBIS; 36, Andre Gallant/Getty Images/The Image Bank, 40, V. Bujarrabal (OAN, Spain), WFPC2, HST, ESA/ NASA ; 41, Royal Observatory, Edinburgh/SPL/Photo Researchers, Inc.; 44 (br), Dr. Christopher Burrows, ESA/STScl/NASA; 44, blt nglo–Australian Telescope Board; 44 (bl), Anglo–Australian Telescope Board; 45, V. Bujarrabal (OAN, Spain), WFPC2, HST, ESA/ NASA ; 46, Bill & Sally Fletcher/Tom Stack & Associates; 47 (br), Dennis Di Cicco/Peter Arnold, Inc.; 47 (bl), David Malin/Anglo–Australian Observatory; 48 (bl), I M House/Getty Images/Stone; 48 (br), Bill &Sally Fletcher/Tom Stack & Associates; 48 (bc), Jerry Lodriguss/Photo Researchers, Inc, 49, NASA/CXC/Smithsonian Astrophysical Observatory; 54, Sam Dudgeon/HRW; 55, John Sanford/Photo Researchers, Inc.; 60 (bl), NASA; 60 (tr), Jon Morse (University of Colorado)/NASA; 61 (r), The Open University; 61 (bkgd), Dutlev Van Ravenswaay/SPL/Photo Researchers, Inc.

Chapter Three 62–63, Anglo–Australian Observatory/Royal Obs. Edinburgh; 64, David Malin/Anglo–Australian Observatory/Royal Observatory, Edinburgh; 72, NASA/Mark Marten/Photo Researchers, Inc.; 73, NASA/TSADO/Tom Stack & Associates; 74, Earth Imaging/Getty Images/Stone; 77, SuperStock; 78 (l), Breck P. Kent/Animals Animals/Earth Scenes; 78 (r), John Reader/Science Photo Library/Photo Researchers, Inc; 80 (bc), Scott Van Osdol/HRW; 84, Sam Dudgeon/HRW; 636, Earth Imaging/Getty Images/Stone; 90 (b), NSO/NASA; 90 (tr), Jon Lomberg/Science Photo Library/Photo Researchers, Inc.; 90 (inset), David A. Hardy/Science Photo Library/Photo Researchers, Inc.; 91 (r), NASA/CXC/SAO/AIP/Niels Bohr Library; 91 (l), Corbis Sygma

Chapter Four 92–93, NASA/CORBIS; 94 (Mercury), NASA; 694 (Venus), NASA/Peter Arnold, Inc; 94 (Earth), Paul Morrell/Getty Images/Stone; 94 (Mars), USGS/TSADO/Tom Stack & Associates; 94 (Jupiter), Reta Beebe (New Mexico State University)/NASA; 98, NASA/Mark S. Robinson; 99, NASA; 100 (b), NASA; 100 (tl), Frans Lanting/Minden Pictures; 101, World Perspective/Getty Images/Stone; 101 (b), 102 (b), 102 (inset), NASA; 103, ESA; 104, NASA/Peter Arnold, Inc.; 105 (t), 105 (br), 105 (Saturn), 105 (Uranus), 105, (Neptune), 105 (Pluto), 106 (t), 107, 108 (t), 110, NASA; 112 (moons), John Bova/Photo Researchers, Inc.; 113, Fred Espenek; 114 (tl), Jerry Lodriguss/Photo Researchers, Inc.; 115 (b), NASA; 116 (t), USGS/Science Photo Library/Photo Researchers, Inc.; 116 (b), World Perspective/Getty Images/Stone; 117 (t), NASA; 118, Bill & Sally Fletcher/Tom Stack & Associates; 121 (bc), Breck P. Kent/Animals Animals/Earth Scenes; 121 (bl), E.R. Degginger/Bruce Coleman Inc.; 121 (br), Ken Nichols/Institute of Meteorites/University of New Mexico; 121 (t), Dennis Wilson/Science Photo Library/Photo Researhers, Inc.; 122, NASA; 123, Ken Nichols/Institute of Meteorites/University of New Mexico; 126, NASA, 127, ESA; 127 (b), NASA; 130 (tr), Richard Murrin; 131 (t), NASA; 131 (bl), Mehau Kulyk/Science Photo Library/Photo Researchers, Inc.

Chapter Five 132, Smithsonian Institution/Lockhead Corportation/Courtesy of Ft. Worth Museum of Science and History; 134, NASA; 135 (tr), Hulton Archive/Getty Images; 138 (bl), Brian Parker/Tom Stack & Associates; 138 (br), Sam Dudgeon/HRW; 140, Aerial Images, Inc. and SOVINFORMSPUTNIK; 141, NASA Marshall/National Space Science and Technology Center; 142 (bl, br), USGS; 144 (bkgd), Jim Ballard/Getty Images/Stone; 145 (bll), JPL/TSADO/Tom Stack & Associates; 145 (bkgd), Jim Ballard/Getty Images/Stone; 146 (bkgd), Jim Ballard/Getty Images/Stone; 146 (b), NASA/JPL/Malin Space Station Systems; 147 (bkgd), Jim Ballard/Getty Images/Stone; 149, JPL/NASA; 150 (bl, br), Bettmann/CORBIS; 151 (t), NASA; 151 (br), Corbis Images; 152 (plane), NASA; 152 (bkgd), Telegraph Colour Library/Getty Images/Taxi; 153, NASA; 154 (b), JPL/NASA; 162 (bl), NASA; 162 (tr), Photo courtesy Robert Zubrin, Mars Society; 163 (tr, b), NASA

Lab Book/Appendix 164, Victoria Smith/HRW; 165, Peter Van Steen/HRW; 166, 167, 168, 170, 171, Sam Dudgeon/HRW; 172, NASA/Getty Images/Stone; 173, Sam Dudgeon/HRW

TEACHER EDITION CREDITS

1E (l), David L. Brown/Tom Stack & Associates; 1E (r), Simon Fraser/Science Photo Library/Photo Researchers, Inc.; 1E (inset), Roger Ressmeyer/Corbis; 1F (tl), MSFC/NASA; 29E (bl), Roger Ressmeyer/CORBIS; 29F (tl), Bill & Sally Fletcher/Tom Stack & Associates; 29E (r), I M House/Getty Images/Stone; 29E (tl), John Sanford/Astrostock; 61E (bl), David Malin/Anglo–Australian Observatory/Royal Observatory, Edinburgh; 61F (tl), Earth Imaging/Getty Images/Stone; 91E (tr), NASA/Mark S. Robinson; 91F (br), Bill & Sally Fletcher/Tom Stack & Associates; 91F (bl), NASA; 131E (l), Hulton Archive/Getty Images; 131E (r), NASA; 131F (plane), NASA; 131F (plane bkgd), Telegraph Colour Library/Getty Images/Taxi

Answers to Concept Mapping Questions

The following pages contain sample answers to all of the concept mapping questions that appear in the Chapter Reviews. Because there is more than one way to do a concept map, your students' answers may vary.

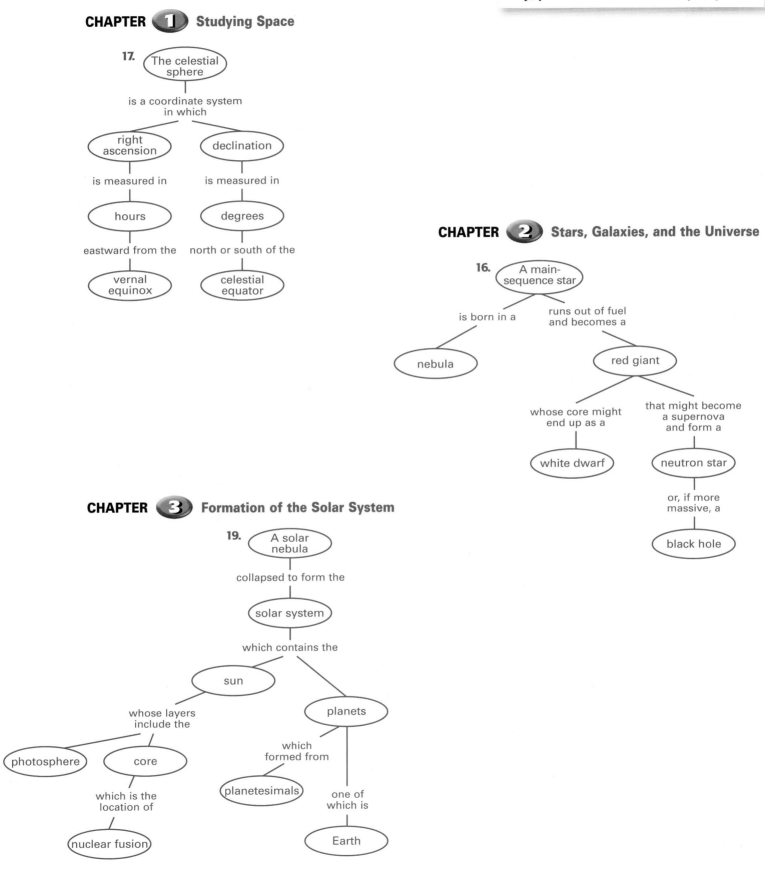

CHAPTER 1 Studying Space

17.
- The celestial sphere
  - is a coordinate system in which
    - right ascension
      - is measured in
        - hours
          - eastward from the
            - vernal equinox
    - declination
      - is measured in
        - degrees
          - north or south of the
            - celestial equator

CHAPTER 2 Stars, Galaxies, and the Universe

16.
- A main-sequence star
  - is born in a
    - nebula
  - runs out of fuel and becomes a
    - red giant
      - whose core might end up as a
        - white dwarf
      - that might become a supernova and form a
        - neutron star
          - or, if more massive, a
            - black hole

CHAPTER 3 Formation of the Solar System

19.
- A solar nebula
  - collapsed to form the
    - solar system
      - which contains the
        - sun
          - whose layers include the
            - photosphere
            - core
              - which is the location of
                - nuclear fusion
        - planets
          - which formed from
            - planetesimals
          - one of which is
            - Earth

CHAPTER 4 A Family of Planets

21.

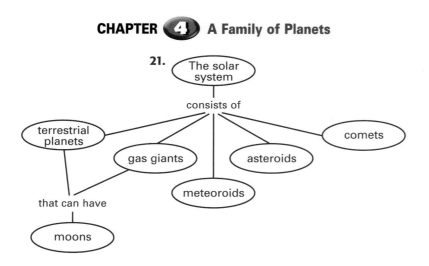

CHAPTER 5 Exploring Space

20.

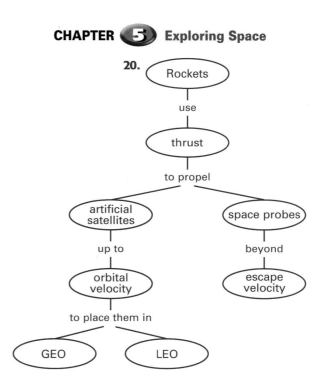